Paediatric Formulation

Paediatric Formulation

Design and Development

Editors

Nunzio Denora
Antonio Lopalco

MDPI • Basel • Beijing • Wuhan • Barcelona • Belgrade • Manchester • Tokyo • Cluj • Tianjin

Editors

Nunzio Denora
Department of
Pharmacy–Pharmaceutical
Sciences, University of Bari
"Aldo Moro"
Italy

Antonio Lopalco
Department of Pharmacy–Drug
Sciences, University of Bari
"Aldo Moro"
Italy

Editorial Office
MDPI
St. Alban-Anlage 66
4052 Basel, Switzerland

This is a reprint of articles from the Special Issue published online in the open access journal *International Journal of Molecular Sciences* (ISSN 1422-0067) (available at: https://www.mdpi.com/journal/ijms/special_issues/paediatr_formul).

For citation purposes, cite each article independently as indicated on the article page online and as indicated below:

LastName, A.A.; LastName, B.B.; LastName, C.C. Article Title. *Journal Name* **Year**, *Volume Number*, Page Range.

ISBN 978-3-0365-0740-8 (Hbk)
ISBN 978-3-0365-0741-5 (PDF)

Contents

About the Editors

Nunzio Denora is an associate professor and research scientist of Pharmaceutical Technology in the Department of Pharmacy-Drug Sciences at the University of Bari. Since 2016, he has been the director of the Inter-University Consortium CIRCSMB research group of Bari with the aim to develop "Inorganic-based Nano Drug Delivery Systems" and a member of the Scientific Committee of the European Paedriatric Translational Research Infrastructure EPTRI. His research is devoted to the development of paediatric formulations, microcarriers, and targeted nanocarriers by conventional and microfluidics techniques.

Antonio Lopalco received a Master's degree in Chemistry and Pharmaceutical Technology from the University of Bari in 2008 and a Ph.D. in Pharmaceutical Technology from the University of Palermo in 2012. After completing postdoctoral work in Pharmaceutical Chemistry at the University of Kansas, he was appointed as associate researcher in that department in 2015. He joined the faculty of the Department of Pharmacy-Pharmaceutical Sciences at the University of Bari in 2016 and worked his way up through the academic ranks to become an associate professor in 2019. Dr. Lopalco's research interests have included the development of targeted drug delivery systems and age-related formulations. Dr. Lopalco is the author or co-author of more than 50 research articles.

International Journal of
Molecular Sciences

MDPI

Editorial

Paediatric Formulation: Design and Development

Antonio Lopalco * and Nunzio Denora *

Department of Pharmacy-Pharmaceutical Sciences, University of Bari Aldo Moro, 70125 Bari, Italy

* Correspondence: antonio.lopalco@uniba.it (A.L.); nunzio.denora@uniba.it (N.D.); Tel.: +39-080-544-2764 (A.L.); +39-080-544-2767 (N.D.)

Received: 28 August 2020; Accepted: 25 September 2020; Published: 27 September 2020

Abstract: The development of medicines designed for children can be challenging since this distinct patient population requires specific needs. A formulation designed for paediatric patients must consider the following aspects: patient population variability; dose flexibility; route of administration; patient compliance; drug and excipient tolerability. The purpose of this Special Issue entitled "Paediatric Formulation: Design and Development" is to provide an update on both state-of-the-art methodology and operational challenges in the design and development of paediatric formulations. It aims at re-evaluating what is needed for more progress in the design and development of age-appropriate treatments for paediatric diseases, focusing on: formulation development; drug delivery design; efficacy, safety, and tolerability of drugs and excipients. This editorial, briefly, summarizes the objects of nine original research and review papers published in this Special Issue.

Keywords: formulation; development; design; children; paediatric; age-related; palatable; taste-masking; acceptable; excipient

1. Introduction

When designing and developing paediatric medicines, the route of administration, dosage form, and dose of the active ingredient (API) are decided on the basis of the expected duration of the therapy, the disease affecting a patient and his/her age, size, physio-pathological condition, API organoleptic and physicochemical properties (taste, aqueous solubility), its pharmacokinetic and pharmacodynamic properties, and stability during manufacture, storage, and use of the chosen formulation [1–4].

The paediatric population is a heterogeneous group ranging from preterm newborn infants to adolescents with wide physiological and developmental differences regarding organ and skin maturation, metabolism, and other factors that impact on the pharmacokinetics and pharmacodynamics of a drug [5].

It is a challenge to formulate one dosage form appropriate for this heterogeneous group of the human population. The goal should be to safely cover as wide an age range as possible with one specific dosage form. The guiding principle for choosing paediatric formulations should be the equilibrium of risks and benefits, taking into account the specific needs of these vulnerable patients. Key aspects in paediatric formulations involve the design and development of novel dosage forms, which should be adjustable for age and size, acceptable and palatable, easy to administer, and, at the same time, safer and effective [2–4].

Current use of medicines for paediatric patients reflects the full range of dosage forms and routes of administration used for adult medicines. There is, however, limited information on the acceptability of different paediatric dosage forms in relation to age and therapeutic needs and on the safety of excipients in relation to the development of the child [2–4,6,7].

Since its establishment, the biopharmaceutical classification system (BCS) has facilitated the development of oral drug formulations designed for adult patients. Theoretically, the BCS tenets could be applied also to paediatrics. Children's peculiarities and physiological differences from adults justify the need for a specific paediatric biopharmaceutics classification system (PBCS). Several scientific works attempted to provisionally classify oral drugs listed on the latest World Health Organization's Essential Medicines List for Children into age appropriate BCS. Validating a PBCS would provide a valuable tool to apply in specific paediatric formulation design by reducing time and costs and avoiding unnecessary paediatric experiments restricted by ethical reasons. Additionally, PBCS could minimize the associated risks to the use of adult medicines on pharmaceutical compound formulations for children. Moreover, developing a PBCS classification might be helpful in the process of harmonizing extemporaneous oral formulations in the hospital setting [8,9].

As a result of the great interest in developing age-appropriate dosage forms for children, this Special Issue, entitled "Paediatric Formulation: Design and Development", was programmed to highlight the need of formulating safer and effective medicines for paediatric patients. A total of nine papers were accepted for publication in this issue: four reviews and five primary data manuscripts, focusing on (1) the design, characterization, and safety evaluation of orodispersable formulations for paediatric tuberculosis pharmacotherapy; (2) development and palatability assessment of ritonavir powder for the paediatric population; (3) development of new praziquantel paediatric formulations in schistosomiasis treatment; (4) tridimensional retinoblastoma cultures as a vitreous seeds model for live-cell imaging of chemotherapy penetration; (5) preparation and characterization of dasatinib/cyclodextrin complex for the potential treatment of paediatric neuromuscular disorders; (6) efficacy, safety, and tolerability of tyrosine kinase inhibitors in the treatment of paediatric chronic myeloid leukaemia; (7) technologies and modern approaches to modify drug release in paediatric dosage forms; (8) challenges in formulating neonatal medicines; (9) efficacy, safety, and tolerability of anti-interleukin-1 (anti-IL-1) treatment in paediatric autoinflammatory diseases.

2. Articles in This Special Issue

Palatable orodispersible film formulations are ideal for patients with swallowing difficulties such as babies and children because they are stable and dissolve rapidly within the oral cavity in the presence of saliva, without the need to chew or drink water. In this Special Issue, Matawo and co-workers [10] investigated the preparation, optimization, and evaluation of a co-polymeric orodispersible pharmaceutical formulation containing pyrazinamide, a model first line antitubercular agent suitable for use in actively infected children. Organoleptic and cell toxicity studies presented the formulation as palatable, easy-to-handle, and biocompatible under applied test conditions. The orodispersible dosage form developed could potentially ease some of the challenges associated with the safe administration of tuberculosis antibiotics in children to aid desirable pharmacotherapeutic outcomes. Moreover, the authors suggested that the carrier matrix designed in this work might be used as is or even modified to accommodate and safely increase the release and/or absorption of other APIs for use in paediatric patients.

In this Special Issue, Morris and colleagues [11] presented an acceptable and age-appropriate dosage form formulation of the inhibitor of human immunodeficiency virus protease, ritonavir, for paediatric patients or patients who might have difficulties in swallowing a tablet. Norvir® oral powder (NOP) was developed to replace the oral solution, which provided reasonable bioavailability but exhibited taste-masking challenges and required the use of solvents with potential paediatric toxicity. In this study, the authors provided an overview of the development of NOP and palatability assessment strategy. In summary, NOP provided dose flexibility, enhanced stability, eliminated solvents, and maintained consistent bioavailability, with reduced bitterness and improved palatability via administration with common food products.

Here, Albertini and co-workers [12] evaluated the association of mechanochemical activation and spray congealing technology for developing a child-friendly praziquantel (PZQ) dosage form, with better product handling and biopharmaceutical properties, compared to mechanochemical activation materials. PZQ is the first line drug for the treatment of schistosome infections and is included in the WHO Model List of Essential Medicines for Children. In this study, they demonstrated that the approach consisting of the association of spray congealing with mechanochemical activation grinding, in the absence of polymeric excipients, was the most favourable, thus, it could be a promising method for designing a new PZQ formulation and a valid option for enhancing the performance of this antischistosomal drug, possibly permitting a significant reduction in therapeutic dose and minimizing the use of excipients in paediatric formulations.

Retinoblastoma is the most common intraocular tumour of childhood [13,14]. This tumour is highly curable if diagnosed in the early stages. Preclinical models could aid in understanding retinoblastoma vitreous seeds behaviour, drug penetration, and response to chemotherapy to optimize patient treatment. Winter et al. [15] described, in this Special Issue, the development of a novel tridimensional in vitro model of retinoblastoma vitreous seeds to assess chemotherapy penetration by means of live-cell imaging. Under the cell culture conditions used, these cells grew as tumourspheres with the ability to interact in a 3D structure. The results showed that tumourspheres provided a valuable model to study in vitro drug penetration and might be useful to optimize drug therapy and improve the efficacy of the treatment in paediatric patients affected by retinoblastoma.

For the first time, in this issue, a new inclusion complex of dasatinib, the first-choice oral drug in the treatment of chronic myeloid leukaemia and the excipient hydroxypropyl-beta-cyclodextrin (HP-β-CD), was described and fully characterized by Cutrignelli and co-workers [16] for the potential treatment of paediatric neuromuscular disorders. The strategy of complexation with HP-β-CD could allow the development of a paediatric oral liquid formulation of the poor water soluble dasatinib, which could be a valid alternative to the one currently present on the market that is solid. Moreover, considering that HP-β-CD is Food and Drug Administration approved for parenteral formulations, the dasatinib/HP-β-CD inclusion complex could also be an interesting tool for the administration of dasatinib by this route.

Recently, tyrosine kinase inhibitors imatinib, dasatinib, and nilotinib have been approved for the treatment of chronic myeloid leukaemia in children, even though the studies that were concerned with efficacy and safety toward this population are still awaiting defined and more accurate data [17–19]. In this scenario, here, Carofiglio and colleagues [20] published a review article pointing out the importance of prospectively validating data extrapolated from adult studies to set a standard therapeutic management for paediatric chronic myeloid leukaemia by employing appropriate formulations on the basis of paediatric clinical trials, which allow a careful monitoring of tyrosine kinase inhibitor-induced adverse effects, especially in growing children exposed to long-term therapy. Limited experience with very young children, the transition of teenagers to adult medicine, and the goal of achieving treatment-free remission for this rare leukaemia are more significant obstacles that require further clinical investigations. Finally, they concluded that in order to carry out a possible and viable therapy with these new anticancer drugs in the paediatric age, a key role is played by developing appropriate formulations specifically customized toward this kind of patient, since these drugs are often available in solid dosage forms, which are difficult to administer in children.

In pharmaceutical technology, the paediatric population still represents the greatest challenge in terms of developing flexible and appropriate drug dosage forms. In their review article published in this issue, Trofimiuk and colleagues [21] elucidated how to modify drug release in paediatric oral dosage forms, discuss the already accessible technologies, and to introduce novel approaches of manufacturing with regard to the paediatric population. Key aspects in modern formulations involve the development of novel formulations when considering chronic diseases that affect children and minimizing the dose frequency. Simultaneously, safety of excipients and child's acceptability should be kept in mind. The limited number of modified release formulations already present on the market arises from the high cost of technologies and lack of relevant clinical trials in the paediatric population. The authors suggested that new regulations and additional funding opportunities, as well as innovative collaborative research initiatives, should be constantly developed.

In this Special Issue, O'Brien at al. [22] published a review that offered insight into those challenges posed by the formulation of medicinal products for neonatal patients in order to support the development of clinically valid products. Neonates cannot be classified as small children when it comes to medicinal products and their formulation development. To design and develop an appropriate formulation for neonates, it is important to understand their physio-pathological status and development as well as route and methods of drug administration. The paper highlighted that a good understanding of the various constraints that could limit the development of an appropriate paediatric formulation would allow the formulator to provide for neonates whilst having due regard for the needs of the older population. If the neonate is considered early in the formulation design process, some delays in clinical trials in this population might be avoided.

The safety profile of treatment is of paramount importance when children with autoinflammatory disorders are managed, since it may affect adherence to treatment and overall clinical efficacy [23–25]. In this regard, in a review paper published in this issue, Bettiol and colleagues [26] aimed at providing current findings on the efficacy, safety, and tolerability of anti-IL-1 agents anakinra and canakinumab in multifactorial autoinflammatory diseases, focusing on the paediatric setting. Recent evidence from both observational studies and clinical trials widely documented the efficacy of IL-1 blocked in the main autoinflammatory diseases, also enlightening a good safety profile with few worries with regard to tolerability. In particular, the observed major side effects of anakinra were skin reactions at the injection-site. These reactions might become so unbearable for paediatric patients that treatment withdrawal might be required. In this regard, convincing young patients to continue therapy could be challenging. Reactions could be mitigated by the application of topical cortisone creams. On the contrary, the overall safety of canakinumab showed an exceptional tolerability, as pointed out by both very few discontinuation rates and injection-site reactions. However, a slightly increased rate of non-serious infections involving the upper respiratory tract was observed. Although these two anti-IL-1 agents currently represent the most effective treatments available in these diseases, and a promising therapeutic tool for managing refractory Kawasaki disease, the development of innovative dosage forms which further reduce side effects in paediatric sceneries is expected.

3. Conclusions

The high number of articles published in this Special Issue entitled "Paediatric Formulation: Design and Development" highlights the significant amount of research being conducted on the development of medicines designed for paediatric patients.

Formulating an appropriate dosage form is a challenge when considering chronic diseases that affect children and minimizing the dose frequency. Key aspects in paediatric formulations involve the design and development of novel dosage forms, which should be adjustable for age, palatable, easy to administer, and, at the same time, safer and effective.

Author Contributions: Writing—original draft preparation; writing—review and editing; visualization; supervision, A.L. and N.D. All authors have read and agreed to the published version of the manuscript.

Funding: This research received no external funding.

Acknowledgments: We would like to thank all the authors who submitted their work for this Special Issue. Particular thanks also to all of the reviewers who participated and enhanced the quality of the articles by seeking clarification for arguments and requesting modifications where necessary.

Conflicts of Interest: The authors declare no conflict of interest.

References

1. Allen, L.V. Dosage form design and development. *Clin. Ther.* **2008**, *30*, 2102–2111. [CrossRef]
2. Lopalco, A.; Denora, N.; Laquintana, V.; Cutrignelli, A.; Franco, M.; Robota, M.; Hauschildt, N.; Mondelli, F.; Arduino, I.; Lopedota, A. Taste masking of propranolol hydrochloride by microbeads of EUDRAGIT® E PO obtained with prilling technique for paediatric oral administration. *Int. J. Pharm.* **2020**, *574*, 118922. [CrossRef]
3. Laquintana, V.; Asim, M.H.; Lopedota, A.; Cutrignelli, A.; Lopalco, A.; Franco, M.; Bernkop-Schnürch, A.; Denora, N. Thiolated hydroxypropyl-β-cyclodextrin as mucoadhesive excipient for oral delivery of budesonide in liquid paediatric formulation. *Int. J. Pharm.* **2019**, *572*, 118820. [CrossRef]

4. Lopalco, A.; Curci, A.; Lopedota, A.; Cutrignelli, A.; Laquintana, V.; Franco, M.; Denora, N. Pharmaceutical preformulation studies and paediatric oral formulations of sodium dichloroacetate. *Eur. J. Pharm. Sci.* **2019**, *127*, 339–350. [CrossRef]
5. Kearns, G.L.; Abdel-Rahman, S.M.; Alander, S.W.; Blowey, D.L.; Leeder, J.S.; Kauffman, R.E. Developmental pharmacology—Drug disposition, action and therapy in infants and children. *N. Engl. J. Med.* **2003**, *349*, 1157–1167. [CrossRef]
6. Perrone, M.; Lopalco, A.; Lopedota, A.; Cutrignelli, A.; Laquintana, V.; Douglas, J.; Franco, M.; Liberati, E.; Russo, V.; Tongiani, S.; et al. Preactivated thiolated glycogen as mucoadhesive polymer for drug delivery. *Eur. J. Pharm. Biopharm.* **2017**, *119*, 161–169. [CrossRef] [PubMed]
7. Perrone, M.; Lopalco, A.; Lopedota, A.; Cutrignelli, A.; Laquintana, V.; Franco, M.; Bernkop-Schnürch, A.; Denora, N. S-preactivated thiolated glycol chitosan useful to combine mucoadhesion and drug delivery. *Eur. J. Pharm. Biopharm.* **2018**, *132*, 103–111. [CrossRef] [PubMed]
8. Shawahna, R. Pediatric Biopharmaceutical Classification System: Using Age-Appropriate Initial Gastric Volume. *AAPS J.* **2016**, *18*, 728–736. [CrossRef]
9. delMoral-Sanchez, J.M.; Gonzalez-Alvarez, I.; Gonzalez-Alvarez, M.; Navarro, A.; Bermejo, M. Classification of WHO Essential Oral Medicines for Children Applying a Provisional Pediatric Biopharmaceutics Classification System. *Pharmaceutics* **2019**, *11*, 567. [CrossRef] [PubMed]
10. Matawo, N.; Adeleke, O.A.; Wesley-Smith, J. Optimal Design, Characterization and Preliminary Safety Evaluation of an Edible Orodispersible Formulation for Pediatric Tuberculosis Pharmacotherapy. *Int. J. Mol. Sci.* **2020**, *21*, 5714. [CrossRef]
11. Morris, J.B.; Tisi, D.A.; Tan, D.C.T.; Worthington, J.H. Development and Palatability Assessment of Norvir®(Ritonavir) 100 mg Powder for Pediatric Population. *Int. J. Mol. Sci.* **2019**, *20*, 1718. [CrossRef] [PubMed]
12. Albertini, B.; Perissutti, B.; Bertoni, S.; Zanolla, D.; Franceschinis, E.; Voinovich, D.; Lombardo, F.; Keiser, J.; Passerini, N. Combining Mechanochemistry and Spray Congealing for New Praziquantel Pediatric Formulations in Schistosomiasis Treatment. *Int. J. Mol. Sci.* **2019**, *20*, 1233. [CrossRef] [PubMed]
13. Abramson, D.H. Retinoblastoma: Saving life with vision. *Annu. Rev. Med.* **2014**, *65*, 171–184. [CrossRef] [PubMed]
14. Chantada, G.; Schaiquevich, P. Management of retinoblastoma in children: Current status. *Paediatr. Drugs* **2015**, *17*, 185–198. [CrossRef] [PubMed]
15. Winter, U.; Aschero, R.; Fuentes, F.; Buontempo, F.; Zugbi, S.; Sgroi, M.; Sampor, C.; Abramson, D.H.; Carcaboso, A.M.; Schaiquevich, P. Tridimensional Retinoblastoma Cultures as Vitreous Seeds Models for Live-Cell Imaging of Chemotherapy Penetration. *Int. J. Mol. Sci.* **2019**, *20*, 1077. [CrossRef]
16. Cutrignelli, A.; Sanarica, F.; Lopalco, A.; Lopedota, A.; Laquintana, V.; Franco, M.; Boccanegra, B.; Mantuano, P.; De Luca, A.; Denora, N. Dasatinib/HP-β-CD Inclusion Complex Based Aqueous Formulation as a Promising Tool for the Treatment of Paediatric Neuromuscular Disorders. *Int. J. Mol. Sci.* **2019**, *20*, 591. [CrossRef]
17. Kalmanti, L.; Saussele, S.; Lauseker, M.; Müller, M.C.; Dietz, C.T.; Heinrich, L.; Hanfstein, B.; Proetel, U.; Fabarius, A.; Krause, S.W.; et al. Safety and efficacy of imatinib in CML over a period of 10 years: Data from the randomized CML-study IV. *Leukemia* **2015**, *29*, 1123–1132. [CrossRef]
18. FDA Approved Dasatinib for Pediatric Patients. Available online: https://www.fda.gov/drugs/resources-information-approved-drugs/fda-approves-dasatinib-pediatric-patients-cml (accessed on 20 January 2020).
19. FDA Approves Nilotinib Pediatric Patients. Available online: https://www.fda.gov/drugs/resources-information-approved-drugs/fda-approves-nilotinib-pediatric-patients-newly-diagnosed-or-resistantintolerant-ph-cml-chronic (accessed on 20 January 2020).
20. Carofiglio, F.; Lopalco, A.; Lopedota, A.; Cutrignelli, A.; Nicolotti, O.; Denora, N.; Stefanachi, A.; Leonetti, F. Bcr-Abl Tyrosine Kinase Inhibitors in the Treatment of Pediatric CML. *Int. J. Mol. Sci.* **2020**, *21*, 4469. [CrossRef]
21. Trofimiuk, M.; Wasilewska, K.; Winnicka, K. How to Modify Drug Release in Paediatric Dosage Forms? Novel Technologies and Modern Approaches with Regard to Children's Population. *Int. J. Mol. Sci.* **2019**, *20*, 3200. [CrossRef]
22. O'Brien, F.; Clapham, D.; Krysiak, K.; Batchelor, H.; Field, P.; Caivano, G.; Pertile, M.; Nunn, A.; Tuleu, C. Making Medicines Baby Size: The Challenges in Bridging the Formulation Gap in Neonatal Medicine. *Int. J. Mol. Sci.* **2019**, *20*, 2688. [CrossRef]

23. Rigante, D.; Lopalco, G.; Vitale, A.; Lucherini, O.M.; Caso, F.; De Clemente, C.; Molinaro, F.; Messina, M.; Costa, L.; Atteno, M.; et al. Untangling the web of systemic autoinflammatory diseases. *Mediat. Inflamm.* **2014**, *2014*, 948154. [CrossRef] [PubMed]
24. Cattalini, M.; Soliani, M.; Rigante, D.; Lopalco, G.; Iannone, F.; Galeazzi, M.; Cantarini, L. Basic Characteristics of Adults with Periodic Fever, Aphthous Stomatitis, Pharyngitis, and Adenopathy Syndrome in Comparison with the Typical Pediatric Expression of Disease. *Mediat. Inflamm.* **2015**, *2015*, 570418. [CrossRef] [PubMed]
25. Vitale, A.; Sota, J.; Rigante, D.; Lopalco, G.; Molinaro, F.; Messina, M.; Iannone, F.; Cantarini, L. Relapsing Polychondritis: An Update on Pathogenesis, Clinical Features, Diagnostic Tools, and Therapeutic Perspectives. *Curr. Rheumatol. Rep.* **2016**, *18*, 3. [CrossRef] [PubMed]
26. Bettiol, A.; Lopalco, G.; Emmi, G.; Cantarini, L.; Urban, M.L.; Vitale, A.; Denora, N.; Lopalco, A.; Cutrignelli, A.; Lopedota, A.; et al. Unveiling the Efficacy, Safety, and Tolerability of Anti-Interleukin-1 Treatment in Monogenic and Multifactorial Autoinflammatory Diseases. *Int. J. Mol. Sci.* **2019**, *20*, 1898. [CrossRef]

International Journal of
Molecular Sciences

MDPI

Article

Optimal Design, Characterization and Preliminary Safety Evaluation of an Edible Orodispersible Formulation for Pediatric Tuberculosis Pharmacotherapy

Nyaradzo Matawo [1], Oluwatoyin A. Adeleke [1,*,†] and James Wesley-Smith [2]

1 Division of Pharmaceutical Sciences, School of Pharmacy, Sefako Makgatho Health Science University, Pretoria 0208, South Africa; nmatawo1@gmail.com

2 Electron Microscope Unit, Sefako Makgatho Health Science University, Pretoria 0208, South Africa; jaime.wesley-smith@smu.ac.za

* Correspondence: oluwatoyin.adeleke@fulbrightmail.org or oluwatoyin.adeleke@smu.ac.za

† Additional address: Laboratory of Parasitic Diseases, Immunobiology Section, National Institute of Allergy and Infectious Diseases, National Institutes of Health, Bethesda, MD 20892, USA.

Received: 26 June 2020; Accepted: 30 July 2020; Published: 10 August 2020

Abstract: The severity of tuberculosis (TB) in children is considered a global crisis compounded by the scarcity of pharmaceutical formulations suitable for pediatric use. The purpose of this study was to optimally develop and evaluate a pyrazinamide containing edible orodispersible film formulation potentially suitable for use in pediatrics actively infected with TB. The formulation was prepared employing aqueous-particulate blending and solvent casting methods facilitated by a high performance Box Behnken experimental design template. The optimized orodispersible formulation was mechanically robust, flexible, easy to handle, exhibited rapid disintegration with initial matrix collapse occurring under 60 s (0.58 ± 0.05 min ≡ 34.98 ± 3.00 s) and pyrazinamide release was controlled by anomalous diffusion coupled with matrix disintegration and erosion mechanisms. It was microporous in nature, light weight (57.5 ± 0.5 mg) with an average diameter of 10.5 mm and uniformly distributed pyrazinamide load of 101.13 ± 2.03 $\%^{w}/_{w}$. The formulation was physicochemically stable with no evidence of destructive drug–excipient interactions founded on outcomes of characterization and environmental stability investigations. Preliminary inquiries revealed that the orodispersible formulation was cytobiocompatible, palatable and remained intact under specific storage conditions. Summarily, an edible pyrazinamide containing orodispersible film formulation was optimally designed to potentially improve TB pharmacotherapy in children, particularly the under 5 year olds.

Keywords: Orodispersible formulation; pyrazinamide; pediatric drug delivery; tuberculosis; design of experiments; children; edible films

1. Introduction

The latest World Health Organization's (WHO) global report estimated that approximately 10 million people developed new tuberculosis (TB) infections that progressed to TB disease with about 1.5–2 million deaths recorded per annum [1]. So far, TB is the deadliest infectious disease globally and millions of people continue to fall sick and die annually. It is amongst the first ten primary causes of death from a single infectious agent worldwide, ranking above HIV/AIDS [1–3] (WHO, 2019; Swaminathan and Rekha, 2010; Kumar et al., 2017;). It remains a global threat with approximately 1.7 billion people having latent TB infection that can turn into active TB disease at any time [1]. Tuberculosis is an airborne, infectious disease that usually affects the lungs (pulmonary TB) leading to severe coughing, fever and chest pains or in some rare cases, other body parts (extra-pulmonary TB). It is caused by *Mycobacterium tuberculosis* also called tubercle bacilli. It is preventable and curable if diagnosed early and managed with the correct medicines [1,4,5].

Generally, TB infection within the pediatric population is considered to be a major cause of morbidity and mortality [6]. According to the latest WHO's global TB report, at least one million children under the age of 15 (accounting for about 11% of the affected population) contract active TB infection with about 230,000 fatalities recorded annually [1] (WHO, 2019). Children may have TB disease at any age, but most often under 5 years old in TB-endemic countries. TB disease is also prevalent amongst children infected with the human immunodeficiency virus (HIV) who are usually at a twenty times greater risk of contracting active TB infection compared to children who are HIV negative [7,8]. Children often contract TB from actively infected adult household members, during birth or when they present with weak immune systems; such as in infants, those infected with HIV or the severely malnourished who are at greater risk of developing TB disease or even dying. Pulmonary TB is the most common in children although extra-pulmonary TB may occur. Pediatric TB is more common in developing countries where there is overcrowding, poverty and malnutrition than in developed states [2]. Treatment and prevention of TB in children is considered neglected regardless of the alarming statistics as there are few scientifically justified studies focusing on: (i) accurate pediatric dosing; (ii) designing desirable drug formulations suitable for use in children of all ages; (iii) developing effective diagnostic tools for this age group as they usually do not manifest any symptoms or signs of disease early; plus (iv) the belief that childhood TB is not important for TB control [9–11].

To date, commonly used pharmaceutical formulations are either liquid dosage forms (e.g., solutions, suspensions), fixed dose dispersible tablets and in most instances, adult tablets are often broken, crushed or mixed with food or water (co-administration) to make pediatric management possible [12–15]. Despite the availability of a few commercialized pediatric preparations, considerable global scarcity still exists, meaning that many children are unable to access these medicines [15–19]. Moreover, studies have shown that co-administration (with food, water etc.) is a common global practice for treating children with TB and that it is performed without appropriate instructions. In most case, caregivers just choose any food or drink without any assessment of its impact on safety and efficacy [13,15,20]. This may potentially lead to inaccurate dosing, resulting in reduced efficacy or adverse effects often caused by under-dosing and over-dosing respectively, disruption of the outer coating leading to physicochemical instabilities, and potential active pharmaceutical ingredient (API) wastage [2,13,21].

The use of alternative dosage forms such as suspensions or solutions can potentially help us overcome some of these challenges but they are also known to be generally less stable even when refrigerated, difficult to taste mask, expensive for safe transportation and have short shelf lives; all of which limit their applicability [2,22]. Dispersible tablets on the other end are deemed more child-friendly but still limited in that they are difficult to administer while in transit or when there is reduced/no access to potable water—like in most underdeveloped and developing countries where TB is endemic. They usually contain additives that are either not safe for use in children or hygroscopic in nature, making them prone to atmospheric moisture/water absorption that can lead to active drug instability, eventual inactivity and possible pharmacotherapeutic inefficacy [17,19,23]. Other potentially applicable delivery systems for children include chewable tablets, which are often more suitable for older children (>3 years) with teeth, and sprinkles, though they are more acceptable for older children that can eat solid food [23,24].

Recent studies show that the most popular age appropriate delivery systems are small sized, solid oral drug delivery systems e.g., minitablets and multi-particulates and orally disintegrating formulations like orodispersible tablets or films [18,25]. Particularly, orodispersible formulations are of choice because of their characteristic advantages such as water free administration, easy to use anywhere and at any time without the need for external help or specialized caregivers, improved stability, easier transportation, cost effectiveness and rapid disintegration when placed within the oral cavity releasing incorporated API for absorption. This definitely allows easy administration to pediatric patients with or without teeth [26]. Orodispersible delivery systems offer advantages such as enhanced pediatric compliance, possibility of local action, dosage accuracy, reduced choking risks, easy handling and portability [27,28]. They also allow rapid onset of action and increase in bioavailability due to rapid dispersion within the mouth and significant pre-gastric absorption, all leading to desirable pharmacotherapeutic efficacy [29]. Furthermore, antitubercular agents are administered at low doses in children so, orodispersible formulations will not be outsized or pose a choking hazard [19,23–25,30,31].

Therefore, this study details the design, optimization and systematic in vitro evaluation of a polymer-based, orodispersible film formulation containing pyrazinamide (PZA) as a potential alternative for flexible pediatric dosing. It is a first line antitubercular agent often used in combination with isoniazid, rifampicin and ethambutol for the treatment of active TB infection [32,33]. PZA is highly bactericidal, and acts by sterilizing slowly metabolizing tubercle bacilli, resulting in low incidence of bacteriological relapse post completion of chemotherapeutic regimen. It facilitates treatment shortening, leading to greater patient compliance [32,34–39]. It is a prodrug which undergoes conversion into active pyrazinoic acid by the bacterial enzyme pyrazinamidase at or below pH 5.6 [33]. Typically, it is administered for the initial 2 months of a 6-month treatment for drug-susceptible infections. PZA is a Class III drug according to the Biopharmaceutics Classification System (BCS) characterized by its high aqueous solubility (15 mg/mL at 25 °C), relatively low permeability (logP = −1.88) and linear absorption over a broad spectrum of doses [36,38] (Becker et al., 2008; Adeleke et al., 2016). The PZA loaded orodispersible matrices were prepared using the solvent casting method [27,40,41]. The PZA loaded formulation was prepared using a combination of pharmaceutical excipients which included copolymer polyvinyl alcohol-polyethylene glycol as a matrix and film forming agent, citric acid as a natural preservative, sodium starch glycolate as a superdisintegrant and xylitol as a sweetener acceptable for pediatric use as documented by Dixit and Puthli [42]. Formulation preparation and optimization were facilitated using a response surface method based on a 4-factor, 3-level Box Behnken experimental design (Minitab® 18 Statistical Software (Minitab LLC, State College, PA, USA), a robust, high performance quadratic template widely applied in the development of viable drug carriers [38,43,44]. The optimized orodispersible film formulation was then physicochemically characterized in vitro by determining its mass, dimensions (inner and outer diameter), disintegration time, drug release and kinetics, drug content, dissolution pH, surface morphology changes, thermal behavior, crystallinity and structural chemical backbone transitioning. Furthermore, we studied the stability of the optimized formulation under common environmental storage conditions, its organoleptic qualities and cytobiocompatibility in vitro.

2. Results and Discussion

2.1. Orodispersible Formulation Variants

Employing the initial one-variable-at-a-time screening together with the systematic 4-factor, 3-level Box Behnken experimental design template, 27 pyrazinamide loaded orodispersible formulations were successfully prepared using the solvent casting method. Through these approaches, the independent variables affecting the response parameters, namely disintegration time (Y_1), dissolution pH (Y_2) and formulation weight (Y_3), were identified. In general, the orodispersible formulations appeared as whitish, dense and bendable, hollow cup-shaped matrices that were light weight (<122 mg) and had average inner and outer diameters of 11 ± 1 mm and 10 ± 0.81 mm respectively. Overall, the 27 orodispersible formulations had average weights ranging between 121.4 ± 8.00 mg and 60.87 ± 3.80 mg, disintegrated with a total matrix structure collapse within 0.20 ± 0.09 to 5.67 ± 0.42 minutes and presented dissolution pHs spanning from 6.59 ± 0.23 to 7.43 ± 0.01 which is relatively close to that of the oral cavity (saliva). Differences observed in the measured response parameters showed that the selected independent variables applied at the varying factors levels and combinations, according to the quadratic design template, had noteworthy effects on the nature of each formulation. Numerical values of response parameters measured for all 27 formulations are presented in Table 1.

Table 1. Response parameters generated based on the quadratic experimental design template.

Formulation	Weight (mg)	Disintegration Time (minutes)	Dissolution pH
F1	91.66 ± 1.55	2.57 ± 0.34	7.33 ± 0.02
F2	65.60 ± 2.58	0.47 ± 0.15	6.69 ± 0.04
F3	106.80 ± 2.00	1.59 ± 0.50	7.04 ± 0.02
F4	100.00 ± 2.38	1.24 ± 0.23	6.96 ± 0.05
F5	105.83 ± 3.95	1.72 ± 0.42	6.78 ± 0.04
F6*	87.86 ± 2.11	1.50 ± 0.09	6.59 ± 0.23
F7	75.07 ± 4.45	1.19 ± 0.06	6.94 ± 0.02
F8	60.87 ± 3.80	0.44 ± 0.04	6.96 ± 0.03
F9*	86.50 ± 2.59	1.45 ± 0.24	6.83 ± 0.06
F10	81.23 ± 1.64	3.46 ±0.51	6.98 ± 0.01
F11	81.50 ± 1.64	5.40 ± 0.10	7.30 ± 0.03
F12	84.77 ± 6.96	1.94 ± 0.62	7.06 ± 0.03
F13	112.06 ± 18.98	3.62 ±2.90	7.23 ± 0.14
F14	121.4 ± 8.00	5.44 ± 0.14	7.01 ± 0.06
F15	101.33 ± 4.51	4.84 ± 0.37	6.70 ± 0.06
F16	82.63 ± 0.06	4.20 ± 0.13	7.29 ± 0.03
F17	82.07 ± 3.04	5.03 ± 1.02	7.07 ± 0.03
F18	73.67 ± 3.16	4.83 ± 1.54	7.05 ± 0.02
F19	77.3 ± 1.42	4.02 ± 0.58	7.02 ± 0.06
F20	77.3 ± 14.03	2.58 ± 0.53	6.95 ± 0.06
F21	109.47 ± 2.39	2.57 ± 1.28	7.33 ± 0.41
F22	98.23 ± 15.06	2.73 ± 1.54	7.00 ± 0.05
F23*	90.30 ± 1.34	1.08 ± 0.10	7.10 ± 0.01
F24	80.97 ± 1.23	1.67 ± 0.57	7.43 ± 0.01
F25	114.97 ± 10.08	5.67 ± 0.42	7.05 ± 0.04
F26	78.40 ± 5.56	0.20 ± 0.09	7.02 ± 0.05
F27	70.37 ± 3.75	0.72 ± 0.34	7.41 ± 0.06

* represents the centrepoint experimental runs.

2.2. Selection and Validation of the Optimized Orodispersible Formulation

The ANOVA analysis revealed that the percentage of polyvinyl alcohol polyethylene glycol, sodium starch glycolate, citric acid and xylitol contained in each formulation variant significantly ($p < 0.05$) impacted the response parameters. Based on the statistical method and constraints applied on the formulation weight, disintegration time and dissolution pH, a formula was developed for the preparation of the optimized orodispersible formulation using the Minitab® 18 Statistical Software. An overall desirability of 0.991 indicating the robustness of the optimization platform and design template was obtained. The optimized formula containing the levels of each independent variable is shown in Table 2. To further confirm the validity and suitability of the optimization process, the optimized formulation was prepared in triplicate following earlier described method and measurements of the formulation weight, disintegration time and dissolution pH were performed experimentally. The magnitude of the observed response parameters measured against that of the predicted values displayed a high degree of correlation, further supporting the precision and robustness of the statistical design employed for generating the desired optimized formulation.

Table 2. Optimized formula and model summary of fitted and experimental outputs as it relates to the experimental design template.

Optimized Formula	Validation of Predicted Outputs with Experimental Values (n = 3)			
Factor Levels	*Responses*	*Desirability Level*	*Predicted*	*Observed*
X_1 = 1.000 g X_2 = 0.483 g X_3 = 0.058 g X_4 = 0.539 g	Y_1 Y_2 Y_3	0.980 1.000 0.994	0.620 min 7.000 65.333 mg	0.583 ± 0.050 min 6.900 ± 0.250 57.500 ± 0.002 mg

Note: X_1 = Polyvinyl alcohol polyethylene glycol (Kollicoat® IR); X_2 = Sodium starch glycolate (Primojel®); X_3 = Citric acid; X_4 = Xylitol, Y_1 = Disintegration time; Y_2 = Disintegration pH; Y_3 = Formulation weight.

2.3. Physical Properties of Optimized Formulation

The optimized orodispersible formulation was thin, flexible making handling possible, whitish in color with a hollow/concave shape as well as inner and outer diameters of 10.00 ± 0.52 mm and 11.00 ± 0.43 mm. The formulation was made up of sodium starch glycolate as a super disintegrate, co-polymer polyvinyl alcohol polyethylene glycol as a matrix and film forming agent, xylitol as a sweetener suitable for pediatrics, citric acid as a natural preservative and pyrazinamide (500 mg) as a model antitubercular agent. The optimized formulation was light weight and small enough for orodispersible applications. It rapidly disintegrated in less than 60 seconds (i.e., 0.58 min ≡ 34.98 s) when placed in simulated saliva at 37 ± 0.1 °C. The dissolution was close to neutral (7.0) and saliva pH (6.8), meaning that the formulation has no potential to irritate the buccal mucosa (Table 2). Digital photographs of the PZA loaded and drug free (placebo) optimized formulations are shown in Figure 1.

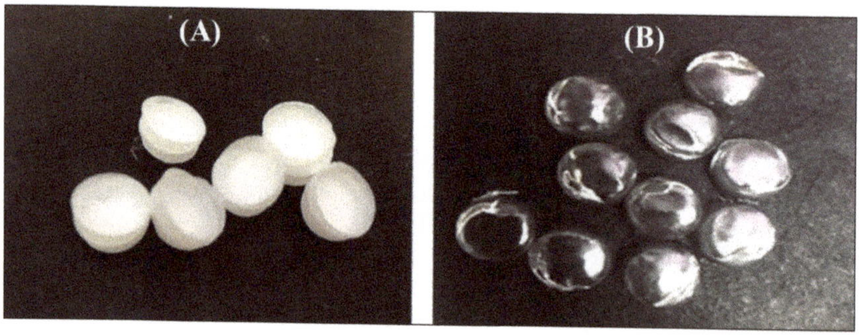

Figure 1. Digital photograph of (**A**) drug loaded and (**B**) placebo orodispersible film formulation.

2.4. Drug Content and Release Behavior

The PZA content of the optimized orodispersible formulation was 25.02 ± 0.71 mg equaling 101.13 ± 2.03%$^{w}/_{w}$, indicating that uniform drug distribution occurred among replicate test samples and drug remained stable during and after preparation. In vitro drug release was carried out under biorelevant conditions to determine the rate at which PZA molecules were released in simulated saliva (pH 6.8) at 37 ± 0.1 °C, mimicking the buccal environment. The generated release profile is illustrated with Figure 2. Figure 2B, which is an expanded form of segments of Figure 2A, shows that drug liberation was initiated under 10 s (0.17 min) with 1.94 ± 0.28% released almost immediately after the formulation came in contact with the dissolution media (onset of matrix disintegration). Subsequently, the amount of drug released continued to increase rapidly, reaching its first peak at 5 min (72.01 ± 11.93%) and then maintained a relatively plateaued profile until 60 min when complete matrix dissolution and drug release (102.50 ± 5.19%) occurred.

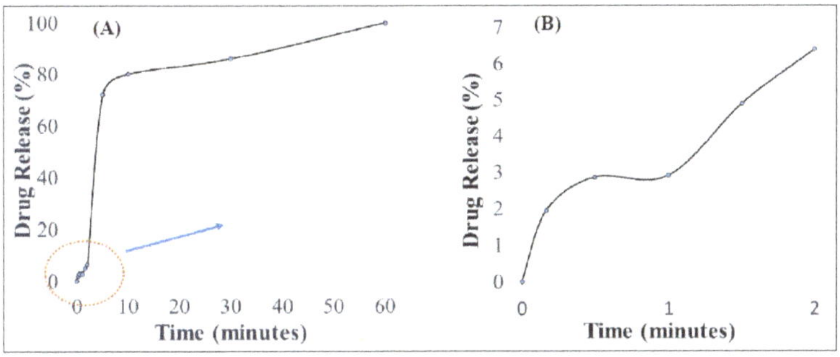

Figure 2. (**A**) In vitro drug release behavior in simulated saliva and (**B**) illustrates an expanded segment of the drug release profile.

The mathematical models applied to represent the release mechanisms of the active drug from the formulation were zero-, first-, second-order as well as Higuchi, Korsmeyer–Peppas and Hixon–Crowell models yielded R values of 0.59, 0.47, 0.33, 0.40, 0.90 and 0.51, respectively. The optimized formulation displayed a good fit to the Korsmeyer–Peppas ($R = 0.90$) and the computed *n-value* was 0.83, meaning that drug release after formulation hydration was controlled by an anomalous drug diffusion process followed by matrix relaxation (disentangling polymer chains or disintegration) and erosion (matrix dissolution) [45].

2.5. Optimized Formulation Characterization

2.5.1. Thermal Behavior Using Differential Scanning Calorimetry and Thermogravimetry

Generated differential scanning calorimetry (DSC) thermograms were employed in the assessment of typifying thermal quantity changes for pure PZA, pharmaceutical excipients (citric acid, xylitol, poly vinyl alcohol polyethylene and sodium starch glycolate), optimized drug loaded and placebo formulations. Measured key thermal quantities include melting point (T_m) and glass transition temperature (T_g). Typical DSC thermograms are shown in Figure 3. The DSC scans of citric acid and xylitol represented in (Figure 3A,B) show sharp endothermic peaks corresponding to their T_m at 153.38 °C and 92.64 °C, respectively, indicating their purity and stable states. Polyvinyl alcohol polyethylene glycol (Kollicoat® IR) thermogram (Figure 3C) depicts multiple broad endothermic and exothermic peaks with a small endothermic peak at 92.3 °C and a more prominent endothermic peak at 212.65 °C as T_m. The presence of two endothermic peaks shows that the Kollicoat® IR is made up of two different natured polymers. The appearance of Tg noted at 44.24 °C can be associated with its amorphous co-polymeric transitioning into a crystalline material. Sodium starch glycolate thermogram (Figure 3D) displayed characteristic exothermic peak at 298.75 °C and glass transition was observed at 46.45 °C. Pure pyrazinamide shows two endothermic peaks at 154.00 °C, corresponding to solid–solid transition and a sharp endothermic peak at 190.64 °C (Figure 3E), which corresponds to the T_m, indicating its α- polymorphic form and purity of PZA [38,46,47]. Thermal peaks identified for each excipient and pure drug are indicative of their purity and stability as individual compound before their inclusion in the orodispersible formulation mix. The placebo thermogram (Figure 3F) presented broad endothermic and exothermic peaks, revealing its semi-crystalline nature which can be related to its crystalline, semi-crystalline and amorphous constituents already mentioned above. Likewise, the optimized PZA loaded orodispersible formulation thermogram (Figure 3G) also displayed a semi-crystalline trend with broad endothermic and exothermic peaks occurring within the melting point region of excipients. This finding further supports its blended pure drug and excipient content. The slight shift of the PZA peak in the mixture may have been influenced by polymeric crystallization occurring during the formation of the orodispersible formulation resulting in the endothermic peak slightly shifting to 193 °C, accounting as the T_m of the newly prepared drug loaded formulation further confirming drug stability within the drug delivery matrix. The disappearance of melting peaks of individual excipients in the drug loaded formulation shows complete solubilization of the drug and excipients within the matrix. It also supports the formation of a new compound, hence the physical transitioning of PZA from crystalline to the amorphous form potentially accounting for the rapidly disintegrating quality of the developed formulation [48]. The obtained DSC thermograms demonstrate the compatibility of the drug (PZA) and pharmaceutical excipients used in the formulation development with no evidence of possible adverse interactions.

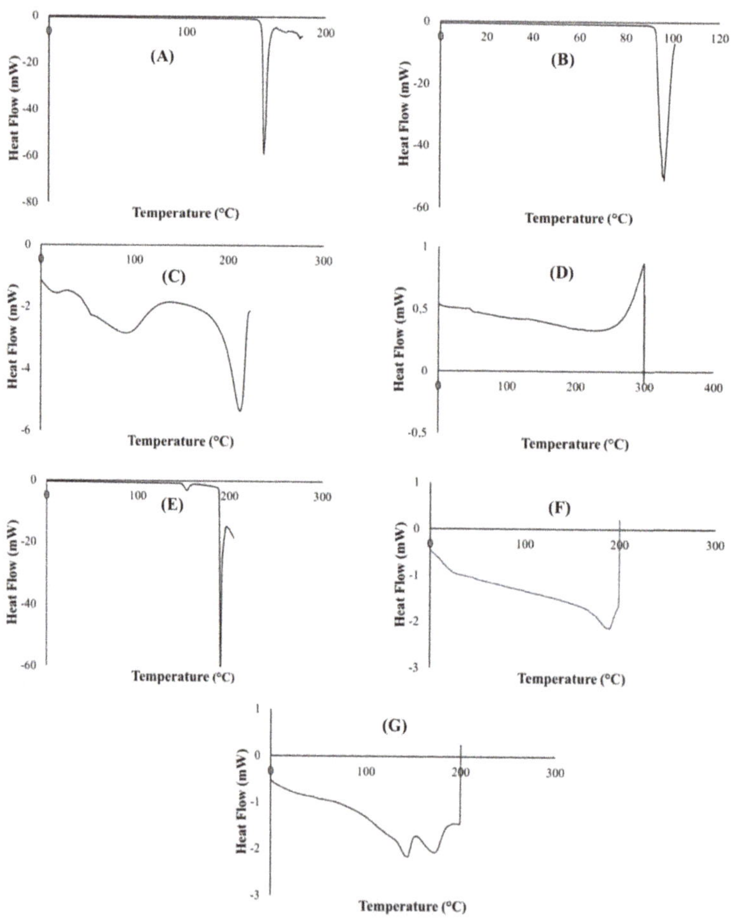

Figure 3. DSC Thermograms of (**A**) citric acid, (**B**) xylitol, (**C**) polyvinyl alcohol polyethylene glycol, (**D**) sodium starch glycolate, (**E**) pyrazinamide, (**F**) placebo and (**G**) drug loaded formulation.

Thermogravimetric curves of pure PZA, pharmaceutical excipients, optimized drug loaded and placebo formulations recorded under a nitrogen saturated atmosphere, purge and heating rates of 25 mL/min and 10 °C/min, respectively, are presented in Figure 4. TGA analysis was conducted for additional investigation of thermal degradation events quantified as percentage weight loss as it relates to temperature and its impact of formulation stability. First, important thermal events were identified for individual excipients and pure PZA. The PZA thermal decomposition T_{onset} was observed at 164 °C and complete weight loss occurred at 199.15 °C (Figure 4A). The onset of thermal decomposition for xylitol (Figure 4B) was noted at 240.59 °C and final decomposition temperature was observed at about 306.73 °C, showing it thermal robustness as T_{onset} was above 200 °C. Citric acid and poly vinyl alcohol polyethylene thermal plots in (Figure 4C,D) commenced decomposition at a T_{onset} of about 185.99 and 236.39 °C, respectively, and complete breakdown only occurred at temperatures greater than 400 °C, confirming thermal stability of these excipients. Sodium starch glycolate (Figure 4E) showed two decomposition events, water loss from the polymer was observed at about 125.01 °C, corresponding to 8% weight loss, followed by final decay at 257.99 °C. Generally, the excipients presented with relatively high, single decomposition temperatures, 199 °C and above, indicating their distinct stability and purity. Thermal decomposition of optimized drug loaded and placebo (Figure 4F,G) began at 156.24 and 200.38 °C, respectively. Complete weight loss followed at 202.72 °C for the PZA loaded formulation and beyond 400 °C for the placebo. Onset temperatures considerably below 200 °C suggests lower thermal stability which, in this case, can be linked to the loss of residual water molecules from the both matrices due to the presence of hydrophilic components in both drug loaded and placebo formulations. Usually, weak drug and hydrophilic polymer interactions break easily around 100 °C [49]. The presence of an additional non-polymeric hydrophilic molecule, PZA, within the drug loaded formulation matrix probably accounts for the reduction in the onset temperature (156.24 °C) compared to that recorded for the placebo (200.38 °C) which contains all other hydrophilic constituent excluding PZA. This can further influence the temperature at which complete weight loss (100%) occurred for both drug loaded (202.72 °C) and placebo formulations (>400 °C). From these observations, it appears that the placebo contains more crystalline domains within its molecular structure and its components are more of a physical blend with less amorphous transitioning happening compared to the drug loaded formulation, where a degree of amorphization seems to occur between the PZA and excipient physical blend as also revealed through the DSC (Figure 3G) and XRD (Figure 5G) analytical outputs. Overall, the TGA thermograms exhibited the thermal stability of both drug and pharmaceutical excipients, either as separate entities or blends in the different formulations, as well as presented no visible trace of any destructive chemical interaction. Additionally, the similarity in weight loss patterns plus relative overlap in final decomposition temperatures for both drug loaded formulation (202.72 °C) and pure PZA (199.15 °C) (Figure 4A,F), can further signify formulation matrix stability and drug intactness which, was also the case with the DSC analyses as the PZA peak remained identifiable in the formulation blend with a slight shift associated with the presence of other excipients (Figure 3G).

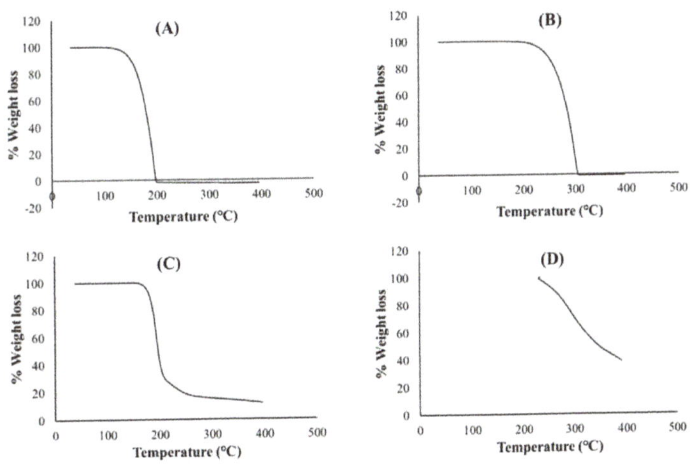

Figure 4. *Cont.*

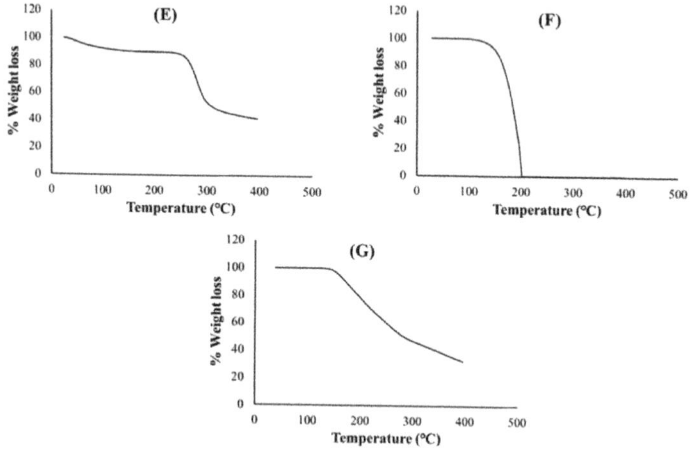

Figure 4. Thermogravimetric curves of (**A**) PZA, (**B**) xylitol, (**C**) citric acid, (**D**) polyvinyl alcohol polyethylene glycol, (**E**) sodium starch glycolate, (**F**) drug loaded formulation and (**G**) placebo.

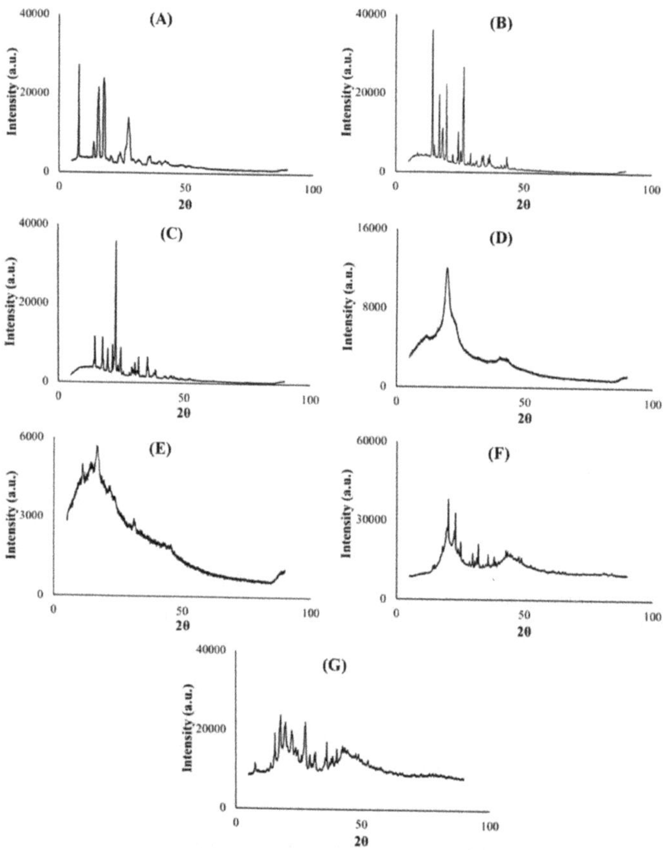

Figure 5. X-ray diffractograms of (**A**) PZA, (**B**) citric acid, (**C**) xylitol, (**D**) polyvinyl alcohol polyethylene glycol, (**E**) sodium starch glycolate, (**F**) placebo and (**G**) drug loaded formulation.

2.5.2. Crystallinity

Powder X-ray diffractometry was used to determine the physical, solid state changes and crystallinity [50] of the model drug PZA, pharmaceutical excipients, optimized drug loaded and placebo formulations. The diffractograms displayed in Figure 5A–C depict the crystalline nature of pyrazinamide, citric acid and xylitol correspondingly with distinct, high intensity peaks (>30,000 a.u.). Co-polymer polyvinyl alcohol polyethylene glycol is semi-crystalline (characteristic of it make-up), typified with broader band, less intense (<16,000 a.u.) crystalline peak observed at 19.2 (2θ) and multiple blunt regions (Figure 5D), while sodium starch glycolate exhibited a low intensity (<6000 a.u.), undefined topography (Figure 5E), which can be related to its amorphous character, largely contributing to the non-crystalline domains within the entire molecular structure of the optimized formulations. The optimized drug and placebo formulation diffractograms are considered a sum of individual diffractions generated by the component excipients and drug [26]. When compared with its drug loaded counterpart, the placebo diffractogram (Figure 5F) also exemplified the presence of recurrent amorphous and higher intensity crystalline regions (>30,000 a.u.) within its matrix which further validates findings documented from the TGA analysis where the placebo was shown to require higher temperatures to exhibit complete decomposition compared to the drug loaded formulation (Figure 4F,G). The optimized PZA formulation diffractogram (Figure 5G) irregularly exhibited slightly broader, sparse and lower intensity peaks (<30,000 a.u.) plus an expanse of blunt regions both associated with its crystalline and amorphous states respectively attributed to the excipients and PZA constituents. The presentation of the amorphous segments within the drug formulation structure appears substantial and may be linked with the rapid disintegration characteristic observed upon hydration, as the amorphous state is known to improve the solubility of drug delivery systems, thereby enhancing dissolution and drug absorption within the body [38,41,49]. The crystalline PZA appeared transformed into a partially amorphous state evidenced by the relative loss of its crystalline peaks intensity, particularly that identified at 8.03 (2θ) which was at about 30,000 a.u. and reduced to 10,000 a.u. in the optimized formulation diffractogram (Figure 5A). These findings also concur with the DSC and XRD analytical outputs.

2.5.3. Structural Interpretation

Fourier transform infrared spectroscopy was performed on all pure excipients, PZA, optimized drug loaded and placebo formulations to determine any drug–excipient interactions, and produced spectra are presented in Figure 6. The analysis was focused on identifying vibrational peaks that depict the presence of particular functional groups in the pure drug and pharmaceutical excipients (Figure 6A–G). Pinpointing xylitol peaks related to O-H stretching were observed at 3354 cm^{-1} and 3284 cm^{-1}; an intense C-H peak at 1418 cm^{-1} (Figure 6A) while C-H and O-H peaks observed at 2916 cm^{-1} and 3438 cm^{-1}, respectively, were specific for sodium starch glycolate xylitol peaks (Figure 6B). Typifying citric acid C-O-H peak was noted at 1372 cm^{-1}, C-O vibration at 3282 cm^{-1} and O-H bending at 1172 cm^{-1} (Figure 6C). Specific peaks documented at 1090 cm^{-1}, 2902 cm^{-1} and 3296 cm^{-1} correspond to C-O, C-H and O-H stretching, respectively; C-H bend at 1448 cm^{-1}, C-H rock at 848 cm^{-1} and -C-C- vibration at 1086 cm^{-1} signify co-polymer polyvinyl alcohol polyethylene glycol—Kollicoat® IR (Figure 6D) [51–53]. Of note are peaks representing N-H stretch at 3148cm^{-1}, 3292 cm^{-1}, 3388 cm^{-1} and 3408 cm^{-1}, C-N (ring, stretching) at 1680 cm^{-1}, C-C ring stretching at 1436 cm^{-1}, C=N at 1162 cm^{-1}, C=O at 786 cm^{-1} and C=C at 1704 cm^{-1}, which are key to its structural make-up were logged for pure PZA (Figure 6E) [54]. The optimized PZA loaded formulation spectra (Figure 6F) displayed distinctive peaks, appearing within identical vibrational frequency range obtained for PZA but with slight shifts due to the presence of polymeric and non-polymeric additives. Peaks at N-H = 3290 cm^{-1}, 3406 cm^{-1} and 3426 cm^{-1}, C-O = 1680 cm^{-1} and 1022 cm^{-1}, C-C (ring, stretching) = 1434 cm^{-1}, typical of the standard PZA chemical backbone, were identified. A presentation of functional groups specific to the pure drug and/or excipients individually reflect in the generated spectra for the optimized drug formulation and exposes the level of structural compatibility amongst these components. This signifies that despite the relative amorphization noted for formulated PZA, typifying structural peaks remained noticeably unaltered, thereby demonstrating the absence of chemically disruptive interactions during the formulation development process. The placebo FTIR spectra (Figure 6G) showed C-O stretching peaks at 1064 and 1008 cm^{-1} including C-H stretching peak seen at 2916 cm^{-1} accounting for the presence of both poly vinyl alcohol polyethylene glycol and sodium starch glycolate, C-O-H peaks were detected at 1420 cm^{-1} and 1374 cm^{-1} representing citric acid and, vibrational frequencies at 3418 cm^{-1}, 3382 cm^{-1} and 3282 cm^{-1} are characteristics of O-H peaks for associated with xylitol, polyvinyl alcohol polyethylene glycol and citric acid chemical backbone structures. These show that the placebo is a homogenous physical blend of the different excipients. Overall, the outcomes of the FTIR analysis indicated that the drug and excipients were well incorporated, compatible and stable with absence of any destructive intermolecular or intramolecular interactions.

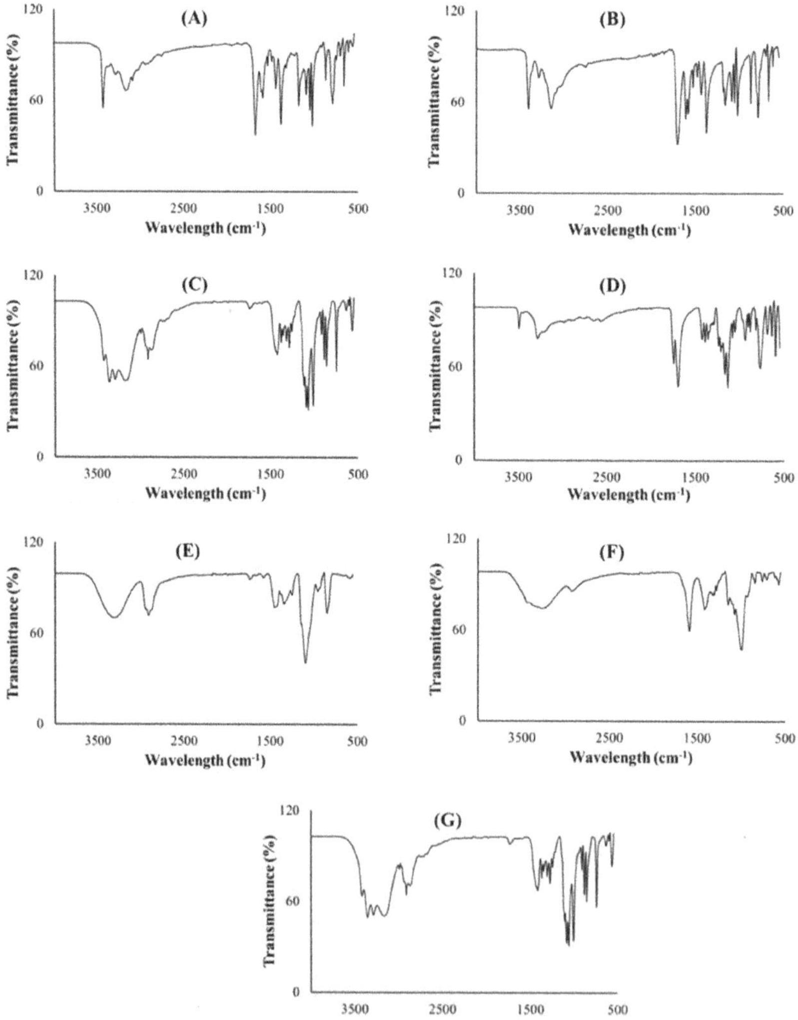

Figure 6. FTIR spectra of (**A**) xylitol, (**B**) sodium starch glycolate, (**C**) citric acid, (**D**) polyvinyl alcohol polyethylene glycol, (**E**) PZA, (**F**) drug loaded formulation and (**G**) placebo.

2.5.4. BET Surface Area and Porosity

The surface areas of optimized drug loaded and placebo formulations were obtained by employing the theory of Brunauer–Emmett–Teller (BET) on nitrogen adsorption isotherms generated from the sample surface measured at 77 K. Typical isotherms are illustrated in Figure 7A,B, respectively. Drug loaded and placebo isotherms show similar trends at relative pressures of 0.2000–0.4000, revealing the existence of micropores in both formulations [55]. Specifically, the micropore volume of the placebo and drug loaded formulations were 0.00001 and 0.00003 cm^3/g, respectively, and the presence of PZA within the matrix seems to increase the pore volume according to the numerical data. The specific BET surface area of drug loaded formulation (Figure 7C) and placebo (Figure 7D) were 0.0015 and 0.0753 m^2/g, while the single point surface area measured 0.0020 and 0.1408 m^2/g, respectively. The surface areas for the drug loaded sample is smaller than that of the placebo. This can be associated with the presence of successfully attached PZA molecules on the optimized formulation matrix. T-plots of the drug loaded (Figure 7D) and placebo (Figure 7E) profiles exhibited similar trends with the same thickness ranging between 0.3000 and 0.5000 nm, which may still be related to the finding that the PZA molecules are well incorporated into the formulation matrix, thus not changing its thickness either PZA-free or PZA-loaded.

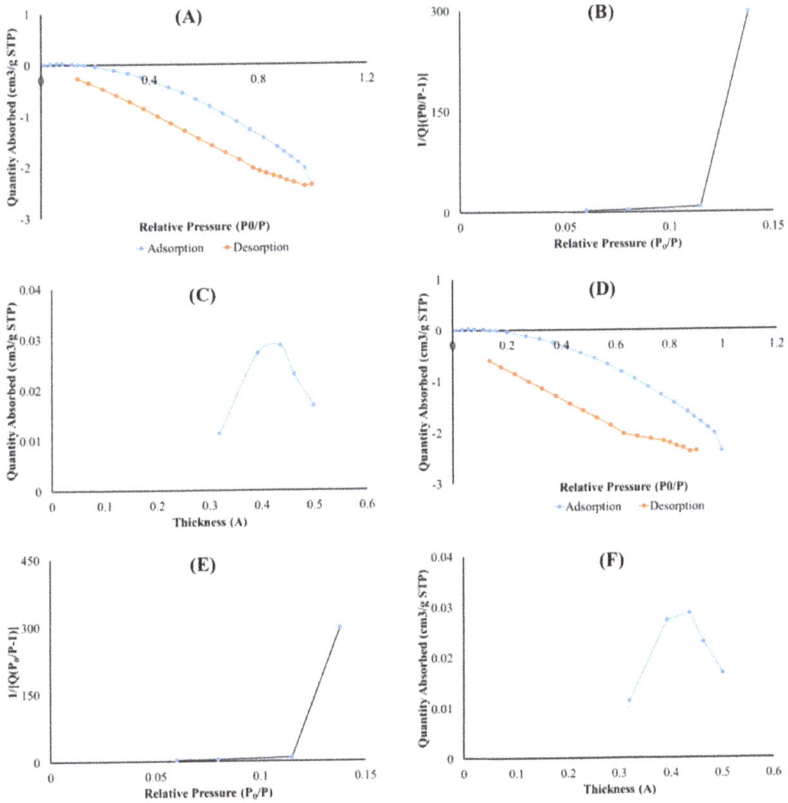

Figure 7. Graph presentation of Isotherm liner plot of (**A**) drug loaded and (**B**) placebo, BET surface area plot for (**C**) drug loaded and (**D**) placebo and t-plot of (**E**) optimized drug loaded and (**F**) placebo.

2.6. Typifying Surface Morphological Features and Transitioning with Hydration

Scanning electron microscopy was used to examine the differences in the surface morphologies of the anhydrous PZA, optimized drug loaded and placebo formulations as shown in Figure 8A–C, respectively. The pure PZA micrograph (Figure 8A) displayed irregular sized, rod shaped particles with defined edges confirming the crystalline nature of the PZA molecules [47,50]. The optimized PZA formulation surface micrograph (Figure 8B) revealed a uniform distribution of the drug molecules embedded throughout the carrier matrix, whereas the placebo (Figure 8C) showed a relatively plane topography with undulating segments confirming the absence of PZA molecules.

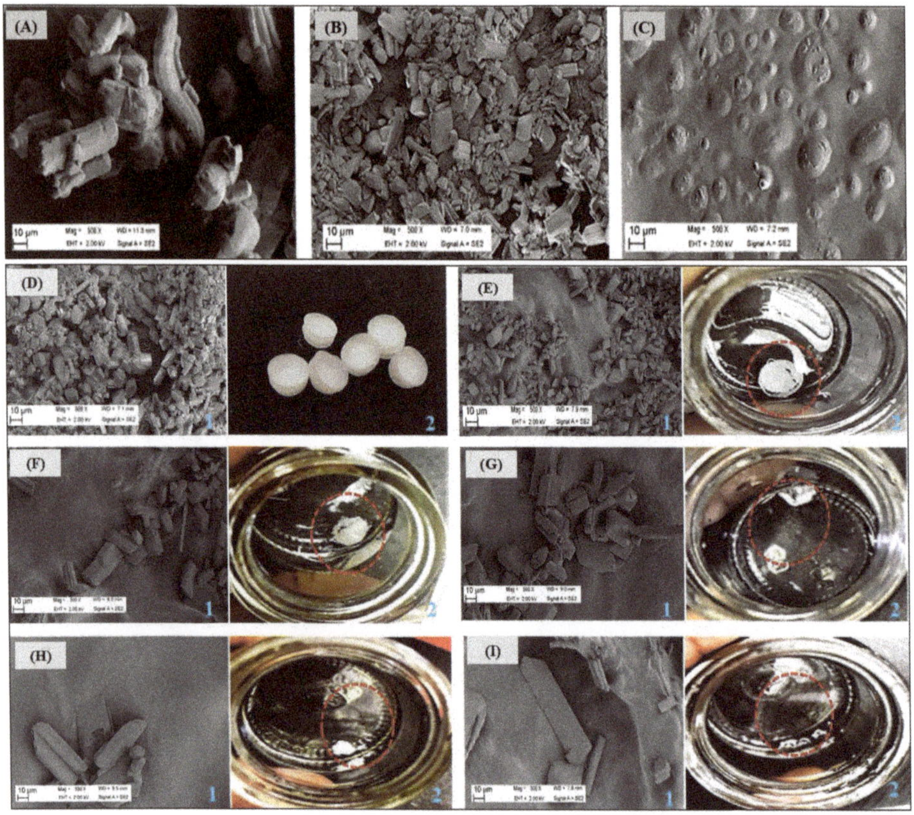

Figure 8. Scanning electron micrographs of (**A**) pure PZA, (**B**) optimized PZA loaded formulation, (**C**) placebo formulations. Changes are visible in the surface topographies of the unhydrated and hydrated optimized drug loaded formulation (**D**) before—0 s and after (**E**)10, (**F**) 30, (**G**) 60, (**H**) 90 and (**I**) 120 s of being in contact with simulated saliva under biorelevant conditions (pH 6.8; 37 ± 0.1 °C). Images were captured as SEM micrographs at magnification 500× (refer to the blue label "1") and high resolution photographs (refer to blue label "2") at each time point. The circular, red dotted lines shown in the photograph images (labelled "2") represent left over formulation fragments after hydration.

Furthermore, time-dependent disintegration patterns of the optimized drug loaded formulation in preheated simulated saliva (pH 6.8; 37 ± 0.1 °C) was investigated using scanning electron microscopy and digital photography. Briefly, images were taken at 0 s (before hydration) and 10, 30, 60, 90 and 120 s after placement in simulated saliva. Captured micrographs and photographs showing changes in formulation surface morphology as it collapsed upon hydration and subsequently released PZA molecules over time are illustrated side-by-side (labelled 1 and 2) in Figure 8D–I. At time 0 s before hydration (Figure 8D), formulation matrix morphology was intact and displayed a wide coverage with PZA molecules (defined edge particles as per micrograph) as also discussed above. As formulation wetting progressed with time, gradual matrix collapse and breakdown occurred, and progressive migration of embedded PZA molecules ensued as a result of matrix disintegration (Figure 8D–I). Initial matrix collapse and drug molecule (visible as embedded particles with defined edges) migration into simulated saliva started at about 10 s (Figure 8E) and was more pronounced at 30 s (Figure 8F), followed by visible matrix erosion, dissolution and continuous outward movement of incorporated drug molecules as time elapsed (Figure 8G–I). These outcomes agree with the low disintegration time (34.98 s) recorded for the drug loaded formulation and is considered an indication of the desired rapid matrix fragmentation, which makes it an attractive and potentially suitable orodispersible delivery system for use in children. In essence, we were able to use the information provided from visualizing microscopic disintegration processes to validate macroscopic level formulation breakdown events.

2.7. Organoleptic Properties of Optimized Drug Loaded Formulation

Preliminary evaluation of organoleptic properties was based on color, texture (general appearance) and acceptability. Results showed that the orodispersible formulations placed on the tongue dispersed quickly—under 60 s in the presence of saliva (required no water for swallowing) and produced a generally satisfying taste as described by the panel. On the average, the volunteers rated the formulation 3.5, implying that the remaining bitterness was minor and mostly considered adequately taste masked/tasteless by them. The panelists also described the formulation color as acceptable, texture as satisfactory and easy to handle. Outcomes of this preliminary qualitative investigation makes the developed orodispersible preparation potentially attractive for pediatric use.

2.8. Cytobiocompatibility Evaluation

In vitro cytotoxicity assay was performed to determine the potential biocompatibility of the optimized drug loaded orodispersible formulation relative to the placebo and pure PZA using hepatocyte cell line (Hep2G) as a model. HepG2 cell line is commonly used to examine the toxicity of antitubercular drugs and respective formulations [56–58]. Cell viability was measured with the 3-(4,5-dimethylthiazol-2-yl) 2,5-diphenyl tetrazolium bromide (MTT) and neutral red (NR) assays at different xenobiotic concentrations ranging from 0.0005—5 mg/mL over a treatment period of 24 h. Cytotoxicity levels were presented as mean percentage cell viability and standard error of mean (SEM) for both MTT and NR assays (Figure 9). For the MTT assay, it was observed that cell viability generally increased as sample concentration decreased (Figure 9A–C). The optimized drug loaded preparation (Figure 9A) showed cell viability 60% and higher for all test concentrations while the placebo formulation displayed lowest cell viability <20% at 5 mg/mL, which consistently increased between 0.5 and 0.005 mg/mL and then insignificantly dropped to 77.09% at the lowest concentration of 0.0005 mg/mL (Figure 9B). Pure pyrazinamide, on the other hand (Figure 9C), expressed highest cell proliferation at lowest concentration 0.0005 mg/mL (96.29%) and the reverse at 5 mg/mL (32.40%). A combination of PZA and excipients in the formulation seemed to promote biocompatibility and cell growth in comparison to outputs captured for the placebo and PZA only. Interestingly, the neutral red analyses showed no cytotoxicity for all three test samples with values majorly greater than 100% for all concentration levels (Figure 9D–F). The only slight decrease in cell viability was noted for PZA at 5 mg/mL and it was statically insignificant. Hence, the NR uptake assessment revealed that drug formulation, placebo and PZA supported cell division and growth, an indication of cytobiocompatibility. These findings may be as a result of the differences in the biochemical reactions of both assays. The MTT assay is based on cellular respiration or mitochondrion cell metabolic activities while the NR analysis measures dye uptake and concentration within the lysosomes thus measuring staining capacity of live cells [3,59]. Consequently, a reduction in MTT-based cell viability can represent a decrease in metabolic activity. The fact that similar trends are not identified for the NR assay at all test concentrations, the cells can be considered to exhibit cytostatic effects at some point. This may mean that the introduction of the test compounds may have inhibited cell growth but does not necessarily promote cell destruction or death—as it is with the case of cytotoxicity. Identified cellular responses resulting from cell exposure to test samples can be termed as dose-dependent and two-phased, an occurrence related with hormesis which, is a two-way adaptive reaction of cells (biological systems) to external stress such as xenobiotics, environmental changes (e.g., pH, temperature) [60]. For both implemented assays, the optimized PZA orodispersible formulation demonstrated no significant reduction in cell viability.

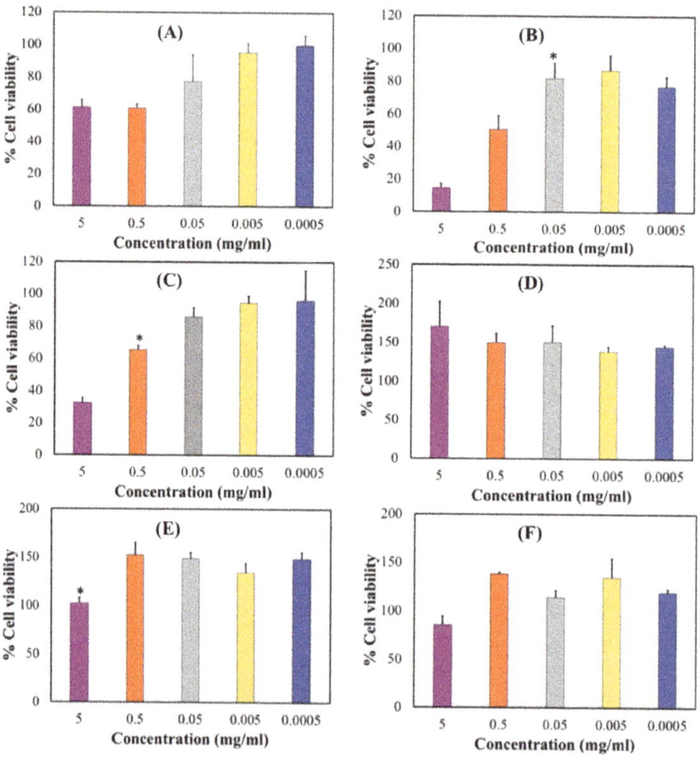

Figure 9. HepG2 cell biocompatibility levels measured by MTT analysis (**A**) drug loaded formulation, (**B**) placebo and (**C**) pyrazinamide and, neutral red assay (**D**) drug loaded formulation, (**E**) placebo and (**F**) pyrazinamide. Results represent mean ± SEM and statistical significance ($p < 0.05$) indicated with an asterisk (*).

2.9. *Drug Formulation Stability under Changing Storage Conditions*

Pyrazinamide formulation stability under varying environmental storage conditions was evaluated over 12 weeks. Briefly, drug formulations were placed in airtight, glass jars containing desiccant bags and stored in: (a) a dark enclosure (23 ± 3 °C/65 ± 5% RH), (b) refrigerator (4 ± 2 °C) and (c) under regular room conditions (24 ± 3 °C/70 ± 5% RH). Tests were performed in triplicate per storage condition and the stability indicators quantified were inner and outer diameters, disintegration time, dissolution pH, weight, and drug content using earlier described methods. Results were reported as average ± standard deviation. Freshly prepared control formulations were tested immediately (time = 0 weeks) and stability indicators recorded in three replicates. At the 12-week time-point, samples stored in dark enclosures (23 ± 3 °C/65 ± 5% RH) and refrigerator (4 ± 2 °C) retained their physical shape, color and showed minimal variation in stability indicators compared to values recorded at the starting point. Samples stored under regular room conditions (24 ± 3 °C/70 ± 5% RH) with fluctuating light exposure were considerably unstable, evidenced by their physical discoloration and documented stability indicators relative to the control samples (Table 3). Summarily, the suggested storage conditions for the optimized PZA containing orodispersible pharmaceutical formulation would be in airtight vessels containing desiccant bags kept away from direct or fluctuating light sources and under ambient or refrigerator settings.

Table 3. Stability indicators recorded under different storage conditions.

Stability Indicators	0 Weeks	Varying Storage Conditions at 12 Weeks		
		I *	II *	III *
Mass (mg)	57.500 ± 0.002	56.100 ± 0.472	58.600 ± 0.321	58.530 ± 1.531
Outer diameter (mm)	11.000 ± 0.426	11.000 ± 0.577	11.000 ± 0.577	11.667 ± 0.577
Inner diameter (mm)	10.000 ± 0.520	10.000 ± 0.577	10.000 ± 0.577	10.667 ± 0.577
Disintegration time (min)	0.583 ± 0.050	0.533 ± 2.510	0.467 ± 3.050	0.483 ± 1.154
Dissolution pH	6.900 ± 0.250	7.290 ± 0.0.015	7.640 ± 0.071	7.633 ± 0.041
% drug content	101.000 ± 2.030	88.000 ± 15.530	84.000 ± 3.020	33.333 ± 10.408
Colour change	None	None	None	Yes

* **Note**: **I**—refrigerator (4 ± 2 °C); **II**—dark enclosure (23 ± 3 °C/65 ± 5% RH); **III**—regular room conditions (24 ± 3 °C/70 ± 5% RH).

3. Materials and Method

3.1. Materials

Pyrazinamide, citric acid, sodium starch glycolate (Primojel®), xylitol, Dulbecco's modified Eagle's medium (DMEM), fetal bovine serum (FBS), L-glutamine, non-essential amino acids, penicillin/streptomycin, disodium hydrogen phosphate, potassium dihydrogen phosphate, sodium chloride, 3-(4,5-dimethylthiazol-2-yl) 2,5-diphenyl tetrazolium bromide (MTT) and neutral red (NR) cell viability assay were purchased from Sigma Aldrich (St. Louis, MO, USA). Copolymer polyvinyl alcohol-polyethylene glycol (Kollicoat® IR) was procured from BASF (Ludwigshafen, Germany). Hepatocyte cell line (HepG2) was purchased from American Tissue Culture Collection (ATCC) (Manassas, VA, USA). All other chemicals employed were of analytical grade and used as received.

3.2. Experimental Design

3.2.1. Constructing the Box Behnken Design Template

The systematic preparation and optimization of the PZA loaded formulation was based on a 4-factor, 3-level Box Behnken experimental design template, a response surface methodology (RSM), constructed utilizing the Minitab® 18 Statistical Software (Minitab LLC, State College, PA, USA). The independent variables were the formulation excipients namely polyvinyl alcohol polyethylene glycol (X_1), sodium starch glycolate (X_2), citric acid (X_3), xylitol (X_4). 3-levels of the independent variables referred to as lower (−1), midpoint (0) and upper (+1) limits were selected for the construction of the design template as represented in Table 1. The dependent variables or responses were parameters key to the performance of the formulation and these included disintegration time (Y_1), dissolution pH (Y_2) and formulation weight (Y_3). Factor level selection for each excipient was set on their ability to produce stable orodispersible formulations, which was based upon the one-variable-at-a-time approach (OVAT) [38,61]. The OVAT approach was implemented by changing one variable per time while keeping the others constant so as to determine the influence exhibited by each excipient. Accordingly, the Box Behnken design template generated 27 possible combinations (F1–F27) with 3 replicates at central points to minimize errors as presented in Table 4 [62]. Model estimation and significance level were executed using the analysis of variance (ANOVA) where p-values below 0.05 indicated statistical significance and correlation coefficient (R) closest to one (>0.9) was selected because of complexities associated with quadratic experimental design templates (Table 3).

Table 4. Independent variables employed for the Box Behnken design template.

Independent Variables	Levels		
	−1	0	+1
X_1: Polyvinyl alcohol polyethylene glycol (g)	1.000	2.000	3.000
X_2: Sodium starch glycolate (g)	0.000	0.500	1.000
X_3: Citric acid (g)	0.025	0.063	0.100
X_4: Xylitol (g)	0.300	0.650	1.000

3.2.2. Preparation of Orodispersible Formulations

Pyrazinamide loaded orodispersible formulations were prepared using the solvent casting technique [27,40,41]. Each formulation consisted of different amounts of polyvinyl alcohol polyethylene glycol, sodium starch glycolate, citric acid and xylitol based on the design template detailed in Table 5. As a result, 27 orodispersible formulations were prepared with each, containing a fixed quantity of pyrazinamide which equaled 500 mg per formulation. Briefly, for every orodispersible formulation variants, all excipients (factor levels) and drug were carefully weighed on a calibrated analytical balance (AS220.R2 Radwag Wagi Electroniczone, Radwag, North Miami Beach, FL, USA) and added to 20 mL deionized water under continuous stirring (Digital Hotplate Stirrer, Model H3760-HSE; Lasec; Ndabeni, Cape Town, South Africa) at 500 rpm over 60 min at 37 ± 0.1 °C until a homogeneous slurry was formed. The homogeneous mixture was left to cure in an airtight and dark environment until all air bubbles were visibly absent. Next, specific amounts (required to produce 20 films per formulation variant) of the cured slurry was filled into specialized, hollow plastic molds and then placed into a Labcon forced air circulation incubator (Model FSIH4, Krugersdorp, Gauteng, South Africa) until dried to constant weight at 25 ± 0.5 °C over 24 h. The resulting drug loaded formulations were then appropriately stored away in airtight, opaque vials for further testing.

Table 5. Box Behnken design template.

Formulation	X_1	X_2	X_3	X_4
F1	3.0000	0.5000	0.0250	0.6500
F2	2.0000	1.0000	0.0625	1.0000
F3	3.0000	1.0000	0.0625	0.6500
F4	1.0000	0.5000	0.1000	0.6500
F5	3.0000	0.5000	0.1000	0.6500
F6 *	2.0000	0.5000	0.0625	0.6500
F7	1.0000	1.0000	0.0625	0.6500
F8	1.0000	0.5000	0.0625	0.3000
F9 *	2.0000	0.5000	0.0625	0.6500
F10	2.0000	0.5000	0.1000	0.3000
F11	2.0000	0.0000	0.0250	0.6500
F12	2.0000	1.0000	0.0625	0.3000
F13	2.0000	0.5000	0.0250	1.0000
F14	1.0000	0.0000	0.0625	0.6500
F15	3.0000	0.5000	0.0625	0.3000
F16	2.0000	1.0000	0.0250	0.6500
F17	2.0000	0.0000	0.0625	1.0000
F18	2.0000	0.0000	0.0625	0.3000
F19	2.0000	0.0000	0.1000	0.6500
F20	2.0000	0.5000	0.1000	1.0000
F21	3.0000	0.5000	0.0625	1.0000
F22	3.0000	0.0000	0.0625	0.6500

Table 5. *Cont.*

Formulation	X_1	X_2	X_3	X_4
F23 *	2.0000	0.5000	0.0625	0.6500
F24	2.0000	0.5000	0.0250	0.3000
F25	2.0000	1.0000	0.1000	0.6500
F26	1.0000	0.5000	0.0625	1.0000
F27	1.0000	0.5000	0.0250	0.6500

Note: * indicates the experimental design center points; X_1 = Polyvinyl alcohol polyethylene glycol (Kollicoat® IR); X_2 = Sodium starch glycolate; X_3 = Citric acid; X_4 = Xylitol. Each orodispersible film variant blend (i.e., F1–F27) contained 20 mL deionized water as solvent, 500 mg pyrazinamide as model drug and produced an average of 20 orodispersible film formulations per variant blend meaning that a single film formulation was loaded with approximately 25 mg pyrazinamide.

3.2.3. Weight Determination for the Matrices

Each prepared orodispersible formulation (F1–F27; Table 2) was weighed using a calibrated analytical balance (AS220.R2; Radwag Wagi Electroniczone, Radwag, North Miami Beach, FL, USA). For each measurement, three independent samples were weighed and mean weight ± standard deviation was calculated and recorded.

3.2.4. In Vitro Disintegration Time and Dissolution pH of the Matrices

The in vitro disintegration time of the 27 experimental design orodispersible formulations was measured utilizing a modified petri dish method [63,64]. Disintegration time represents the specific period when formulation matrix collapse begins [38]. The disintegration time was determined visually using a dual-display digital stopwatch (Fotronic Corporation, Melrose, MA, USA). In this case, each sample was placed in 5 mL of pH 6.8 simulated saliva solution contained in a glass vial and placed in the shaking water bath (ST 30, NÜVE, Akyurt, Ankara, Turkey) maintained at 37 ± 0.1 °C and 10 rpm to mimic the oral cavity [26]. The vial was swirled after every 10 seconds and physical appearance of the formulation was consistently observed for any dimensional changes [28]. The simulated saliva was prepared by dissolving 2.38 g disodium hydrogen phosphate, 0.19 g potassium dihydrogen phosphate and 8.00 g sodium chloride in a liter of distilled water [65]. In vitro disintegration time was recorded at the point when the sample started breaking apart. Thereafter, test samples were allowed to dissolve completely to form a homogenous solution and dissolution pH recorded using a pH meter (GmbH 8603, Mettler Toledo, Sonnenbergstrasse, Schwerzenbach, Switzerland) [40,66]. All the measurements were done in three replicates.

3.3. *Formulation Optimization*

The main objective of the statistical design approach was to develop an optimal pyrazinamide loaded orodispersible formulation. After generating a full quadratic polynomial regression which connected dependent with independent variables from the Box-Behnken design template, experimental outputs were fitted within set limits for predicting the optimal orodispersible formulation. Selection and analyses of optimized levels were performed using the Minitab® 18 statistical software by simultaneously applying specific constraints on the dependent variables namely, disintegration time, dissolution pH and formulation weight, as presented in Table 6. Accuracy and efficiency of the statistical optimization process was measured using the desirability function in which case a value closest to one is indicative of precision. To validate the experimental design approach, the optimized orodispersible formulation was prepared in triplicate, dependent variables measured and obtained values were compared to the predicted values. Thereafter, more optimized drug loaded and placebo formulations were prepared for additional in vitro characterization and testing.

Table 6. Model summary of optimization constrains and statistical significance of the selected response parameters.

Dependent Variables	Constrain	Lower	Upper	Goal	R	*p*-Value
Y_1: Disintegration time (minutes)	Minimize	0.500	0.700	0.600	0.910	0.002
Y_2: Disintegration pH	Target	6.900	7.100	7.000	0.940	0.001
Y_3: Formulation weight (mg)	Minimize	50.00	70.00	60.00	0.901	0.020

3.4. Physical Properties of the Optimized Orodispersible Formulation

3.4.1. Weight Determination

The optimized formulation weight was measured in triplicate using a calibrated analytical balance as previously described.

3.4.2. Measurement of Inner and Outer Diameter

The inner and outer diameter of the optimized formulation was manually measured in triplicate using a centimeter calibrated precision ruler.

3.4.3. Disintegration Time and Dissolution pH

The time elapsed at the onset of in vitro disintegration and the media pH after complete formulation dissolution was quantified using methods already detailed above.

3.5. Drug Content Analysis

Pyrazinamide loaded and placebo optimized formulations of about 12 × 10 mm dimension were separately dissolved in 100 mL of simulated saliva contained in an Erlenmeyer flask. The resulting aqueous mixture was placed on a digital hotplate magnetic stirrer (Model H3760-HSE; Lasec; Ndabeni, Cape Town, South Africa) set at 37 ± 0.1 °C and 500 rpm. The samples were visually monitored until a complete clear solution was formed. Subsequently, 1 mL of the clear solution was appropriately diluted in simulated saliva and passed through the 0.45 µm nylon syringe filter (Whatman®, GD/X syringe filters, Sigma Aldrich, Johannesburg, South Africa). The placebo formulation was also subjected to the same dilution and filtration processes as the drug loaded samples and used as blank measurements to nullify background absorbance associated with included excipients. Filtrates collected from both drug loaded and placebo samples were then separately analyzed by measuring absorbance using a UV/VIS spectrophotometer (Nanocolour® UV/VIS, Macherey Nagel, Separations, Bellville, Cape Town, South Africa) set at a λ_{max} of 268 nm, specific for PZA [26]. The final absorbance measurements obtained from this differential computation were fitted into a linear calibration curve ($y = 654.34\ x$; $R^2 = 0.96$) to obtain the actual and percentage PZA content of the optimized formulation. All quantifications were performed using three replicate samples.

3.6. Evaluation of In Vitro Drug Release Kinetics

The in vitro drug release experiment was carried out on three separate optimized formulations. Each sample was separately enclosed in lidded glass vials containing 5 mL simulated saliva and the entire contrivance was immersed into a shaking water bath at 37 ± 0.1 °C under gentle agitation of 10 rpm, mimicking the buccal environment. Thereafter, 2 mL sample was collected and replaced with an equal volume of freshly prepared, temperature equilibrated simulated saliva (37 ± 0.1 °C) at different time intervals (10, 30, 60, 90 s and 2, 5, 10, 30, 60 min). The samples were then diluted, filtered using 0.45 µm Whatman® nylon syringe and analyzed with a Nanocolour® UV/VIS spectrophotometer at λ_{max} = 268 nm to detect drug absorbance which was eventually translated into percentage drug release values employing a linear polynomial equation ($y = 654.34\ x$; $R^2 = 0.96$). Furthermore, obtained drug release profile was analyzed employing model dependent methods namely zero-, first-, second-order as well as Higuchi and Korsmeyer–Peppas and Hixon–Crowell [67]. The model of best-fit optimally describing the mechanism of drug release from the optimized orodispersible formulation was selected based on the coefficient of determination (R^2) closest to one. All mathematical fitting was performed using the KinetDS, version 3.0 open source software.

3.7. *Physicochemical Characterization*

3.7.1. Differential Scanning Calorimetry (DSC)

The thermal properties of PZA, all excipients used, optimized drug loaded and placebo were evaluated and compared using a differential scanning calorimeter (DSC, Q2000 DSC, TA Instruments, New Castle, DE, USA). Approximately 6 mg of each sample was placed into a flat bottomed standardized aluminum pan which was directly transferred into the calorimeter for testing purposes. For referencing, an empty aluminum pan was included for each measurement as needed. All test samples were analyzed three times at 10 °C/min^{-1}, temperature range between −65 °C and 300 °C under an inert nitrogen gas flow rate of 25 mL/min. The thermograms obtained were recorded and analyzed.

3.7.2. Thermogravimetric Analysis (TGA)

The drug model PZA, excipients, optimized drug loaded and placebo formulations were assessed using a thermogravimetric analyzer (TGA Q500 V20.13 Build 39, TA Instruments, USA). About 8 mg of each sample was separately placed into platinum pans, heated at a temperature range of 10–400 °C, flow rate of 5 °C/min and maintained under constant nitrogen and air flow set at 40 mL/min and 60 mL/min respectively. The percentage weight loss during each heating cycle was recorded using the TGA universal analysis software. Measurements were completed in triplicate and results expressed as the mean of the three readings.

3.7.3. Evaluation of Structural Transitions

A Fourier transform infrared (FTIR) spectrophotometer (Perkin Elmer Spectrum 100 Series, Beaconsfield, UK) equipped with the Spectrum V 6.2.0 software was utilized for the characterization of PZA, all excipients, optimized drug loaded and placebo formulation samples. The FTIR spectra of each sample were recorded in the transmission mode at a frequency range of 550–4000 cm^{-1}. Each spectrum was an average of 32 scans combined in order to achieve a satisfactory signal-to-noise ratio. In all cases, spectra resolution was maintained at 8 cm^{-1} and the gauge force at 150. The compatibility of the samples was checked and FTIR spectra documented for further analysis.

3.7.4. Surface Area and Porosity Analyses

The surface area and porosity of optimized drug loaded and placebo formulations were quantified utilizing the Brunauer–Emmett–Teller (BET) analyzer (Micromeritics TRISTAR II 3020, Micromeritics, Norcross, GA, USA) employing nitrogen adsorption mechanisms. About 0.3 g of each sample was degassed under a vacuum environment overnight at 40 °C. The specific surface area for each specimen was calculated using the BET method with experimental points fixed at a relative pressure of 0.01–1.

3.7.5. X-ray Diffraction (XRD)

The differences in the crystalline structures of PZA, excipient, drug loaded and placebo formulations were identified using an X'Pert Pro Powder X-ray diffractometer (PANalytical, Westborough, MA, USA). Anode material used was copper based, machine divergence slit was set at 0.38 mm and measurements were performed using a reflection-transmission spinner. Measurement operations were carried out using 1.54 Cu K-alpha (1 and 2) radiation, 45 kV generator voltage and 40 mA tube current. Continuous scanning was performed at 0.026 scan step size and 126.99 s/step between 5° and 90° (2θ).

3.8. *Surface Conformational Transitions of Dry and Hydrated Formulations*

First, the surface morphology of pure PZA, drug loaded and placebo formulations were viewed using the Zeiss Supra 55 SM Scanning Electron Microscope (SEM) (Carl Zeiss, Germany) at a 2 kV accelerating voltage. The samples were cut into small pieces, mounted on aluminum stubs using double sided adhesive carbon tape and then sputter coated with approximately 15 nm chromium using a Quorum T150 ES coater (East Sussex, UK) before imaging.

Afterwards, the changes in the surface geometry of the optimized drug loaded formulation upon hydration under biorelevant conditions, similar to that earlier described for the disintegration analysis, were studied to further corroborate previously observed disintegration and drug release patterns. At predetermined time intervals (10, 30, 60, 90, 120 s), photographs of the observed physical changes were captured, then remnants of the disintegrating formulation were carefully collected and dried to constant weight with a Labcon forced air circulation incubator at 25 ± 0.5 °C. Subsequently, dried remnants collected at the different time points were processed as described above and mounted for viewing on a Zeiss Supra 55 SM Scanning Electron Microscope. Photomicrographs for both dry (whole) and hydrated samples were taken at 500× magnification.

3.9. *Preliminary Organoleptic Evaluation*

A single blinded approach was used to evaluate taste acceptability and physical appearance of optimized drug loaded orodispersible formulation (each containing 25 mg PZA) by human volunteers ($n = 5$) [68,69]. Each volunteer was requested to allow formulations disperse in their mouths and to record the taste of each formulation on the provided chart after some seconds (under a minute) before removing formulation remnants from their mouths without ingestion. All panelists were provided with potable water to thoroughly rinse their mouths of any formulation residue using drinkable water before evaluating another sample (each panelist assessed 3 samples). The bitterness and quality attributes were evaluated using a 4-point hedonic scale with 1 point = very bitter, 2 points = moderate to bitter, 3 points = slightly bitter and 4 points = tasteless/taste masked). An average numerical value indicating the overall acceptability of the formulation was computed [68,69].

3.10. *In Vitro Cytotoxicity Assay*

The PZA, PZA loaded and placebo optimized formulation were employed as samples for investigating the cytobiocompatibility using Hepatocyte cell line (Hep G2 also referred to as ATCC® HB-8065™) was obtained from the American Type Culture Collection (Manassas, VA, USA). Two colorimetric assays were employed to quantify cell viability of the samples namely 3-(4,5-dimethylthiazol-2-yl) 2,5-diphenyl tetrazolium bromide (MTT) and neutral red (NR) cell viability assay.

3.10.1. Cell Culturing and Sample Preparation

The hepatocyte cell line was cultured in Dulbecco's modified Eagle's medium (DMEM), 10% fetal bovine serum (FBS), 1% L-glutamine, 1% non-essential amino acids (NEAA), and 1% penicillin/streptomycin. Tissue culture flasks (75 cm^2) were used to grow the cell in an incubator maintained at 37 °C in 5% of carbon dioxide. The cells were harvested and passaged when they were confluent. Assay samples were dissolved in Dulbecco's modified Eagle's medium (DMEM) with serial dilutions (5, 0.5, 0.005, 0.0005 mg/mL) and prepared samples evaluated using MTT and NR assay detailed below.

3.10.2. MTT Cell Viability Assay

A modified technique outlined by Mosmann (1983) and Vistica (1991) was used for MTT viability assay [57,70]. HepG2 cells were seeded at a density of 40,000 cells/mL in a 96-well plate. The cells were left to be attached overnight, then they were exposed at different concentrations (µg/mL) of the samples. The spent medium was aspirated after 24 h of incubation at 37 °C and substituted in simple DMEM by a 0.5 mg/mL MTT. After another 3 h of incubation at 37 °C, the medium was removed and 200 μL DMSO dissolved the purple formazan crystals. A microplate reader (SpectraMax® Paradigm® Multi-Mode Detection Platform, Molecular Devices LLC, San Jose, CA, USA) measured the absorbance at 540 nm.

3.10.3. Neutral Red Cell Viability Assay

The media was aspired after the 24 h of incubation with the sample products, then adding 20 μL of the neutral red solution (Sigma Aldrich) to each well. The culture was incubated in a humidified chamber at 37 °C for 3 h (5% carbon dioxide). The cells were washed with pre-warmed PBS after incubation, followed by 200 μL of neutral red solubilization solution and left at room temperature for 10 min. A micro plate reader (SpectraMax ® Paradigm® multi-mode detection platform) was used to measure absorbance readings at 540 nm. The cytotoxicity of both the MTT and NR results are reported as a percentage according to the following calculation:

$$\% \text{ Cell viability} = \text{Sample} - \text{Blank/Control} - \text{Blank}) \times 100. \quad (1)$$

3.11. Stability Studies

Environmental stability studies were performed on drug loaded formulations over a period of 12 weeks utilizing selected settings that were intended to simulate everyday use. Generic protocols set by the International Conference on Harmonization (ICH) were considered for the environmental conditions used for this preliminary investigation [71,72]. Samples were kept in airtight, glass jars containing desiccant bags and stored: (a) in a dark enclosure (23 ± 3 °C/65 ± 5% RH), (b) refrigerator (4 ± 2 °C) and (c) under room conditions (24 ± 3 °C/70 ± 5% RH) and tested in triplicate. Formulation weight, disintegration time, drug content uniformity, dissolution pH, inner and outer diameter were selected as indicators for determining the influence of set storage conditions on the physical and chemical stability of these samples. All stability indicators quantified at the end of 12 weeks were compared to measurements conducted at the point when the formulations were freshly prepared (time = 0 weeks).

4. Conclusions

Palatable orodispersible film formulations are ideal for patients with swallowing difficulties such as pediatrics because they are stable and dissolve rapidly within the oral cavity in the presence of saliva, without the need to chew or drink water. This current investigation details the successful preparation, optimization and evaluation of an edible, co-polymeric orodispersible pharmaceutical formulation containing pyrazinamide, a model first line antitubercular agent suitable for use in actively infected children. The orodispersible formulation was manufactured by blending polymeric and non-polymeric excipients with drug molecules in an aqueous milieu coupled with the solvent casting approach. The production and optimization processes were facilitated by a one-variable-at-a-time and high performance Box Behnken experimental sesign approaches. The optimized orodispersible formulation was hollow-shaped, uniformly whitish in color, mechanically robust and bendable enough to withstand safe handling. It disintegrated rapidly (34.98 ± 3.00 s) under biorelevant conditions, maintained a close to neutral surrounding pH of 6.90 ± 0.25 and total matrix dissolution and drug release, an indication of complete drug absorption, occurred at approximately 60 min. Drug release from the optimized formulation followed the Korsmeyer–Peppas mathematical model, showing that drug liberation was controlled by anomalous diffusion coupled with matrix disintegration and erosion mechanisms. Pyrazinamide molecules were well incorporated into the formulation matrix and displayed a high loading capacity (25.02 ± 0.71 mg ≡ 101.13 ± 2.03 $\%^{w}/_{w}$). According to the WHO, a pediatric patient requires an average dose of 35 mg/kg, meaning that multiple films (relative to body weight) may be needed per child; an approach not unusual in TB management with oral or water dispersible tablets.

This may therefore be more usable in under 5-year-old children and, should not pose any choking hazards considering the rapidly disintegrating characteristic of the fabricated pyrazinamide films which does not necessitate the use of water for swallowing. Captured SEM micrographs and digital photographs showed that the drug formulation matrix was micro-structured and also confirmed its quick disintegration sequence. The orodispersible drug preparation was thermodynamically and environmentally stable under specific storage conditions based on findings from physicochemical characterization (TGA, DSC, FTIR, XRD, BET analyses) and stability testing processes. Preliminary organoleptic and cell toxicity enquiries presented the drug formulation as palatable, easy-to-handle and biocompatible under applied test conditions. In conclusion, the orodispersible pharmaceutical formulation developed herein can potentially ease some of the current global challenges associated with the safe administration of TB antibiotics in pediatric patients to aid desirable pharmacotherapeutic outcome. Besides, the carrier matrix designed in this study may be used as is or even modified to accommodate and safely improve the release/absorption of other antitubercular agents for use in children.

Author Contributions: Conceptualization, O.A.A.; Methodology, O.A.A., N.M., J.W.-S.; Software, O.A.A.; Formal Analysis, N.M., O.A.A., J.W.-S.; Investigation, N.M., O.A.A., J.W.-S.; Resources, O.A.A., J.W.-S.; Data Curation, O.A.A., N.M.; Writing—Original Draft Preparation N.M.; Writing—Review & Editing O.A.A., N.M., J.W.-S.; Visualization, O.A.A., N.M.; Validation, O.A.A., N.M., J.W.-S.; Supervision, O.A.A., J.W.-S.; Project Administration, O.A.A.; Funding Acquisition, O.A.A. All authors have read and agreed to the published version of the manuscript.

Funding: This work was funded by the South African National Research Foundation (Grant Number: 113143) and Sefako Makgatho Health Sciences University-DHET Research Development Grant (Grant Number: D112-RDG).

Acknowledgments: The authors thank the National Centre for Nano-structured Materials, Council for Scientific and Industrial Research, Pretoria, South Africa and the Preclinical Drug Development Platform, North-West University, Potchefstroom, South Africa for sample characterization and cytotoxicity testing respectively. We are grateful to Chantelle Baker from the SMU Electron Microscopy Unit, Pretoria, South Africa for her valuable advice and discussions.

Conflicts of Interest: The authors declare no conflict of interest. The funders had no role in the design of the study; in the collection, analyses, or interpretation of data; in the writing of the manuscript, or in the decision to publish the results.

References

1. World Health Organization. 2019. Available online: https://apps.who.int/iris/bitstream/handle/10665/329368/9789241565714-eng.pdf?ua=1 (accessed on 1 May 2020).
2. Swaminathan, S.; Rekha, B. Paediatric tuberculosis: Global overview and challenges. *Clin. Infect. Dis.* **2010**, *50*, S184–S194. [CrossRef] [PubMed]
3. Kumar, V.; Sharma, N.; Maitra, S.S. In vitro and in vivo toxicity assessment of nanoparticles. *Int. Nano Lett.* **2017**, *7*, 243–256. [CrossRef]
4. Fogell, N. Tuberculosis: A disease without boundaries. *Tuberculosis* **2015**, *95*, 527–531. [CrossRef] [PubMed]
5. Izudi, J.; Semakula, D.; Sennono, R.; Tamwesigire, I.K.; Bajunirwe, F. Treatment success rate among adult pulmonary tuberculosis patients in sub-Saharan Africa: A systematic review and meta-analysis. *BMJ Open* **2019**, *9*, 029400. [CrossRef] [PubMed]
6. Peña, M.J.M.; García, B.S.; Baquero-Artigao, F.; Pérez, D.M.; Pérez, R.P.; Echevarría, A.M.; Amador, J.T.R.; Durán, D.G.-P.; Noguera-Julian, A. Tuberculosis treatment for children: An update. *Anales de Pediatría* **2018**, *88*, 52.e1–52.e12. [CrossRef]
7. Graham, S.M. Treatment of paediatric TB: Revised WHO guidelines. *Paediatr. Respir. Rev.* **2011**, *12*, 22–26. [CrossRef]
8. World Health Organization. 2018. Available online: https://www.who.int/tb/publications/global_report/gtbr2018_main_text_28Feb2019.pdf (accessed on 3 May 2019).
9. Newton, S.M.; Brent, A.J.; Anderson, S.; Whittaker, E.; Kampmann, B. Paediatric tuberculosis. *Lancet Infect. Dis.* **2008**, *8*, 498–510. [CrossRef]
10. Tsai, K.S.; Chang, H.L.; Chien, S.T.; Chen, K.L.; Chen, K.H.; Mai, M.H.; Chen, K.T. Childhood tuberculosis: Epidemiology, diagnosis, treatment, and vaccination. *Paediatr. Neonatol.* **2013**, *54*, 295–302. [CrossRef]

11. Whittaker, E.; López-Varela, E.; Broderick, C.; Seddon, J.A. Examining the Complex Relationship between Tuberculosis and other Infectious Diseases in Children: A Review. *Front. Paediatr.* **2019**, *7*, 233. [CrossRef]
12. Pouplin, T.; Phuong, P.N.; Van Toi, P.; Pouplin, J.N.; Farrar, J. Isoniazid, pyrazinamide and rifampicin content variation in split fixed-dose combination tablets. *PLoS ONE* **2014**, *9*, e102047. [CrossRef]
13. Piñeiro Pérez, R.; Santiago García, B.; Fernández-Llamazares, C.M.; Baquero Artigao, F.; Noguera Julian, A.; Mellado Peña, M.J. The challenge of administering anti-tuberculosis treatment in infants and pre-school children. pTBred Magistral Project. *Anales De Pediatría* **2016**, *85*, 4–12. [CrossRef] [PubMed]
14. World Health Organization. 2016. Available online: https://apps.who.int/iris/handle/10665/250441 (accessed on 7 May 2019).
15. TB Alliance. 2020. Available online: https://www.tballiance.org/sites/default/files/assets/TB-Alliance_Child-Friendly-Medicines_pamphlet_2020.pdf (accessed on 12 February 2020).
16. Sosnik, A.; Carcaboso, Á.M.; Glisoni, R.J.; Moretton, M.A.; Chiappetta, D.A. New old challenges in tuberculosis: Potentially effective nanotechnologies in drug delivery. *Adv. Drug Deliv. Rev.* **2010**, *62*, 547–559. [CrossRef] [PubMed]
17. Svensson, E.M.; Yngman, G.; Denti, P.; McIlleron, H.; Kjellsson, M.C.; Karlsson, M.O. Evidence-based design of fixed-dose combinations: Principles and application to paediatric anti-tuberculosis therapy. *Clin. Pharmacokinet.* **2018**, *57*, 591–599. [CrossRef] [PubMed]
18. Apis, V.; Landi, M.; Graham, S.M.; Islam, T.; Amini, J.; Sabumi, G.; Mandalakas, A.M.; Meae, T.; du Cros, P.; Shewade, H.D.; et al. Outcomes in children treated for tuberculosis with the new dispersible fixed-dose combinations in Port Moresby. *Public Health Action* **2019**, *9*, S32–S37. [CrossRef] [PubMed]
19. Suárez-González, J.; Santoveña-Estévez, A.; Soriano, M.; Fariña, J.B. Design and optimization of a child-friendly dispersible tablet containing Isoniazid, Pyrazinamide and Rifampicin for treating Tuberculosis in paediatrics. *Drug Dev. Ind. Pharm.* **2020**, *46*, 309–317. [CrossRef]
20. Martir, J.; Flanagan, T.; Mann, J.; Fotaki, N. Recommended strategies for the oral administration of paediatric medicines with food and drinks in the context of their biopharmaceutical properties: A review. *J. Pharm. Pharmacol.* **2017**, *69*, 384–397. [CrossRef]
21. Craig, S.R.; Adams, L.V.; Spielberg, S.P.; Campbell, B. Paediatric therapeutics and medicine administration in resource-poor settings: A review of barriers and an agenda for interdisciplinary approaches to improving outcomes. *Soc. Sci. Med.* **2009**, *69*, 1681–1690. [CrossRef]
22. Ivanovska, V.; Rademaker, C.M.; van Dijk, L.; Mantel-Teeuwisse, A.K. Paediatric drug formulations: A review of challenges and progress. *Paediatrics* **2014**, *134*, 361–372. [CrossRef]
23. Singh, J.; Philip, A.K.; Pathak, K. Optimization Studies on Design and Evaluation of Orodispersible Peadiatric Formulation of Indomethacin. *AAPS Pharm. Sci. Tech.* **2008**, *9*, 60–66. [CrossRef]
24. World Health Organization. 2017. Available online: https://www.who.int/tb/publications/global_report/gtbr2017_main_text.pdf?ua=1 (accessed on 25 April 2020).
25. Lopez, F.L.; Ernest, T.B.; Tuleu, C.; Gul, M.O. Formulation approaches to paediatric oral drug delivery: Benefits and limitations of current platforms. *Expert Opin. Drug Deliv.* **2015**, *12*, 1727–1740. [CrossRef]
26. Preis, M.; Woertz, C.; Schneider, K.; Kukawka, J.; Broscheit, J.; Roewer, N.; Breitkreutz, J. Design and evaluation of bilayered buccal film preparations for local administration of lidocaine hydrochloride. *Eur. J. Pharm. Biopharm.* **2014**, *86*, 552–561. [CrossRef]
27. Slavkova, M.; Breitkreutz, J. Orodispersible drug formulations for children and elderly. *Eur. J. Pharm. Sci.* **2015**, *75*, 2–9. [CrossRef] [PubMed]
28. Senta-Loys, Z.; Bourgeois, S.; Pailler-Mattei, C.; Agusti, G.; Briançon, S.; Fessi, H. Formulation of orodispersible films for paediatric therapy: Investigation of feasibility and stability for tetrabenazine as drug model. *J. Pharm. Pharmacol.* **2017**, *69*, 582–592. [CrossRef] [PubMed]
29. Ibrahim, H.K.; El-Setouhy, D.A. Valsartan orodispersible tablets: Formulation, in vitro/in vivo characterization. *AAPS Pharm. Sci. Tech.* **2010**, *11*, 189–196. [CrossRef] [PubMed]
30. Singh, B.; Bhatowa, R.; Tripathi, C.B.; Kapil, R. Developing micro-/nanoparticulate drug delivery systems using "design of experiments". *Int. J. Pharm. Investig.* **2011**, *1*, 75. [CrossRef]
31. Bala, R.; Khanna, S.; Pawar, P. Design optimization and in vitro-in vivo evaluation of orally dissolving strips of clobazam. *J. Drug Deliv.* **2014**, *2014*, 1–15. [CrossRef] [PubMed]

32. Zhang, Y.; Mitchison, D. The curious characteristics of pyrazinamide: A review. *Int. J. Tuberc. Lung Dis.* **2003**, *7*, 6–12. [PubMed]
33. Ando, H.; Mitarai, S.; Kondo, Y.; Suetake, T.; Sekiguchi, J.I.; Kato, S.; Mori, T.; Kirikae, T. Pyrazinamide resistance in multidrug-resistant Mycobacterium tuberculosis isolates in Japan. *Clin. Microbiol. Infect.* **2010**, *16*, 1164–1168. [CrossRef] [PubMed]
34. Zhang, Y.; Shi, W.; Zhang, W.; Mitchison, D. Mechanisms of pyrazinamide action and resistance. *Microbiol. Spectr.* **2013**, *2*, 1–12.
35. Hu, L.; Sun, C.; Song, A.; Chang, D.; Zheng, X.; Gao, Y.; Jiang, T.; Wang, S. Alginate encapsulated mesoporous silica nanospheres as a sustained drug delivery system for the poorly water-soluble drug indomethacin. *Asian J. Pharm. Sci.* **2014**, *9*, 183–190. [CrossRef]
36. Becker, C.; Dressman, J.B.; Amidon, G.L.; Junginger, H.E.; Kopp, S.; Midha, K.K.; Shah, V.P.; Stavchansky, S.; Barends, D.M. Biowaiver monographs for immediate release solid oral dosage forms: Pyrazinamide. *J. Pharm. Sci.* **2008**, *97*, 3709–3720. [CrossRef] [PubMed]
37. Mitchison, D.A.; Fourie, P.B. The near future: Improving the activity of rifamycins and pyrazinamide. *Tuberculosis* **2010**, *90*, 177–181. [CrossRef] [PubMed]
38. Adeleke, O.A.; Monama, N.O.; Tsai, P.C.; Sithole, H.M.; Michniak-Kohn, B.B. Combined atomistic molecular calculations and experimental investigations for the architecture, screening, optimization, and characterization of pyrazinamide containing oral film formulations for tuberculosis management. *Mol. Pharm.* **2016**, *13*, 456–471. [CrossRef] [PubMed]
39. Chirehwa, M.T.; McIlleron, H.; Rustomjee, R.; Mthiyane, T.; Onyebujoh, P.; Smith, P.; Denti, P. Pharmacokinetics of pyrazinamide and optimal dosing regimens for drug-sensitive and-resistant tuberculosis. *Antimicrob. Agents Chemother.* **2017**, *61*, 00490–17. [CrossRef]
40. Adeleke, O.A.; Tsai, P.C.; Karry, K.M.; Monama, N.O.; Michniak-Kohn, B.B. Isoniazid-loaded orodispersible strips: Methodical design, optimization and in vitro-in silico characterization. *Int. J. Pharm.* **2018**, *547*, 347–359. [CrossRef]
41. Zhu, Y.; You, X.; Huang, K.; Raza, F.; Lu, X.; Chen, Y.; Dhinakar, A.; Zhang, Y.; Kang, Y.; Wu, J.; et al. Effect of taste masking technology on fast dissolving oral film: Dissolution rate and bioavailability. *Nanotechnology* **2018**, *29*, 304001. [CrossRef]
42. Dixit, R.P.; Puthli, S.P. Oral strip technology: Overview and future potential. *J. Control. Release* **2009**, *139*, 94–107. [CrossRef]
43. Sadoun, O.; Rezgui, F.; G'Sell, C. Optimization of valsartan encapsulation in biodegradables polyesters using Box-Behnken design. *Mater. Sci. Eng. C* **2018**, *90*, 189–197. [CrossRef]
44. Ranch, K.M.; Maulvi, F.A.; Naik, M.J.; Koli, A.R.; Parikh, R.K.; Shah, D.O. Optimization of a novel in situ gel for sustained ocular drug delivery using Box-Behnken design: In vitro, ex vivo, in vivo and human studies. *Int. J. Pharm.* **2019**, *554*, 264–275. [CrossRef]
45. Ciro, Y.; Rojas, J.; Alhajj, M.J.; Carabali, G.A.; Salamanca, C.H. Production and Characterization of Chitosan–Polyanion Nanoparticles by Polyelectrolyte Complexation Assisted by High-Intensity Sonication for the Modified Release of Methotrexate. *Pharmaceuticals* **2020**, *13*, 11. [CrossRef]
46. Cherukuvada, S.; Thakuria, R.; Nangia, A. Pyrazinamide polymorphs: Relative stability and vibrational spectroscopy. *Cryst. Growth Des.* **2010**, *10*, 3931–3941. [CrossRef]
47. Eedara, B.B.; Tucker, I.G.; Das, S.C. Phospholipid-based pyrazinamide spray-dried inhalable powders for treating tuberculosis. *Int. J. Pharm.* **2016**, *506*, 174–183. [CrossRef] [PubMed]
48. Alshehri, S.M.; Park, J.B.; Alsulays, B.B.; Tiwari, R.V.; Almutairy, B.; Alshetaili, A.S.; Morott, J.; Shah, S.; Kulkarni, V.; Majumdar, S.; et al. Mefenamic acid taste-masked oral disintegrating tablets with enhanced solubility via molecular interaction produced by hot melt extrusion technology. *J. Drug Deliv. Sci. Technol.* **2015**, *27*, 18–27. [CrossRef] [PubMed]
49. Reddy, K.A.; Karpagam, S. Cellulose orodispersible films of donepezil: Film characterization and drug release. *Pharm. Chem. J.* **2017**, *51*, 707–715. [CrossRef]
50. Eedara, B.B.; Rangnekar, B.; Sinha, S.; Doyle, C.; Cavallaro, A.; Das, S.C. Development and characterization of high payload combination dry powders of anti-tubercular drugs for treating pulmonary tuberculosis. *Eur. J. Pharm. Sci.* **2018**, *118*, 216–226. [CrossRef] [PubMed]

51. Demitri, C.; Del Sole, R.; Scalera, F.; Sannino, A.; Vasapollo, G.; Maffezzoli, A.; Ambrosio, L.; Nicolais, L. Novel superabsorbent cellulose-based hydrogels crosslinked with citric acid. *J. Appl. Polym. Sci.* **2008**, *110*, 2453–2460. [CrossRef]
52. Santos, M.G.; Bozza, F.T.; Thomazini, M.; Favaro-Trindade, C.S. Microencapsulation of xylitol by double emulsion followed by complex coacervation. *Food Chem.* **2015**, *171*, 32–39. [CrossRef]
53. Salehi, S.; Boddohi, S. Design and optimization of Kollicoat® IR based mucoadhesive buccal film for co-delivery of rizatriptan benzoate and propranolol hydrochloride. *Mater. Sci Eng. C* **2019**, *97*, 230–244. [CrossRef]
54. Gunasekaran, S.; Sailatha, E. Vibrational analysis of pyrazinamide. *Indian J. Pure Appl. Phys.* **2009**, *47*, 259–264.
55. Tang, Q.; Xu, Y.; Wu, D.; Sun, Y. A study of carboxylic-modified mesoporous silica in controlled delivery for drug famotidine. *J. Solid State Chem.* **2006**, *179*, 1513–1520. [CrossRef]
56. Fotakis, G.; Timbrell, J.A. In vitro cytotoxicity assays: Comparison of LDH, neutral red, MTT and protein assay in hepatoma cell lines following exposure to cadmium chloride. *Toxicol. Lett.* **2006**, *160*, 171–177. [CrossRef] [PubMed]
57. Mosmann, T. Rapid colorimetric assay for cellular growth and survival: Application to proliferation and cytotoxicity assays. *J. Immunol. Methods* **1983**, *65*, 55–63. [CrossRef]
58. Elje, E.; Mariussen, E.; Moriones, O.H.; Bastús, N.G.; Puntes, V.; Kohl, Y.; Dusinska, M.; Rundén-Pran, E. Hepato (Geno) Toxicity assessment of nanoparticles in a HepG2 liver spheroid model. *Nanomaterials* **2020**, *10*, 545. [CrossRef] [PubMed]
59. Van Tonder, A.; Joubert, A.M.; Cromarty, A.D. Limitations of the 3-(4, 5-dimethylthiazol-2-yl)-2, 5-diphenyl-2H-tetrazolium bromide (MTT) assay when compared to three commonly used cell enumeration assays. *BMC Res. Notes* **2015**, *8*, 47. [CrossRef] [PubMed]
60. Adeleke, O.A.; Hayeshi, R.K.; Davids, H. Development and evaluation of a reconstitutable dry suspension containing isoniazid for flexible paediatric dosing. *Pharmaceutics* **2020**, *12*, 286. [CrossRef] [PubMed]
61. Jana, A.; Maity, C.; Halder, S.K.; Mondal, K.C.; Pati, B.R.; Mohapatra, P.K.D. Tannase production by Penicillium purpurogenum PAF6 in solid state fermentation of tannin-rich plant residues following OVAT and RSM. *Appl. Biochem. Biotechnol.* **2012**, *167*, 1254–1269. [CrossRef]
62. Fachel, F.N.S.; Medeiros-Neves, B.; Dal Prá, M.; Schuh, R.S.; Veras, K.S.; Bassani, V.L.; Koester, L.S.; Henriques, A.T.; Braganhol, E.; Teixeira, H.F. Box-Behnken design optimization of mucoadhesive chitosan-coated nanoemulsions for rosmarinic acid nasal delivery-In vitro studies. *Carbohydr. Polym.* **2018**, *199*, 572–582. [CrossRef]
63. Garsuch, V.; Breitkreutz, J. Comparative investigations on different polymers for the preparation of fast-dissolving oral films. *J. Pharm. Pharmacol.* **2010**, *62*, 539–545. [CrossRef]
64. Brniak, W.; Maślak, E.; Jachowicz, R. Orodispersible films and tablets with prednisolone microparticles. *Eur. J. Pharm. Sci.* **2015**, *75*, 81–90. [CrossRef]
65. Koland, M.; Charyulu, R.N.; Vijayanarayana, K.; Prabhu, P. In vitro and in vivo evaluation of chitosan buccal films of ondansetron hydrochloride. *Int. J. Pharm. Investig.* **2011**, *3*, 164. [CrossRef]
66. Tomar, A.; Sharma, K.; Chauhan, N.S.; Mittal, A.; Bajaj, U. Formulation and evaluation of fast dissolving oral film of dicyclomine as potential route of buccal delivery. *Int. J. Drug Dev. Res.* **2012**, *4*, 408–417.
67. Wójcik-Pastuszka, D.; Krzak, J.; Macikowski, B.; Berkowski, R.; Osiński, B.; Musiał, W. Evaluation of the Release Kinetics of a Pharmacologically Active Substance from Model Intra-Articular Implants Replacing the Cruciate Ligaments of the Knee. *Materials* **2019**, *12*, 1202. [CrossRef] [PubMed]
68. Saxena, S.; Gautam, S.; Sharma, A. Microbial decontamination of honey of Indian origin using gamma radiation and its biochemical and organoleptic properties. *J. Food Sci.* **2010**, *75*, M19–M27. [CrossRef] [PubMed]
69. Khadra, I.; Obeid, M.A.; Dunn, C.; Watts, S.; Halbert, G.; Ford, S.; Mullen, A. Characterisation and optimisation of diclofenac sodium orodispersible thin film formulation. *Int. J. Pharm.* **2019**, *561*, 43–46. [CrossRef] [PubMed]
70. Vistica, D.T.; Skehan, P.; Scudiero, D.; Monks, A.; Pittman, A.; Boyd, M.R. Tetrazolium-based assays for cellular viability: A critical examination of selected parameters affecting formazan production. *Cancer Res.* **1991**, *51*, 2515–2520. [PubMed]

71. Markens, U.; Asia-Pacific, R.H. Conducting stability studies-recent changes to climatic zone IV. *Life Sci. Tech. Bull.* **2009**, *13*, 1–4.
72. Williams, H.E.; Bright, J.; Roddy, E.; Poulton, A.; Cosgrove, S.D.; Turner, F.; Harrison, P.; Brookes, A.; MacDougall, E.; Abbott, A.; et al. A comparison of drug substance predicted chemical stability with ICH compliant stability studies. *Drug Dev. Ind. Pharm.* **2018**, *45*, 379–386. [CrossRef]

International Journal of
Molecular Sciences

MDPI

Article

Development and Palatability Assessment of Norvir® (Ritonavir) 100 mg Powder for Pediatric Population

John B. Morris [1,*], David A. Tisi [2], David Cheng Thiam Tan [1] and Jeffrey H. Worthington [2]

1 AbbVie Inc., North Chicago, IL 60064, USA; cheng.tan@abbvie.com
2 Senopsys LLC, Woburn, MA 01801, USA; david.tisi@senopsys.com (D.A.T.); jeff.worthington@senopsys.com (J.H.W.)
* Correspondence: John.B.Morris@abbvie.com; Tel.: +1-847-938-4996

Received: 25 February 2019; Accepted: 3 April 2019; Published: 6 April 2019

Abstract: Norvir® (ritonavir) is a Biopharmaceutical Classification System Class IV compound with poor solubility in water (~5 µg/mL) and limited oral bioavailability. Early stage development efforts were focused on an oral solution (OS) which provided reasonable bioavailability but exhibited taste-masking challenges and required the use of solvents with potential pediatric toxicity. Norvir® oral powder, 100 mg (NOP) was developed to replace OS. The objective of this study is to provide an overview of the development of NOP and palatability assessment strategy. Palatability of NOP was assessed using the flavor profile method: (1) As an aqueous suspension dose/response and (2) evaluation with foods. The dose/response sensory analysis indicated that NOP has strong intensity bitterness and burnt aromatics (3 on the 0–3 flavor profile scale) at the clinical dose (100 mg/10 mL) and the recognition threshold was determined to be 0.3 mg/10 mL. To improve palatability, 100 mg/10 mL NOP aqueous suspension was evaluated with foods. Consuming foods high in fat and/or sugar content after NOP administration successfully reduced bitterness to a 1.5 intensity. In summary, NOP provides dose flexibility, enhanced stability, eliminated solvents, and maintains consistent bioavailability, with reduced bitterness and improved palatability via administration with common food products.

Keywords: Norvir®; ritonavir; poorly soluble compound; pediatric; palatability assessment; bioavailability; flavor profile

---

1. Introduction

Norvir® (ritonavir) is an inhibitor of human immunodeficiency virus (HIV) protease and is indicated in combination with other antiretroviral (ARV) agents for the treatment of HIV-1 infection [1,2]. Due to the CYP3A inhibitory capabilities of ritonavir, it is also co-administered at lower doses as a pharmacokinetic (PK) enhancer to increase exposures of other HIV protease inhibitors (PIs) [3]. When used as a PK enhancer, ritonavir is most commonly administered at 100 to 200 mg once or twice daily [4]. As a PK enhancer, ritonavir has become a mainstay in the management of both treatment-naïve and treatment-experienced patients and is typically no longer prescribed as a sole protease inhibitor in antiretroviral regimens today [5–7]. The PK enhancement often allows for a reduction of pill burden, dosing frequency, and food restrictions, while maintaining efficacy [6,8].

Ritonavir is practically insoluble in water (~5 µg/mL), however, this solubility can be enhanced to 1.2 mg/mL at approximately pH 1 (0.1 N hydrochloric acid solution) [9]. Due to the lack of aqueous solubility, ritonavir showed essentially no bioavailability in an animal model (dog) when administered as an unformulated solid in a capsule [9]. Attempts to enhance the oral bioavailability of ritonavir from a solid dosage form by incorporation of surfactants, acids, and other wetting agents failed to increase the bioavailability to greater than ~4%. Similar results were obtained by substituting ritonavir base with salt derivatives of ritonavir. However, bioavailability of 37% in dogs was achieved with a solution formulation containing 5.0 mg ritonavir per mL in a solvent system consisting of 20:30:50 ethanol:propylene glycol:water [2]. Therefore, early formulation development of ritonavir was focused on oral solutions (OS) because of the poor oral bioavailability observed with solid oral dosage forms and the inability to find a solvent system that was compatible with hard or soft gelatin capsules.

OS development efforts concentrated on increasing the drug loading while maintaining oral bioavailability. It was concluded that: a) The relative bioavailability of ritonavir in liquid formulations is inversely proportional to the drug concentration and b) the presence of a surfactant, such as Cremophor EL, enhances the bioavailability of ritonavir in liquid formulations. OS also has a short shelf-life of six months at ambient storage to ensure ritonavir remains in solution (physical stability) while maintaining acceptable levels of degradation (chemical stability).

In addition to bioavailability and stability challenges, OS is known to possess multiple aversive sensory attributes: Basic taste, aromas, trigeminal irritation, and textures, collectively known as "flavor". These include bitterness and aromatic off-notes from the active pharmaceutical ingredient (API) and trigeminal irritation from the solvents. To address these aversive attributes, an identifying flavor system comprised of sweeteners and aromas (peppermint and caramel) was developed.

In a preliminary study, the sweetened/flavored formulation somewhat reduced aversive sensory attributes, but overall, the formulation remained relatively low in palatability. Five dose-administration approaches to further ameliorate the sensory effects of OS were subsequently evaluated, first by a trained adult sensory panel using the flavor profile method to identify the most promising one(s). The results are summarized in Table 1. Chasing liquid dose administration with foods was identified as the most promising approach. Six products were selected for confirmation testing with patients based on differences in form, flavor strength, and mastication characteristics: SnackWells® Fudge Brownie, Freshen-Up® Peppermint Gum, Toast Crackers with Peanut Butter, Oats'n Honey Crunch® Granola Bar, Riesen® Chocolate Chews, and Nutella® Spread. Using a 7-point intensity scale, 74 OS patients rated the strength of the "medicine flavor" remaining in the mouth two minutes after taking the OS (mean score 4.95) and chasing with these six foods. All of the food products tested reduced the intensity of the aversive flavor compared to the OS alone, with mean "medicine flavor" ranging from a score of 4.18 to 1.50. An ideal formulation would not require external vehicles, but when faced with significant formulation technical challenges, this chaser approach was effective [10].

Table 1. Results of approaches to ameliorate the flavor impact of Norvir® 80 mg/mL oral solution.

Approach	Example Products	Results
Pre-coat mouth to dull sensory receptors	Peppermint Patties (trigeminal cooling); Orange Sherbet (thermal cooling)	No reduction in active pharmaceutical ingredient (API) bitterness or burning mouthfeel.
Mix with food/beverages to dilute sensory effects	Chocolate milk (50/50)	No reduction in API bitterness or burning mouthfeel. Produced a larger volume of an equally bitter solution.
Chase with foods/beverages to wash out aversive sensory attributes	Iced products, milk-based products, fruit juices, savory products, candies, cereals, chewing gums	Solid products reduced the aversive attributes more than liquids. Those more strongly flavored, requiring longer mastication and promoting salivation performed best.
Pre-coat mouth (prime) and chase with foods/beverages	Iced products, milk-based products, fruit juices, savory products, candies	Liquids were ineffective in reducing the aversive attributes. The same solids that performed best as chasers worked as primer/chaser.
Dose with oral syringe	N/A	Produced burning in throat and esophagus and did not reduce bitterness.

The development of an amorphous solid dispersion (ASD) formulation for ritonavir was initiated following the introduction of the Kaletra® (lopinavir/ritonavir) tablet in 2005. Experience gained with solid dispersion technology enabled the successful development of the Norvir® 100 mg tablet containing ritonavir ASD which achieved the desired bioavailability and acceptable ambient chemical and physical stability [11]. The Norvir® 100 mg tablet provides significant benefits to patients and physicians, primarily through non-refrigerated storage compared to the OS, offering more robust stability required for storage under global climatic conditions [12]. However, the need for a liquid formulation remains for pediatric patients and adults who are unable to swallow the tablets. To address this gap, AbbVie developed a new powder formulation, Norvir® oral powder, 100 mg (NOP), to provide a more suitable formulation for the pediatric population and with the intent to replace the marketed OS. The objective of this study is to provide an overview of the development and palatability assessment of the age appropriate NOP.

Development Overview of NOP

The NOP development program was initiated, primarily to mitigate or eliminate the risk of potential toxicities associated with ethanol and propylene glycol solvents in OS, which contains 43.2% (*v/v*) ethanol and 26.0% (*v/v*) propylene glycol [4,13]. Other excipients such as colorants, flavoring agents, and preservatives found in the OS were also removed from NOP. NOP also facilitates dose preparation and administration, aligned with current and future needs for pediatric patients.

NOP is manufactured by milling the ritonavir ASD extrudate intermediate and filling the resulting powder into sachets. For dose preparation, NOP is suspended in liquid vehicles or sprinkled on soft foods. The dispersion of NOP produces a supersaturated aqueous solution of ritonavir drug substance that maintains the bioavailability achieved with OS [14].

Key design targets for the development of the NOP are:

1. Reduction/removal of the ethanol and propylene glycol solvents;
2. Flexibility to accommodate doses for pediatric patients;
3. An acceptable dosage form for pediatric patients or patients who may have difficulty swallowing a tablet;
4. Storage stability to achieve an acceptable commercial shelf life under long term storage conditions;
5. Bioavailability that maintains comparable pharmacokinetic profiles and exposures to the commercial oral solution;
6. Offer opportunities to improve palatability.

Early formulation development efforts evaluated both uncoated and coated powders. To mask the inherent bitter taste of ritonavir, a methacrylic acid–ethyl acrylate copolymer coating (enteric coating, insoluble in acidic media, and soluble above pH 5.5 to allow dissolution in the intestine) was applied to the uncoated powder. Exposure to various pH environments during dose preparation and administration was taken into consideration when assessing impact on the coated powder and potential for drug release and associated aversive attributes as evaluated in healthy adult volunteers. When exposed to various pHs, the uncoated powder provided a more homogeneous suspension compared to the coated powder. This is an important feature to ensure complete dose administration of the amorphous drug suspension for pediatric doses. The uncoated powder formulation achieves the majority of the key design targets, but still had opportunity to improve palatability for dose administration.

The coated and uncoated powders were evaluated for ritonavir pharmacokinetics and palatability as an aqueous suspension administered within approximately 5 min after suspending. In order to evaluate the potential impact of the dissolved aqueous soluble excipients on the ritonavir pharmacokinetics and palatability, uncoated powder was pre-dispersed and held for approximately 30 min prior to administration to allow dissolution of soluble excipients and suspension of the amorphous drug particles prior to administration. The pharmacokinetic results of these three ritonavir regimens (coated, uncoated, and uncoated pre-dispersed powders) showed comparable bioavailability. The bioavailability for all three regimens ranged from 80%–90% relative to OS. The palatability results from this study indicated that no taste-masking benefit was gained from coated powder formulation as compared to the uncoated powder and OS.

The initial milled ritonavir ASD extrudate intermediate had a rather broad particle size distribution with a significant fraction of fine particles (<100 μm) and a particle shape that is not ideal for polymeric coating. Coating for conventional tablets, round pellets, or mini-tablets can be effective in reducing or eliminating API taste, resulting in "taste-neutral" formulations [15–17].

However, coating for irregularly shaped particulates/granules often results in an imperfect barrier film coat, exposing API in the oral cavity where it can be perceived. The milling and coating processes were not optimized, and it is possible that the finer irregular particles may have had incomplete coating with additional ritonavir particles embedded in the outer polymeric coating layer (Figure 1) leading to potential premature dissolution or exposure to taste receptors during administration.

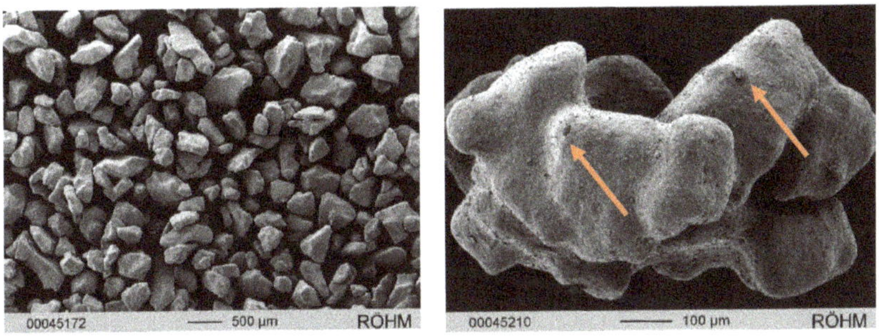

Figure 1. Ritonavir particle shape and small ritonavir particles embedded in eudragit coating. (Orange arrows point to the ritonavir particles).

To further investigate the pharmacokinetic behavior of a larger particle size, still meeting the recommended size for sprinkle products, a uniform 2-mm ritonavir extrudate particulate was coated with the methacrylic acid–ethyl acrylate copolymer (Figure 2) [18]. Although the coated particulate showed improvements in flavor as measured by a trained adult sensory panel, the pharmacokinetic results showed a reduced relative bioavailability of approximately 50% for the coated particulate relative to the OS. The delayed dissolution and drug release profile had a negative impact relative to the known narrow absorption window of ritonavir in the upper intestine. Given the challenges of achieving sufficient coating for taste-masking purposes, without negatively impacting bioavailability, development of NOP was focused on using uncoated powder.

Figure 2. Coated ritonavir extrudate 2-mm particulate.

2. Results and Discussion

2.1. NOP Palatability Assessment, Part 1: Dose/Response Sensory Analysis

All five strengths of NOP aqueous suspension were characterized by a strong and lingering bitter basic taste and secondary aromatic off-notes described as "burnt" (polyethylene, wax, and hair). As shown in Figure 3 bitterness of the five strengths spanned the upper half of the flavor profile supra-threshold intensity range, the most challenging from a taste-masking perspective.

The bitterness of NOP is very strong at the 100mg clinically relevant dose and all strengths would be patient-perceptible (≥1 intensity) initially and for varying lengths of time in the aftertaste. In the absence of a taste-masking system, it would be necessary to reduce NOP strength to about 0.3 mg/10 mL in order for the bitterness to be imperceptible to patients (i.e., <1 intensity).

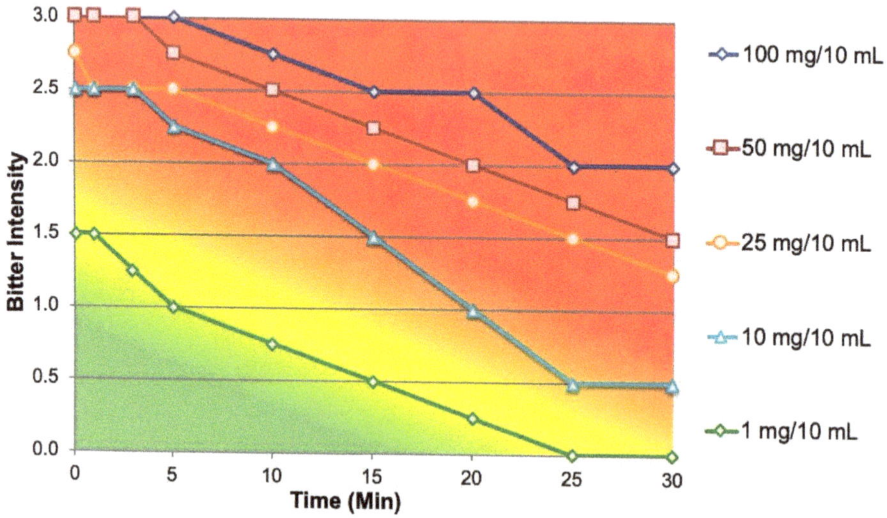

Figure 3. Bitterness dose/response of Norvir® oral powder (NOP).

The dose/response results for the burnt aromatics are shown in Figure 4. The four highest strengths of NOP (10, 25, 50, and 100 mg/10 mL) had moderate-to-strong intensity burnt aromatic off-notes. In the absence of an appropriate aroma masking system, it would be necessary to reduce NOP strength to about 1 mg/10 mL for the burnt aromatics to unrecognizable to patients.

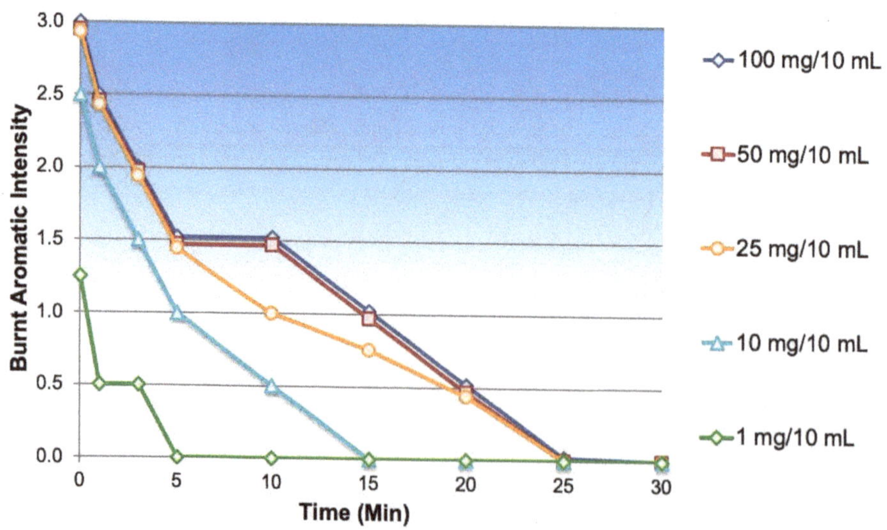

Figure 4. Burnt aromatics dose/response of NOP.

2.2. NOP Palatability Assessment, Part 2: Evaluation with Foods

As shown in Figure 5, and consistent with available information, the foods produced varied effects on bitterness reduction, ranging from $<\frac{1}{2}$-unit (sweet potato chaser and chocolate milk mix-in) to a $1\frac{1}{2}$-unit reduction in perception of bitterness being associated with food chasers high in fat and/or sugar content with strong flavor intensity (peanut butter, Nutella®). The 100-mg clinically relevant dose of NOP is shown for comparison.

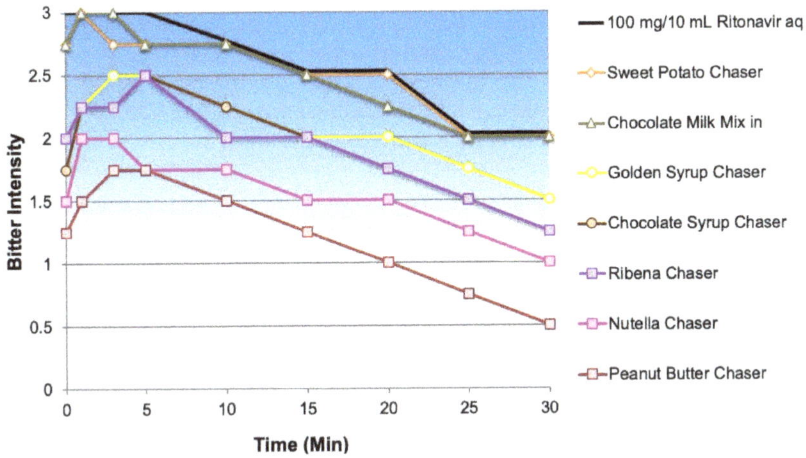

Figure 5. Bitterness profile for NOP with food chasers.

The foods also had varied effects on the burnt aromatic off-notes of the NOP (Figure 6). All of the chasers produced greater reduction in the burnt aromatics than the chocolate milk mix-in (minimal reduction) with several at or below the threshold for perception (≤1). The poor performance of chocolate milk may have been due to its administration as a mix-in, which would have allowed for an extended time for hydration, potentially releasing volatile aromatics.

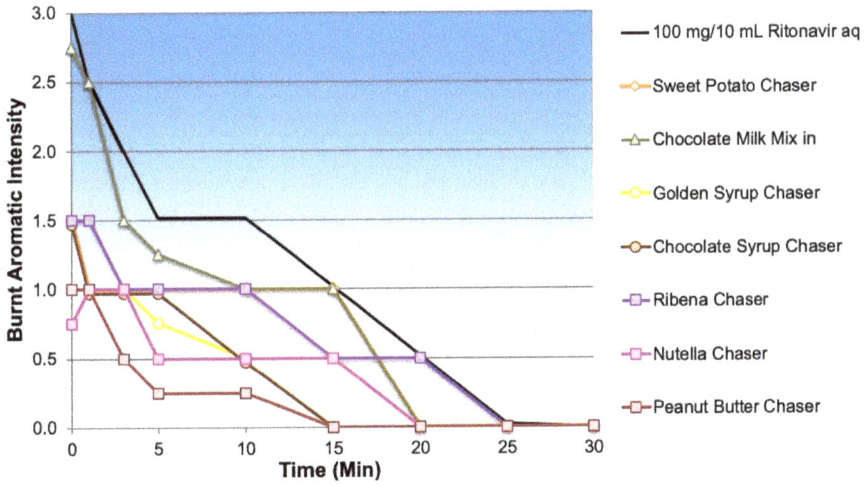

Figure 6. Burnt aromatics profile for NOP with food chasers.

During the bioavailability studies, palatability of NOP in various beverages (water, infant formula, and chocolate milk) as well as admixed into a soft food (applesauce or vanilla pudding) was compared to that of the OS through study participant questionnaires. Only the NOP mixed in chocolate milk showed a modest improvement in overall palatability compared to the OS. The bioavailability study also demonstrated that administration of NOP with infant formula, chocolate milk, applesauce, or pudding was bioequivalent to administration in water [19]. This observation suggests that NOP may be administered with a wide variety of vehicles (soft foods and liquids) with limited impact on bioavailability.

Based on these results, the foods were ranked in descending order of their overall ability to reduce the aversive sensory attributes of the drug product when administered as chasers. The composition and physical characteristics of foods can be important determinants of their ability to "mask" the aversive sensory attributes of drug products. Table 2 summarizes the fat, sugar content, and water activity of the model foods. Water activity is a measure of relative vapor pressure of water molecules in the headspace above a food versus vapor pressure above pure water from 0 to 1 (pure water). This type of food science-based categorization system will ensure that the most varied food types are tested with a minimum of overlap between samples. The best performing foods may then be screened for chemical compatibility, resulting in specific dosing recommendations that are both effective at reducing the aversive sensory attributes of the drug product and efficacious.

Table 2. Composition and physical characteristics of model foods selected as dosing vehicle or chaser.

Food	Brand	Quantity	Fat (g/5 g)	Sugar (g/5 g)	Water Activity (a_w)
Peanut Butter	Jif™ Natural; Creamy	5 g	2.4	0.5	0.251
Hazelnut Spread	Nutella™	5 g	1.5	3.5	0.335
Black Currant Concentrate	Ribena™; Concentrate	5 g	0	2.5	0.912
Golden Syrup	Lyle's™	5 g	0	5	0.545
Chocolate Syrup	Hershey's™; Regular Syrup	5 g	0	2.5	0.830
Sweet Potato Puree	Gerber™; 1st Foods	5 g	0	0.2	0.995
Chocolate Milk (Mix-In)	Nesquik™; Lowfat Chocolate Milk	10 mL	0.1	0.6	0.990

3. Materials and Methods

3.1. Materials

The NOP was supplied as sachets containing 100 mg milled ritonavir extrudate powder for addition to water for preparation of an aqueous suspension.

3.2. Methods

Samples were evaluated using the flavor profile method of sensory analysis, an internationally recognized and approved open-source method [20]. Flavor profile is used to identify, characterize, and quantify the sensory attributes of the samples. Flavor profile measures the perceived strength or intensity of the attributes, the order in which they appear; and a description of all flavors—basic tastes, feeling factors, and aromatics—remaining at specified time intervals in the aftertaste. Per the methodology, 4–6 panelists evaluated each sample and collectively arrived at a consensus judgement of the attribute intensity using chemical reference standards to establish the intensity scale. Sensory characteristics above a slight intensity on the flavor profile scale (>1) are clearly perceptible to consumers/patients; this intensity is known as the recognition threshold. Therefore, in order not to be perceived, negative attributes (e.g., bitterness or irritation) should be below this threshold. The flavor profile terms are as shown in Figure 7.

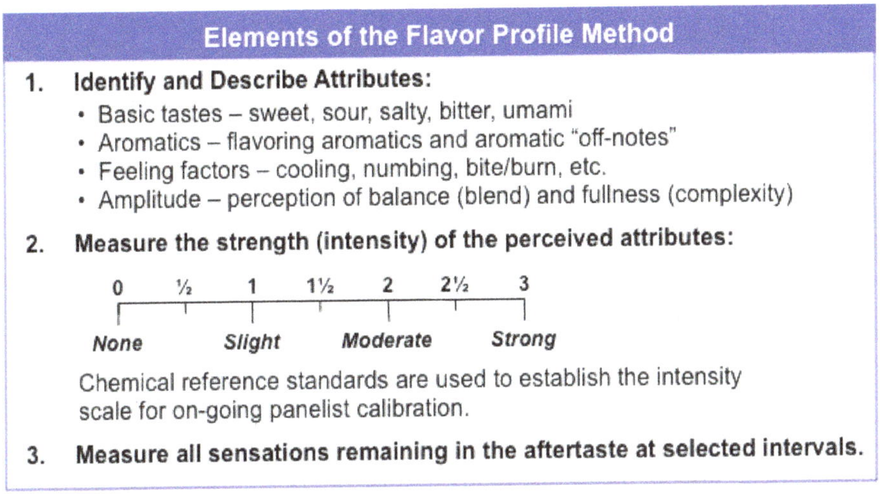

Figure 7. Flavor profile definitions.

A two-part study was conducted with healthy adult sensory panelists (subjects). Part 1 was a dose/response sensory analysis of NOP to determine the maximum concentration that is patient perceptible. Part 2 was an assessment of the sensory performance of NOP when administered with model foods. The study was conducted in accordance with good clinical practice.

Part I: Dose/Response Sensory Analysis of NOP

The NOP was prepared by milling ritonavir extrudate intermediate material. To characterize NOP across a range of concentrations, five strengths (1, 10, 25, 50, and 100 mg/10 mL) were evaluated, the highest being clinically relevant.

A 10-mL aliquot of sample was dispensed into individual 1-ounce plastic cups using a graduated oral syringe and distributed to each panelist. Starting at the same time, each panelist poured the sample into their mouth, swished the contents around the oral cavity for 10 s, and expectorated. The panelists independently evaluated and recorded the sensory characteristics at nine discrete time points (0, 1, 3, 5, 10, 15, 20, 25, and 30 min) using the flavor profile method. Each sample was evaluated twice to generate the final flavor profile. Multiple sessions were required to complete the evaluations to limit exposure to the drug active and minimize sensory fatigue.

Part 2: Evaluation of NOP with Model Foods

There are several approaches that should be considered when improving the "palatability" of drug products with extremely aversive sensory attributes, e.g., bitterness. For NOP, taste-masking was not achievable via a formulation strategy, so two potential alternative approaches were considered, including dosing with food (the term "food" also includes beverages) and cleansing the palate with a food immediately following dose administration (using a food "chaser") [21]. The principle of food "chasers", consumption of a food immediately following dose administration, was explored based on historical experience evaluating various approaches with common food products to reduce the bitterness intensity with OS.

The selection of food products is based on multiple factors including patient age, ease of preparation, and availability. Choices for infants and toddlers often include applesauce, yogurt, and formula [22]. However, limited consideration is given to the composition and chemical properties of the foods, e.g., fat, sugar moisture content, pH, water activity, flavor impact (intensity and duration). Seven products were selected in part based on their compositional diversity from a food science perspective as well as availability and fit with the target demographics and geography (e.g., pediatrics in Africa). These products were peanut butter, hazelnut spread, black currant concentrate, golden syrup, chocolate syrup, sweet potato puree, and chocolate milk. Nutrient data was determined from the commercial nutrition facts labels and water activity was measured on an AquaLab Model 4TEV water activity meter.

The commercial food products were evaluated as dosing vehicles for NOP ("mix-in") or following administration of the NOP ("chaser") to help cleanse the palate. The sensory panelists first evaluated each of the model food products alone (i.e., without NOP) to develop a flavor profile of the native product.

For dosing with model liquid food (mix-in), the panelists were provided a sachet containing NOP and a 1-oz sample cup containing 10 mL of the model liquid food. Starting at the same time, panelists emptied the sachet in to the model food and mixed. After mixing, panelists took the sample into the mouth and agitated for 10 s and expectorated the liquid. The panelists independently evaluated and recorded the sensory characteristics at nine discrete time points (0, 1, 3, 5, 10, 15, 20, 25, and 30 min) using the flavor profile method. The process was repeated for a second evaluation of each sample to generate the final flavor profile for the sample. Multiple sessions were required to complete the evaluations to limit exposure to the drug active and minimize sensory fatigue.

For evaluating the effects of "chasers", the panelists were provided a sachet containing NOP, a 1-oz sample cup containing 10 mL water, and a disposable spoon containing 5 g of the tested food chaser. Starting at the same time, panelists emptied the sachet into the water and mixed, forming a suspension. After mixing, starting at the same time, the panelists sipped the NOP suspension into the oral cavity, agitated in the mouth for 10 s, and expectorated the liquid. Immediately following expectoration, panelists took the food chaser from the spoon into the oral cavity, agitated in the mouth for 10 s, and swallowed the food chaser. The panelists independently evaluated and recorded the sensory characteristics at the same nine time intervals using the flavor profile method. As before, the process was repeated for a second evaluation of each sample to generate the final flavor profile for the sample.

4. Conclusions

The NOP formulation is an acceptable and age-appropriate dosage form to replace the OS for pediatric patients or patients who may have difficulties in swallowing a tablet. The majority of key design targets were achieved for the development program and suitable palatability assessments (flavor profile method) were executed. The development and taste assessment work conducted for the NOP complements the extensive work previously performed for OS. While it was not feasible to substantially improve the palatability of the formulation itself, there may be widely available options for patients to reduce the lingering bitterness.

NOP may be added to soft food (applesauce, vanilla pudding) or suspended in a liquid (water, infant formula, chocolate milk) as a convenient method for dose administration, though the effects on bitterness reduction were determined in previous studies to be modest. In this study, a variety of beverages and soft foods and consumed immediately after NOP dosing (peanut butter, hazelnut chocolate spread, black currant fruit drink concentrate, golden syrup, and chocolate syrup) were shown to reduce the intensity and duration of the bitter aftertaste, most notably peanut butter and hazelnut chocolate spread. The variety of beverages and soft foods as vehicles, as well as those taken immediately after dosing, offers patients more choices according to their individual taste preferences.

In summary, the NOP pediatric formulation provides dosing flexibility, enhanced stability and commercial shelf life under long term global climatic (30 °C/75% RH) storage conditions to support use in tropical climates of Africa where more than 95% of children with HIV live, and absence of propylene glycol and ethanol. It maintains consistent bioavailability when administered with a wide variety of vehicles and improved palatability when common food products were employed as "chasers" following dose administration.

Author Contributions: Conception and design, J.B.M. and J.H.W.; Collection and assembly of data, J.B.M., D.A.T. and J.H.W.; Investigation, D.C.T.T.; Manuscript writing; all authors, Final approval of the manuscript; all authors. Accountable for all aspects of the work; all authors.

Funding: This research received no external funding.

Conflicts of Interest: John B. Morris and David Cheng Thiam Tan are employees of AbbVie Inc. and may own AbbVie stock options. David A. Tisi and Jeffrey H. Worthington are employees of Senopsys LLC. AbbVie sponsored and funded the study; contributed to the design, participated in the collection of data. AbbVie and Senopsys participated in the interpretation of data, in writing, reviewing, and approval of the final publication.

References

1. Hsu, A.; Granneman, G.R.; Bertz, R.J. Ritonavir. Clinical pharmacokinetics and interactions with other anti-HIV agents. *Clin. Pharmacokinet.* **1998**, *35*, 275–291. [CrossRef] [PubMed]
2. Kempf, D.J.; Marsh, K.C.; Denissen, J.F.; McDonald, E.; Vasavanonda, S.; Flentge, C.A.; Green, B.E.; Fino, L.; Park, C.H.; Kong, X.P.; et al. ABT-538 is a potent inhibitor of human immunodeficiency virus protease and has high oral bioavailability in humans. *Proc. Natl. Acad. Sci. USA* **1995**, *92*, 2484–2488. [CrossRef] [PubMed]
3. Kumar, G.N.; Rodrigues, A.D.; Buko, A.M.; Denissen, J.F. Cytochrome P450-mediated metabolism of the HIV-1 protease inhibitor ritonavir (ABT-538) in human liver microsomes. *J. Pharmacol. Exp. Ther.* **1996**, *277*, 423–431. [PubMed]
4. *Norvir (Ritonavir) [Package Insert]*; AbbVie Inc.: North Chicago, IL, USA, 2017.
5. Kempf, D.J.; Marsh, K.C.; Kumar, G.; Rodrigues, A.D.; Denissen, J.F.; McDonald, E.; Kukulka, M.J.; Hsu, A.; Granneman, G.R.; Baroldi, P.A.; et al. Pharmacokinetic enhancement of inhibitors of the human immunodeficiency virus protease by coadministration with ritonavir. *Antimicrob. Agents Chemother.* **1997**, *41*, 654–660. [CrossRef] [PubMed]
6. Moyle, G.J.; Back, D. Principles and practice of HIV-protease inhibitor pharmacoenhancement. *HIV Med.* **2001**, *2*, 105–113. [CrossRef] [PubMed]
7. Sham, H.L.; Kempf, D.J.; Molla, A.; Marsh, K.C.; Kumar, G.N.; Chen, C.-M.; Kati, W.; Stewart, K.; Lal, R.; Hsu, A. ABT-378, a highly potent inhibitor of the human immunodeficiency virus protease. *Antimicrob. Agents Chemother.* **1998**, *42*, 3218–3224. [CrossRef] [PubMed]
8. Klein, C.E.; Chiu, Y.-L.; Awni, W.; Zhu, T.; Heuser, R.S.; Doan, T.; Breitenbach, J.; Morris, J.B.; Brun, S.C.; Hanna, G.J. The tablet formulation of lopinavir/ritonavir provides similar bioavailability to the soft-gelatin capsule formulation with less pharmacokinetic variability and diminished food effect. *JAIDS J. Acquired Immune Defic. Syndromes* **2007**, *44*, 401–410. [CrossRef] [PubMed]
9. Law, D.; Schmitt, E.A.; Marsh, K.C.; Everitt, E.A.; Wang, W.; Fort, J.J.; Krill, S.L.; Qiu, Y. Ritonavir-PEG 8000 amorphous solid dispersions: In vitro and in vivo evaluations. *J. Pharm. Sci.* **2004**, *93*, 563–570. [CrossRef] [PubMed]
10. European Medicines Agency (EMA). *Guideline on Pharmaceutical Development of Medicines for Paediatric Use*; EMA: London, UK, 2013.
11. Tho, I.; Liepold, B.; Rosenberg, J.; Maegerlein, M.; Brandl, M.; Fricker, G. Formation of nano/micro-dispersions with improved dissolution properties upon dispersion of ritonavir melt extrudate in aqueous media. *Eur. J. Pharm. Sci.* **2010**, *40*, 25–32. [CrossRef] [PubMed]
12. Law, D.; Krill, S.L.; Schmitt, E.A.; Fort, J.J.; Qiu, Y.; Wang, W.; Porter, W.R. Physicochemical considerations in the preparation of amorphous ritonavir-poly(ethylene glycol) 8000 solid dispersions. *J. Pharm. Sci.* **2001**, *90*, 1015–1025. [CrossRef] [PubMed]
13. Strickley, R.G. Solubilizing excipients in oral and injectable formulations. *Pharm. Res.* **2004**, *21*, 201–230. [CrossRef] [PubMed]
14. Kanzer, J.; Hupfeld, S.; Vasskog, T.; Tho, I.; Hölig, P.; Mägerlein, M.; Fricker, G.; Brandl, M. In situ formation of nanoparticles upon dispersion of melt extrudate formulations in aqueous medium assessed by asymmetrical flow field-flow fractionation. *J. Pharm. Biomed. Anal.* **2010**, *53*, 359–365. [CrossRef] [PubMed]
15. Strickley, R.G.; Iwata, Q.; Wu, S.; Dahl, T.C. Pediatric drugs–a review of commercially available oral formulations. *J. Pharm. Sci.* **2008**, *97*, 1731–1774. [CrossRef] [PubMed]
16. Xu, M.; Heng, P.W.; Liew, C.V. Evaluation of coat uniformity and taste-masking efficiency of irregular-shaped drug particles coated in a modified tangential spray fluidized bed processor. *Expert Opin. Drug Deliv.* **2015**, *12*, 1597–1606. [CrossRef] [PubMed]

17. Douroumis, D. Orally disintegrating dosage forms and taste-masking technologies; 2010. *Expert Opin. Drug Deliv.* **2011**, *8*, 665–675. [CrossRef] [PubMed]
18. Food and Drug Administration (FDA). *Guidance for Industry Size of Beads in Drug Products Labeled for Sprinkle;* FDA: Silver Spring, MD, USA, 2012.
19. Keane, P. *The Flavor Profile. ASTM Manual on Descriptive Analysis Testing for Sensory Evaluation;* ASTM International: West Conshohocken, PA, USA, 1992; pp. 2–15.
20. Zajicek, A.; Fossler, M.J.; Barrett, J.S.; Worthington, J.H.; Ternik, R.; Charkoftaki, G.; Lum, S.; Breitkreutz, J.; Baltezor, M.; Macheras, P.; et al. A report from the pediatric formulations task force: Perspectives on the state of child-friendly oral dosage forms. *AAPS J.* **2013**, *15*, 1072–1081. [CrossRef] [PubMed]
21. Ternik, R.; Liu, F.; Bartlett, J.A.; Khong, Y.M.; Thiam Tan, D.C.; Dixit, T.; Wang, S.; Galella, E.A.; Gao, Z.; Klein, S. Assessment of swallowability and palatability of oral dosage forms in children: Report from an M-CERSI pediatric formulation workshop. *Int. J. Pharm.* **2018**, *536*, 570–581. [CrossRef] [PubMed]
22. Salem, A.H.; Chiu, Y.L.; Valdes, J.M.; Nilius, A.M.; Klein, C.E. A novel ritonavir paediatric powder formulation is bioequivalent to ritonavir oral solution with a similar food effect. *Antivir. Ther.* **2015**, *20*, 425–432. [CrossRef] [PubMed]

International Journal of
Molecular Sciences

MDPI

Article

Combining Mechanochemistry and Spray Congealing for New Praziquantel Pediatric Formulations in Schistosomiasis Treatment

Beatrice Albertini [1,*], Beatrice Perissutti [2], Serena Bertoni [1], Debora Zanolla [2], Erica Franceschinis [3], Dario Voinovich [2], Flavio Lombardo [4,5], Jennifer Keiser [4,5] and Nadia Passerini [1,*]

1 Department of Pharmacy and BioTechnology, University of Bologna, Via S. Donato 19/2, 40127 Bologna, Italy; serena.bertoni4@unibo.it
2 Department of Chemical and Pharmaceutical Sciences, University of Trieste, P.le Europa 1, 34127 Trieste, Italy; bperissutti@units.it (B.P.); debora.zanolla@gmail.com (D.Z.); vojnovic@units.it (D.V.)
3 Department of Pharmaceutical and Pharmacological Sciences, University of Padova, via Marzolo 5, 35131 Padova, Italy; erica.franceschinis@unipd.it
4 Helminth Drug Development Unit, Department of Medical Parasitology and Infection Biology, Swiss Tropical and Public Health Institute, Socinstr.57, CH-4051 Basel, Switzerland; flavio.lombardo@unibas.ch (F.L.); jennifer.keiser@swisstph.ch (J.K.)
5 Universität Basel, Petersplatz 1, P.O. Box, CH-4001 Basel, Switzerland
* Correspondence: beatrice.albertini@unibo.it (B.A.); nadia.passerini@unibo.it (N.P.)

Received: 18 February 2019; Accepted: 8 March 2019; Published: 12 March 2019

Abstract: Praziquantel (PZQ) is the first line drug for the treatment of schistosome infections and is included in the WHO Model List of Essential Medicines for Children. In this study, the association of mechanochemical activation (MA) and the spray congealing (SC) technology was evaluated for developing a child-friendly PZQ dosage form, with better product handling and biopharmaceutical properties, compared to MA materials. A 1:1 by wt PZQ—Povidone coground—was prepared in a vibrational mill under cryogenic conditions, for favoring amorphization. PZQ was neat ground to obtain its polymorphic form (Form B), which has an improved solubility and bioactivity. Then, activated PZQ powders were loaded into microparticles (MPs) by the SC technology, using the self-emulsifying agent Gelucire® 50/13 as a carrier. Both, the activated powders and the corresponding loaded MPs were characterized for morphology, wettability, solubility, dissolution behavior, drug content, and drug solid state (Hot Stage Microscopy (HSM), Differential Scanning Calorimetry (DSC), X-Ray Powder Diffraction Studies (PXRD), and FT-IR). Samples were also in vitro tested for a comparison with PZQ against *Schistosoma mansoni* newly transformed schistosomula (NTS) and adults. MPs containing both MA systems showed a further increase of biopharmaceutical properties, compared to the milled powders, while maintaining PZQ bioactivity. MPs containing PZQ Form B represented the most promising product for designing a new PZQ formulation.

Keywords: poorly water soluble drug; solubility enhancement; grinding; spray congealing; neglected tropical diseases; polymorph

1. Introduction

Schistosomiasis is one of today's foremost neglected tropical diseases (NTDs) and a disease of poverty affecting more than 200 million people, worldwide [1]. NTDs are a group of 17 endemic diseases that prevail in less developed areas where large numbers of people have little or no access to adequate health care, clean water, housing, transport, and information [2]. In particular, Schistosomiasis is a tropical and subtropical disease caused by one of the six different species of the trematode worm *Schistosoma*, with the great majority of cases either infected with *Schistosoma haematobium*, *S. japonicum* or *S. mansoni* [3,4]. In the absence of effective vaccines for helminthic infections, praziquantel (PZQ) is the first line drug used in endemic countries, for the treatment and prevention therapy of schistosome infections [1,5], and is included in the WHO Model List of Essential Medicines for Children [6]. However, there is still a need for discovering alternative treatments and superior formulations [1], especially for pre-school-age children and infants, which require more appropriate dosage forms, in terms of ease of dose adjustment and swallowing [7]. In fact, PZQ is available as a 600 mg film-coated tablet, and a high dose (maximum 40 mg/kg bodyweight) is required. More recently, 150 mg tablets have become commercially available [8].

From the biopharmaceutical point of view, several approaches have been undertaken to reduce the PZQ therapeutic dose, due to its solubility enhancement [7,9,10]. Recently we have proposed some strategies involving the mechanochemical activation of the drug, proving the amelioration of the PZQ biopharmaceutical properties (solubility, intrinsic dissolution rate (IDR)) [11–13]. In particular, in the first study, PZQ was coground with different pharmaceutical polymers, using a lab-scale vibrational mill and an explorative analysis of formulation variables (drug-polymer wt. ratio and polymer type), and process-related parameters (type of grinding media, grinding time, and frequency) was carried out with the help of an experimental screening design. The most promising sample, in terms of drug solubility enhancement, was a coground composite with crospovidone, with a 50%-by-weight drug content, permitting a 4.6-fold solubility improvement, in comparison to the starting drug. This product displayed a high amorphous character, as it was able to maintain the in vitro bioactivity of the PZQ, against the *S. mansoni*, and appeared to be chemically and physically stable, over six ageing months [11]. Considering that this system revealed an incomplete drug recovery (88.01%), a second research experience encompassed the use of the cryomilling technique (using liquid nitrogen as a cryogenic media), to prevent chemical degradation of the drug during comilling, and to improve activation efficiency. The most promising polymers selected from the former experience (linear and cross-linked povidone) were then used to successfully prepare cryo-composites with a high PZQ content. Cryomilling PZQ with linear povidone led to a significant decrease of drug degradation (PZQ recovery ranged from 95.2% to 98.9%), along with a remarkable reduction of the crystalline content (ranging from 47% to 0%). From both experiences it was clear that the drug, when comilled in presence of polymers, was more prone to amorphization and a certain amount of degradation. Conversely, PZQ that was ground by itself, did not show this propensity to degradation while maintaining the tendency to physical transformations [13]. In addition, when subjected to neat grinding in adequate conditions, PZQ evidenced to form an anhydrous polymorphic variety with a very high efficiency. The obtained new polymorph, named Form B, displayed a two-fold increased water solubility and IDR, in comparison to PZQ, and a better activity than the raw PZQ [12]. Therefore, all these attempts testified to the possibility of forming different activated crystal modifications of PZQ (amorphous/nanocrystalline or polymorphous) with ameliorated biopharmaceutical properties, by means of a mechanochemical process.

With the final ambitious aim of developing a child-friendly PZQ dosage form, a further solubility improvement and a better product handling with respect to the milled systems (drug and coground mixture), was then pursued. In the development of pediatric formulation, an important concern is the appropriate choice of excipients. The EMA "Guideline on pharmaceutical development of medicines for pediatric use" suggests to avoid any excipients that are potentially toxic or unsuitable for children [14]. In this study, the association of the mechanochemical activation with the formation of a PZQ solid dispersion by the spray congealing technology was evaluated. Raw PZQ, its physical mixture with polyvinylpyrrolidone (50:50 *w*/*w*) and activated materials, such as the new polymorphic form, Form B, and the cryo-coground of PZQ:PVP (50:50 *w*/*w*) were, thus, loaded into spray congealed microparticles. The 1:1 drug to polymer wt. ratio was selected, since it showed a significant improvement in solubility performance than a 1:2 ratio [11]. As microparticle carrier, Gelucire® 50/13 was selected, since it formerly demonstrated its great ability to improve the biopharmaceutical properties of several active ingredients—silybum Marianum dry extract [15], glibenclamide [16], carbamazepine [17], melatonin [18], and piroxicam [19]. No safety concerns regarding the selected excipients (PVP and Gelucire® 50/13) were documented.

Both, the activated materials and the corresponding loaded microparticles were then fully characterized with regards to morphology (SEM and particle size), wettability, solubility, dissolution behavior, drug content, and drug solid state (Hot Stage Microscopy (HSM), Differential Scanning Calorimetry (DSC), X-Ray Powder Diffraction Studies (PXRD), and FT-IR). The results were then compared to those obtained from the raw materials (PZQ and physical mixture). Finally, the loaded microparticles were tested, in vitro, in comparison to PZQ, against the *S. mansoni* newly transformed schistosomula (NTS) and adults.

2. Results and Discussion

2.1. Analysis of the Activated Materials

The characteristics of the activated materials were analyzed, prior to their addition to the microparticles. Table 1 summarizes the results of the HPLC assay and of the thermal analysis. Regarding the HPLC analysis, all impurities were absent, both in the standard solution (raw PZQ) and in Form B, prepared by neat grinding, indicating that milling process of the pure drug induced a physical transformation but not the formation of impurities. Thus, the PZQ recovery was 100%. In the case of cryo-coground sample (obtained by vibrating for 60 min at 20 Hz), the PZQ recovery slightly decreased and the total amount of impurity was about 1.11%. Further, analogous to our previous recent study [12], the presence of an additional compound, indicated as "X" impurity, which has not been reported in the Ph-. Eur. PZQ monography or in that of the USP, was detected. This compound, having a molecular ion m/z 147, was identified in a previous research work [20]. Impurity X always showed a higher concentration than impurity A, which never exceeded the 0.2% (the acceptance criteria of the USP 36-NF 31). Conversely, impurity B was never detected. All ground samples were, hence, in accordance with the E.P. criteria. This result displayed that the milling process realized in the suitable conditions did not induce a significant chemical degradation of PZQ; therefore, both of the activated materials were considered suitable for their loading into microparticles.

Table 1. Characteristics of the mechanochemically activated materials, in comparison to the raw Praziquantel (PZQ) and PZQ:PVP physical mixture.

Samples	HPLC Assay: Impurity Retention Time (min) (Content, %)			PZQ Recovery (%)	Solubility (mg/L)	Endothermic Peak (°C)	Residual Crystallinity (%)
	Impurity A	Impurity B	Impurity X				
PZQ	-*	-*	-*	100	140.30 ± 9.26	143.51 ± 0.35	100
PZQ-PVP PM	-*	-*	-*	99.98 ± 0.04	151.78 ± 27.22	142.31 ± 0.22	100
PZQ-PVP CC	3.4–3.5 (0.18)	-*	4.0–4.2 (0.93)	98.89 ± 0.08	267.40 ± 11.00	129.36 ± 0.43	27.70
PZQ Form B	-*	-*	-*	99.55 ±0.05	284.61 ± 4.67	112.50 ± 0.32	n.a. **

* not present; ** not applicable since a new polymorphic form was obtained.

As previously stated, one of the main aims of the milling procedure for a poorly bioavailable drug, is its solubility enhancement. The results of the solubility test, reported in Table 1, clearly showed a 2-fold improvement of the solubility, in comparison to raw PZQ. In particular, both the cryo-comilled sample and Form B was found to be significantly more soluble than raw PZQ and the physical mixture, although no significant difference between the two activated samples was found.

SEM pictures, presented in Figure 1, show that starting PZQ had an acicular habitus (Figure 1C), while Form B had a very different habit, consisting of agglomerates of long and very thin whiskers (Figure 1B). In the CC sample (Figure 1A), compared to primary particles, the particles exhibited a significant change in the shape and surface morphology, all showing irregular blocky-shaped particles. In fact, the PVP particles their lost typical spherical shape and drug crystal were no longer recognizable during cryo-cogrinding. These phenomena suggest that the raw materials were intimately and homogenously combined in the CC, and PZQ particles were completely covered by the PVP macromolecules.

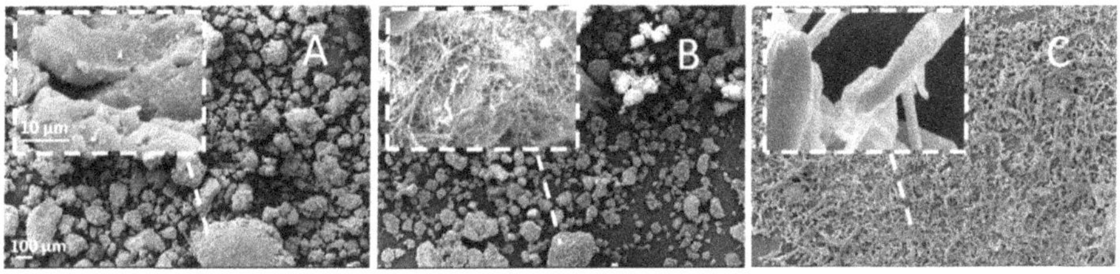

Figure 1. SEM images of PZQ:PVP CC (**A**), PZQ Form B (**B**), and raw PZQ (**C**) (magnification 150×; in the top left frame a 5.0 K× magnified image).

Laser light scattering analysis revealed that the original PZQ had a mono-modal particle size distribution with a median diameter of 23.90 μm. PZQ Form B had a $d_{(0.1)}$ of 20.90 μm, $d_{(0.5)}$ of 77.44 μm, and a $d_{(0.9)}$ 139.14 μm. PVP particles showed the following size distribution $d_{(0.1)}$ of 25.50 μm, $d_{(0.5)}$ of 76.20 μm, and a $d_{(0.9)}$ 130.78 μm. When the PZQ was physically mixed with PVP in a 1:1 wt. proportion, a multimodal size distribution was obtained, with peaks in correspondence to those of the components. The CC system showed a variation of the particle size, with respect to the untreated mixture of powders, reaching a median diameter of 243.99 μm, in agreement with the SEM pictures reported in Figure 1.

Then, the thermal analysis was conducted to check the solid state transformation of PZQ, after the milling process. Table 1 reports the temperatures of the PZQ endotherm and the calculated residual crystallinity of the analyzed samples; while Figure 2 shows the corresponding DSC curves. The DSC scan of PZQ showed only a single endothermic peak at about 143.5 °C (ΔH= 91.9 ± 5.32 J/g), in agreement with the melting point and enthalpy of fusion of the racemic form of the drug [7]. In the physical mixture, this event was anticipated by the dehydration of the hygroscopic polymer, as PVP K30 showed only a large dehydration peak between 50 and 100 °C. The CC sample evidenced a lowering of the PZQ thermal event, due to the intense physical disruption of the crystalline structure—the endothermic peak was lower than the original racemic PZQ (see Table 1), in agreement with a very scarce residual crystallinity and with the prevalent presence of the nanocrystals of PZQ. In fact, the lower the nanocrystal size, the lower was its melting temperature and enthalpy [21]. In the case of the PZQ Form B, the endothermal event corresponding to its melting point appeared at about 112.1 °C (ΔH= 55.9 ± 1.5 J/g), according to Zanolla et al. [12].

Hot stage microscopy analysis clearly showed physical changes in the samples, during the temperature scan (Figure 3). PZQ showed elongated acicular crystals of different sizes, which completed their fusion at 144 °C. A slight PZQ melting point reduction to 140 °C was visible by analyzing the physical mixture (due to the dilution effect). The HSM pictures of the PZQ:PVP CC and PZQ Form B confirm the fusion at about 130 °C and 112 °C, respectively.

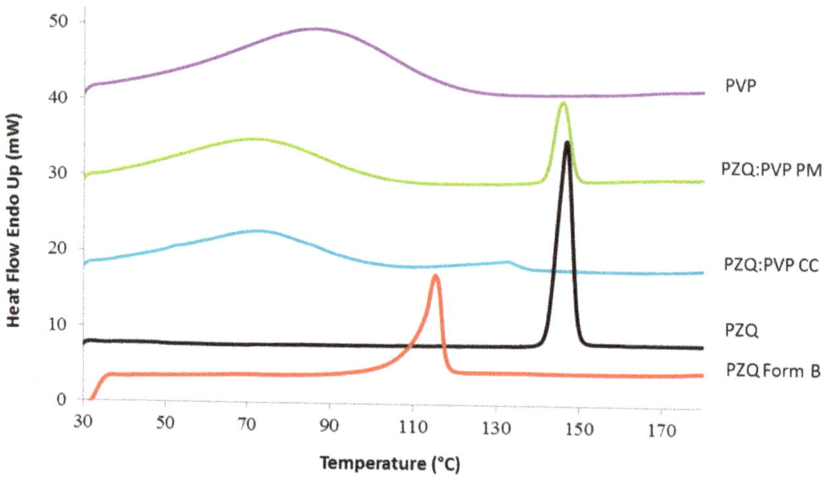

Figure 2. Differential Scanning Calorimetry (DSC) curve overlay of raw PZQ, PVP K30, PZQ:PVP PM, and activated materials (PZQ:PVP CC and PZQ Form B).

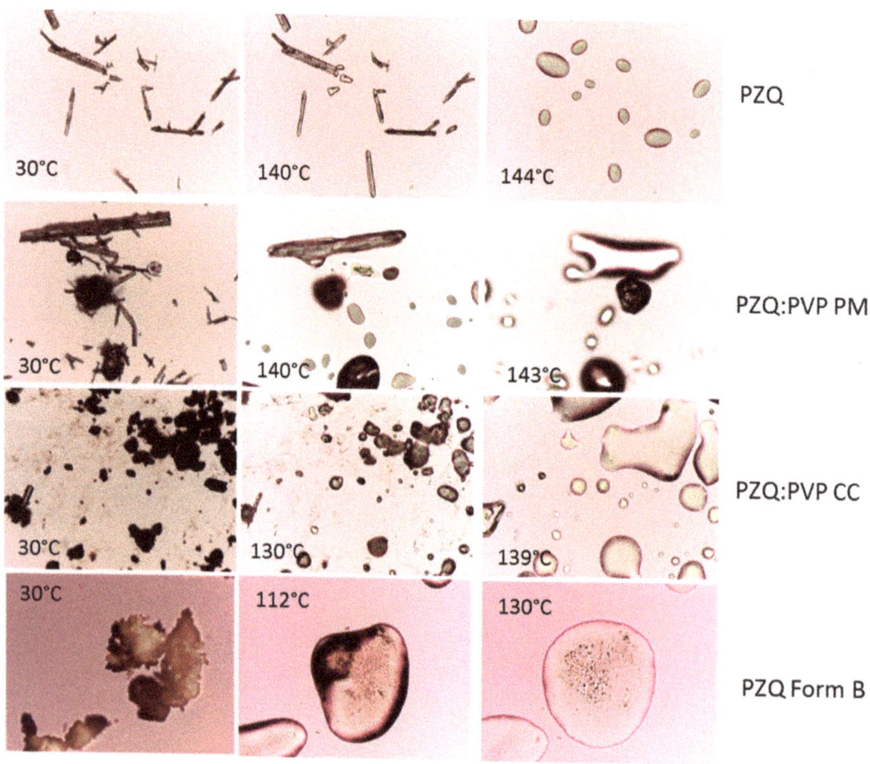

Figure 3. Hot Stage Microscopy (HSM) analysis of raw PZQ, PVP K30, physical mixture, and activated materials (magnification 10×).

FTIR was then performed to detect structural modifications and possible interactions among the components in milled samples (Figure 4). The FTIR spectrum of PZQ shows characteristic peaks at

2929 cm^{-1} and 2852 cm^{-1}, due to the C–H and C–H_2 stretching vibration and an intense multiband at 1650–1600 cm^{-1} due to the region of amide stretching vibrations, with two separate and equal spikes at 1623 cm^{-1} and at 1646 cm^{-1}, corresponding to the C=0 joined to the cyclohexyl and to the heterocyclic carbonyl stretching vibrations, respectively [20]. The IR spectrum of PM appears to be the sum of the pattern of the individual components. The FTIR spectrum of the CC sample shows broader and less intense signals with respect to raw PZQ's. Particularly evident is the difference with the starting drug in correspondence of the stretching of the amide groups, which are involved in molecular packing in the crystal form [22,23] the two original sharp peaks were replaced in the coground spectrum with a broader band. In spectrum of Form B, as already noticed [12], the frequency difference in ν(CO) between the stretching of carbonyl groups is lower than those of original PZQ, due to the presence of anti-conformers in Form B, instead of the syn ones of commercial PZQ crystal structure. In fact, Form B exhibits two signals at 1630 and 1639 cm^{-1} due to the symmetric stretching mode of the carbonyl group joined to the cyclohexyl and to that of the heterocyclic carbonyl, respectively.

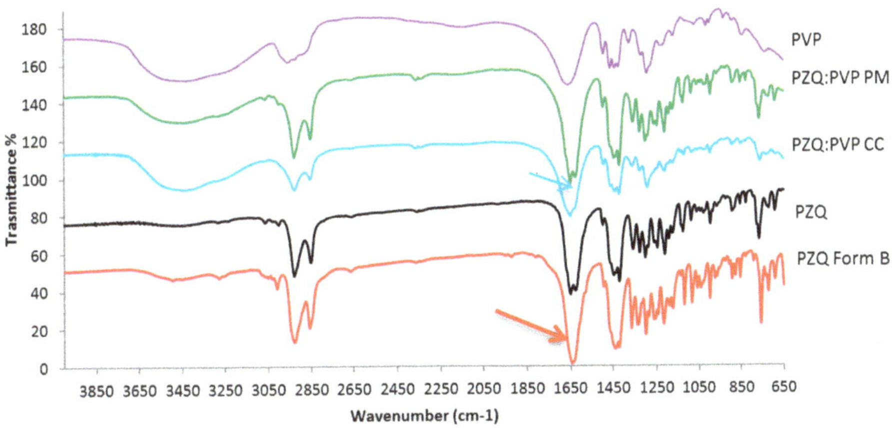

Figure 4. FT-IR spectra of raw PZQ, PVP K30, PZQ:PVP PM and activated materials (PZQ:PVP CC and PZQ Form B). The arrows show the carbonyl stretching vibration.

2.2. Evaluation of the Activated PZQ-Loaded Spray-Congealed Microparticles

To improve the biopharmaceutical properties of PZQ, the activated materials were loaded into hydrophilic microparticles and their characteristics were compared to microparticles containing either raw PZQ or the PZQ:PVP PM. Four batches of MPs (A–D) were then prepared; their composition is reported in Table 2. The real amount of non-degraded PZQ in the cryo-comilled samples was first assayed, in order to prepare the PM and then to calculate the amount of sample to load into the microparticles. After cryo-comilling the 50:50 w/w PZQ:PVP mixture, the effective drug content was about 46% (Table 2).

Table 2. List of the analyzed samples, their relative composition and effective drug content.

	Samples (Abbreviation)			Composition (%, *w/w*) PZQ	PVP	Gelucire 50/13	Real Drug Content (%)	Encapsulation Efficiency (%)
Powders	Activated materials	PZQ:PVP cryocomilled	(PZQ:PVP CC)	50	50	-	46.0 ± 1.2	-
		Milled PZQ (Form B)	(PZQ Form B)	100	-		100	-
	Raw materials	PZQ:PVP physical mixture	(PZQ:PVP PM)	50	50	-	46	-
		Raw PZQ	(PZQ)	100	-	-	100	-
Microparticles	MPs	PZQ:PVP cryo-comilled	(MPsA)	15	15	70	13.75 ± 0.14	91.7
		Milled PZQ_Form B	(MPsB)	15	-	85	13.54 ± 0.83	90.3
		PZQ:PVP physical mixture	(MPsC)	15	15	70	14.35 ± 0.18	95.7
		Raw PZQ	(MPsD)	15	-	85	15.16 ± 0.11	101.0

It is worth noting that the raw PZQ, the PZQ:PVP PM, and the PZQ:PVP CC remained suspended in the molten Gelucire® 50/13. Conversely, the addition of PZQ Form B to the molten excipient formed a clear solution, suggesting the formation of a solid solution for MPs B, rather than a solid dispersion, as supposed for MPs A, C, and D, containing cryo-comilled, PM and raw PZQ, respectively. Microparticle size analysis shows that all MPs ranged between 75–500 µm, with different distribution in sizes (Figure 5).

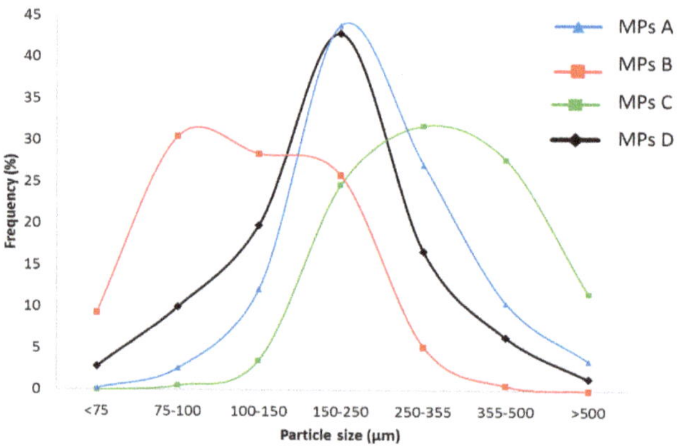

Figure 5. Particle size distribution of the microparticle formulations.

Both MPs A and D displayed a Gaussian curve with the 150–250 µm as a prevalent fraction, while MPs C was found to be bigger, with a broader distribution. MPs B were on average smaller—more than 90% of MPs B had dimensions lower than 250 µm. This different trend was probably due to lower viscosity of the solution, with respect to the suspensions sprayed at the same pressure. In fact, the viscosity of the fluid increased from 100 mPa·s of molten carrier to 960 mPa·s of mixture MPs D, in the following order: MPs B< MPs D<MPs A<MPs C. Therefore, the viscosity of the fluid was not only influenced by the amount of powder included in the molten carrier but also by the particle morphology (size and shape) and by the drug solid state. In fact, the MPs D fluid, containing bigger acicular particles was more viscous than the MPs B fluid, having small dispersed particles and a certain amount of molecularly dispersed drug. MPs A and MPs C fluids differed from the dimensions of the dispersed particles and the viscosity increased proportionally to the particle size.

All microparticles displayed a drug loading, very close to the theoretical one, with encapsulation efficiencies higher than 90%, for all batches (Table 2). As regards to the solubility tests, the main goal was to verify the effect of the Gelucire®-based MPs on the PZQ solubility and, second, to assess the effect of the combination of a mechanochemical activation and spray congealing. The results, reported in Figure 6, clearly demonstrated that the adopted manufacturing and formulation strategies were very useful to enhance the solubility of the BCS class II drug. MPs A increased the drug solubility by four and two times, compared to PZQ and PZQ:PVP CC, respectively. Nevertheless, the more pronounced solubility increase occurred for MPs B, as their solubility was five times greater than the raw PZQ and more than twice that of PZQ Form B. Thus, the solubility of these MPs containing the new polymorph was found to be the highest among the samples. The statistical comparison revealed that solubility of MPs B was significantly higher than that of MPs loaded with PZQ (MPs D), while there were no significant differences among the other MPs.

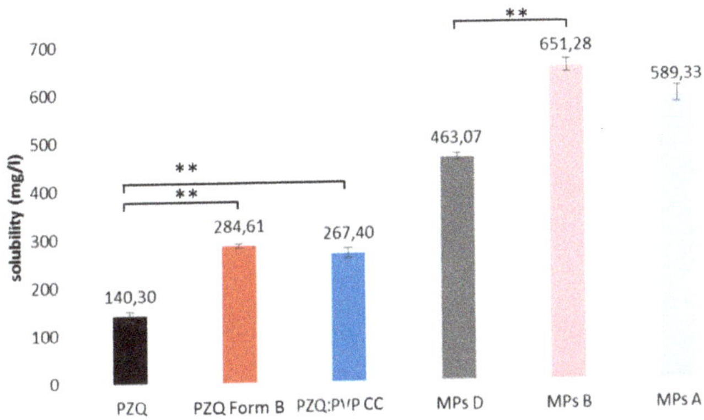

Figure 6. Comparison of the solubility data of the analyzed samples (**indicates significant difference).

In agreement with previous findings [15], the surface of both MPs B and D appeared to be quite rough (Figure 7), which is a typical feature of Gelucire® 50/13-based microparticles obtained by the spray congealing technique. The morphological analysis also revealed that the MPs D sample mainly consists of spherical non-aggregated microparticles, and only a few elongated particles could be detected. On the contrary, the MPs B sample, was characterized by highly spherical particles with polydispersed dimensions and with a remarkable presence of aggregates of very small particles. This was a further confirmation that the addition of micronized particles into the fluid to be atomized, favored the formation of smaller droplets.

The dissolution test was then performed to highlight the difference between the starting samples and the MPs formulations, in the PZQ dissolution rates, and to identify the best formulation approach. Since the dissolution rate varied, depending on the size, wettability, and drug solubility, any factor that might increase one of these properties might have improved the overall bioavailability.

Figure 8 shows the results of the analyzed samples. Analyzing the first 20 min, both ground samples displayed a slight increase of the dissolution rate, relative to the pure PZQ, presumably due to the modification of the crystalline state of the PZQ, both in the CC (semi-crystalline state) and in the milled PZQ (polymorphic form B). In fact, the wettability of Form B meant that it was quite similar to that of the pure PZQ—mean contact angles were 68.77 ± 1.46° and 81.73 ± 1.87°, respectively. However, looking at the whole dissolution profile, the differences were not significant (f_2 = 62.5), in comparison to the pure drug, and the extent of drug dissolved was equal to the raw PZQ (about 52%). This behavior was unexpected, considering their higher solubility, in comparison to the raw PZQ, and it could be explained by taking into account the difference in mean particle size and in particle aggregation—the remarkably higher particle diameter in both ground samples could be a reason behind this dissolution of performance.

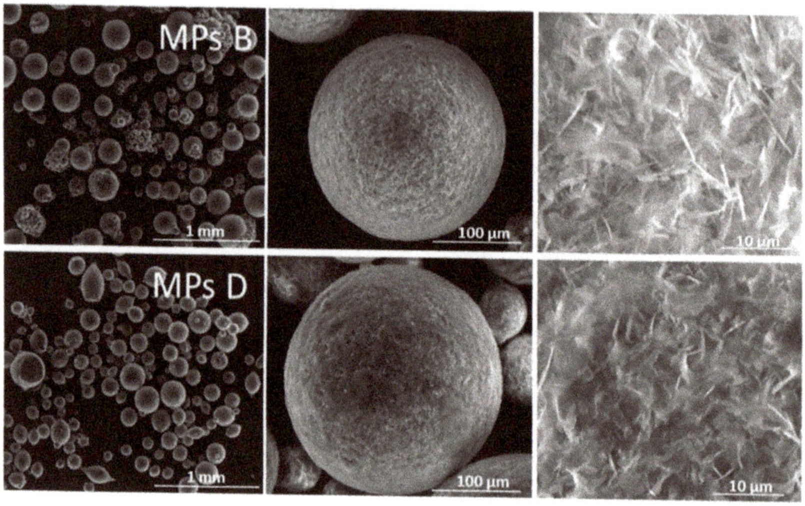

Figure 7. Environmental Scanning Electron Microscopy (ESEM) pictures of MPs B (containing Form B) and MPs D (containing PZQ) (magnification 60×, 500×, and 4.0 K×, in the order from left to right).

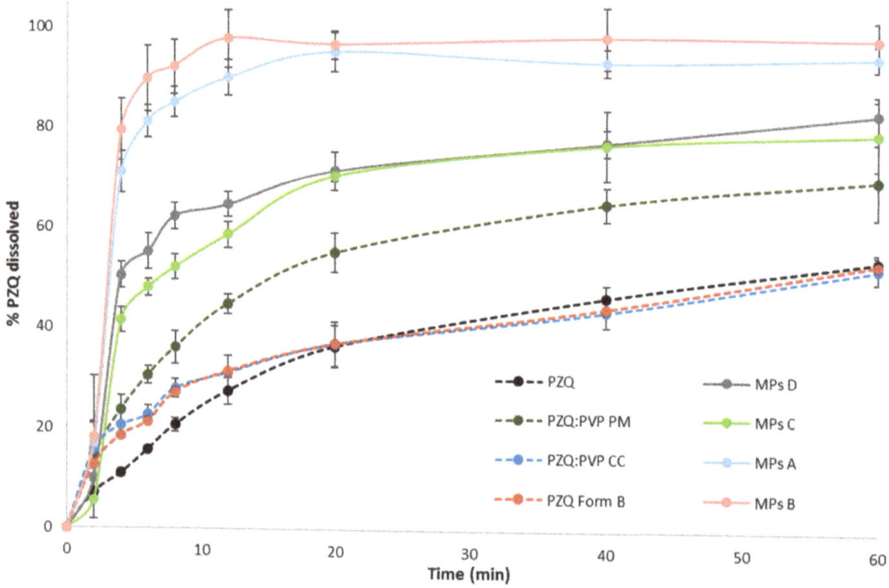

Figure 8. Dissolution profiles of the activated materials and microparticles, in comparison to raw PZQ and a PZQ:PVP physical mixture.

Conversely, a significant improvement of the dissolution rate could be obtained from the physical mixture—after 20 min the 55% of the drug was dissolved as compared to 35% of the raw PZQ (f_2 = 42.6), while after 60 min, the amount of drug solubilized was about 66%. Comparing the PZQ:PVP PM with the CC, the difference between the dissolution profiles was borderline significant (f_2 = 49.7). Although the PVP was present in both systems, in the physical mixture, the drug and excipient were simply blended and, thus, did not interact with each other [24]; unlike in the CC system, the milling process led to the formation of bigger aggregates, decreasing the dissolution rate.

Considering the microparticles, all formulations showed a great improvement in drug dissolution rate, compared to the raw PZQ and ground PZQ, confirming that the spray congealing technology enabled the formation of a system with excellent biopharmaceutical properties. Previous paper evidenced that Gelucire 50/13, being a mixture of mono-, di-, and triglycerides, and PEG 1500 esters of stearic acid (HLB 13), it was able to promote micelles formation, consequently, increasing the dissolution rate of Glibencamide [16]. Moreover, it was reported that Gelucire 50/13 microspheres release the entrapped piroxicam via the formation of a lyotropic liquid crystalline phase, which allowed the dissolution of the drug particles in a finely divided, high surface area, and a well-wetted state [19]. In the case of the PZQ-Gelucire 50/13 system (MPs D), the dissolution profiles were significantly different from raw PZQ (f_2 = 24.3), confirming its ability in enhancing the dissolution drug dissolution rate. Hence, although the MPs had bigger particle sizes than raw PZQ, the greater solubility of microparticles D (Figure 5) represented the main driving force in the dissolution rate enhancement. MPs A and B displayed the highest dissolution profiles, without being different from each other (f_2 >50), while being different from MPs D, with f_2 values of 34.6 and 27.2, respectively. In the case of MPs A containing the cryo-coground system, the combination of the drug in the nano-crystalline state and amorphous form, derived from the milling process, with the dispersion into Gelucire 50/13 contributed to the increase of the dissolution performance. For MPs B, in addition to the high solubility of the polymorphic form (Figure 5), the high dissolution rate of the drug could be attributed to the solubilization of Form B, into the molten carrier, during the MP preparation, forming a single-phase system. Thus, the solid state of the activated drug might be changed.

In order to better elucidate the drug solid state into the MPs, thermal analysis by means of DSC and HSM were performed. DSC studies of microparticles (graph not shown) revealed the presence of the carrier melting peak in all formulations and the disappearance of the PZQ endotherm, apart from MPs C, which displayed a broad peak at about 130 °C. Since PZQ might have completely solubilized in the molten Gelucire, during the DSC scan, and its melting point might not have been detectable any longer (as observed in previous studies [17]), an additional HSM analysis was carried out (Figure 9). In the case of the PZQ-loaded microparticles (MPs D), at 45 °C, the fusion of Gelucire started at 48 °C and the PZQ acicular crystals were easily recognizable and dispersed in the melted carrier. Afterward, the gradual dissolution of the crystals took place and had completed at about 110 °C, with a clear and transparent appearance of the sample. Therefore, PZQ was still present in the MPs D in the form of acicular crystals, a characteristic that could not be deduced from the DSC thermogram. The disappearance of the drug melting peak in their DSC trace was mainly due to the dissolution of the drug in the melted Gelucire, during the analysis, rather than due to a possible amorphization or change in its solid state, during the spray congealing process. This also meant that PZQ maintained its starting acicular habitus, previously reported in [11,12], even after the spray congealing.

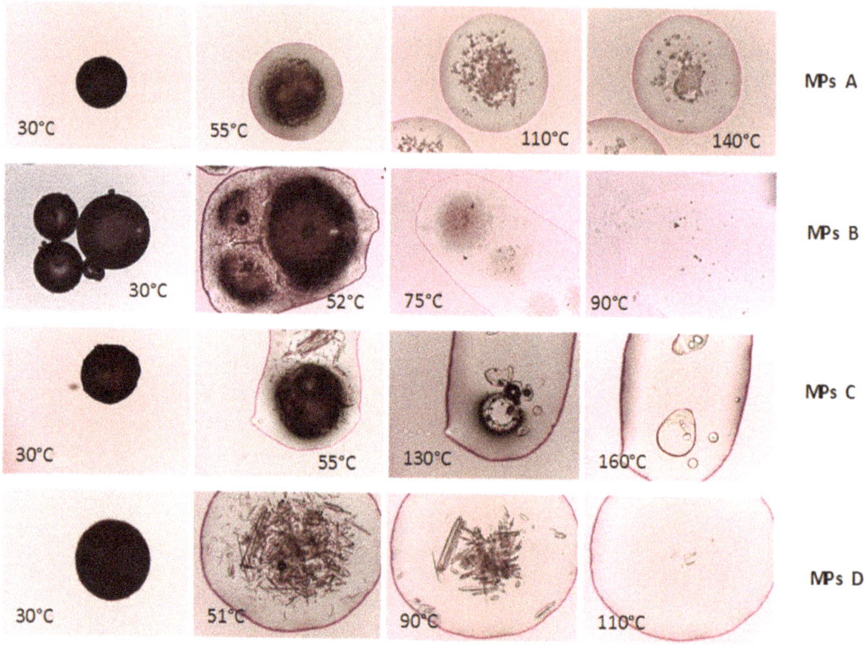

Figure 9. HSM images of microparticle formulations.

The microscopic analysis of MPs A, containing the cryo-coground, revealed that after the fusion of the Gelucire, the activated system started its melting at 55 °C (black mass reduction), and only at 150 °C, the fusion was completed. At this temperature, the PVP polymer (Tg = 150–180 °C) also appeared in a rubbery state, forming a system with a biphasic appearance, as clearly shown in the picture at 140 °C. These images confirmed the presence of a certain amount of PZQ, in the nanocrystalline form, within the cryo-comacinate (Figure 2). The observation of MPs C, during heating, revealed that after the fusion of Gelucire, both needle PZQ crystals and round PVP particles could be clearly recognized. Additionally, in this case, the disappearance of the PZQ crystals was gradual and started from 50–60 °C, depending on the crystal size and completed at 110 °C, while the PVP completed its phase transition at a higher temperature (160 °C). Finally, the images of MPs B, containing the PZQ polymorphic variety, highlighted the presence of dark spots with a crystal lattice appearance inside the molten droplets, at about 52 °C. Their amount and their dimensions were dramatically reduced, in comparison to what was seen for MPs D, attesting that the residual limited crystalline content of Form B was in the form of a very fine dispersion in the molten Gelucire. Increasing the temperature, they progressively solubilized, until their complete disappearance (at about 90° C, at a temperature lower than the Form B melting point, 112 °C).

FTIR spectra of the MPs (Figure 10) were dominated by the Gelucire 50/13 bands, due to its high content in the samples. In particular, in agreement with Brubach et al. [25], particularly evident were the bands in the 1200–1100 cm^{-1} range, while carbonyl stretching vibration appeared at 1715–1738 cm^{-1}·s. This permitted the detection of MPs A, MPs C, and MPs D in the two previously mentioned carbonyl stretching vibrations of PZQ, at 1623 cm^{-1} and 1646 cm^{-1}, which remained unchanged. On the other side, MPs B also displayed two distinct peaks at 1628 cm^{-1} and 1640 cm^{-1}, as is typical of the PZQ Form B [12].

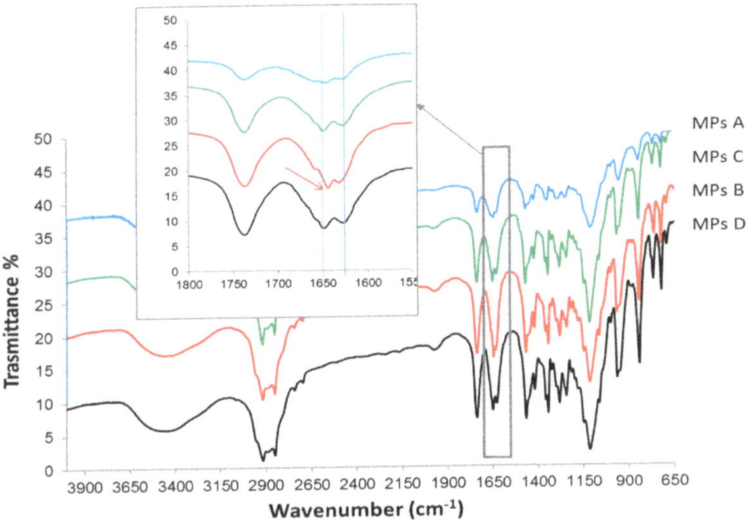

Figure 10. FT-IR spectra of the microparticle formulations.

To gather more information about the solid state of PZQ in MPs B, PXRD analysis was performed, and the spectra of both MPs B and D and their corresponding physical mixture (PM) were compared. PXRD pattern of the samples are reported in Figure 11. These results attested that in the microparticle MPs D, the main reflections of the starting PZQ were still visible, confirming the presence of the crystalline drug in this sample. The comparison with the corresponding physical mixture (PM (MPs D)), had an identical composition, revealing that was likely that during the microparticles formation, a slight amorphization or drug dissolution in the Gelucire 50/13 happened, anyway, though PZQ mainly remained in its original crystalline state. Analyzing MPs B and the relative physical mixture, the disappearance of the signal at about 4° of 2θ, in both samples, could be clearly noticed, which is typical of the pure PZQ [11,12]. In addition, the signal attributable to the polymorphic form B was less intense (as evidenced by the arrows), demonstrating the massive molecular dispersion/amorphization of Form B in the carrier. Only a very limited amount of Form B crystals was still present inside the microparticles. The diffractograms of the MPs B showed no evident change in the pattern, compared to the fresh samples, suggesting the stability of the PZQ amorphous form, within the microparticles. Additionally, the dissolution profile of the MPs B remained unchanged, after 1 year of storage, thus, confirming the stability of the pharmaceutical performance of this formulation.

Finally, the in vitro antischistosomal activity of the compounds was tested on NTS and adult *S. mansoni*. The IC_{50} values on NTS were 3.16 μg/mL for MPs B, 2.4 μg/mL for the Form B and 2.58 μg/mL for the PZQ. On *S. mansoni* adult worms we determined an IC_{50s} of 0.48 μg/mL for MPs B, and slightly lower values of 0.23 μg/mL for the milled PZQ and 0.13 μg/mL for the standard PZQ. Therefore, MPs was found to show a very similar bioactivity to that of PZQ. This meant that the microparticles permitted the release of the active drug and maintenance of the activity against *S. mansoni* adult worms.

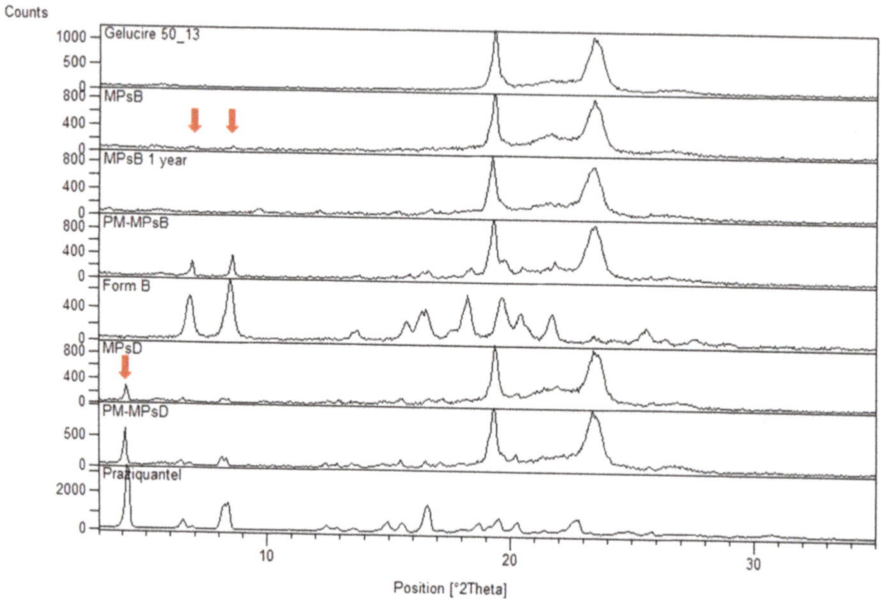

Figure 11. X-Ray Powder Diffraction Studies (PXRD) analysis of Gelucire 50/13, PZQ raw, PZQ Form B, MPs D (red arrow indicates a typical PZQ reflection), and MPs B (fresh samples and after 1 year storage) (red arrows indicate residual Form B signals) and their corresponding PM.

3. Material and Methods

3.1. Materials

Praziquantel (PZQ) Ph. Eur. grade ((11bRS)-2-(Cyclohexylcarbonyl)-1,2,3,6,7,11b-hexahydro-4-H-pyrazino[2,1-a]isoquinolin-4-one) was kindly donated by FATRO S.p.A., Ozzano Emilia, Bologna, Italy. PZQ impurities: Impurity A (2-Benzoyl-1,2,3,6,7,11b-hexahydro-4-H-pyrazino[2,1-a]isoquinolin-4-one) and impurity B (2-Cyclohexanecarbonyl-2,3,6,7-tetrahydro-pyrazino[2,1-a]isoquinolin-4-one) were of Ph. Eur. grade and were purchased from Endotherm Gmbh (Saarbruecken, Germany). Povidone (Kollidon K30, PVP K30) was supplied by BASF (Ludwigshafen, Germany) while Gelucire 50/13 was kindly supplied by Gattefossè (Milan, Italy).

3.2. Preparation of Activated Materials by Neat Grinding

These experiments were performed in a vibrational mill-Retsch MM400 (Retsch GmbH, Haan, Germany), which was equipped by two screw-type zirconium oxide jars, each with a capacity of 35 mL. A ceramic material like zirconium oxide was selected, due to its high density (5.9 g/cm^3). Three zirconium oxide spheres of 15 mm (weighing 10.72 g) were used as the milling media.

In particular, to obtain the PZQ crystalline polymorphic form, Form B, PZQ was ground by itself. The vibrational frequency was set at 20 Hz, for 240 min, without interruption. The amount of powder to be introduced in the milling jar was determined to be 0.800 g per jar, and no cooling was provided to the grinding jar, during room temperature milling. These process parameters were necessary to obtain the new polymorphic form, Form B, which were selected on the basis of our previous work [11]. A 20 Hz vibrational frequency was applied for a duration of 4 h. The experiment was performed, twice, to obtain enough material for both the experimental analysis and the spray congealing process.

For the preparation of the coground system, PZQ and PVP were manually gently mixed in an agate mortar, in a 1:1 drug-to-polymer weight ratio, for the standardized time of 3 min (batch size ranging about 1g). On the basis of previous experiences [11], the amount of powder to be introduced in the milling jar was determined to be 1.072 g per jar, and a vibrational frequency and a milling time of 60 min were set. Prior to milling, the jars containing the samples were immersed in liquid nitrogen for 1 min; re-cooling of the milling jars with liquid nitrogen for 1 min, was performed, every 15 min of milling. The experiment was repeated four times, to obtain enough material for both the experimental analysis and the spray congealing process.

Post milling, all samples were collected and stored in glass vials in the dark, in a desiccator, over anhydrous calcium chloride, at 25 °C, for further characterization and processing.

For comparison purposes, the properties of the raw PZQ and binary physical mixtures (prepared in the same agate mortar, by manually mixing PZQ and PVP), were investigated.

3.3. Preparation of Microparticles by Spray Congealing

Four batches of microparticles (indicated as MPs A–MPs D) using Gelucire® 50/13 as the carrier were produced by the spray-congealing technology, using an external-mix two-fluid atomizer, called Wide Pneumatic Nozzle (WPN) [18]. Gelucire® was heated up to about 10 °C, above its melting point. The active material (whose amount was determined in order to have a final percentage of PZQ equal to 15% *w*/*w*) was added to the molten carrier, as a powder, and magnetically stirred. The suspension or solution obtained was then loaded into the feeding tank of the spray congealing apparatus. The temperature of the feeding tank of the nozzle and the inlet air pressure were set at 60 °C and 3 bar, respectively. The atomized molten droplets hardened during the fall into a cylindrical cooling chamber, which was held at room temperature. Finally, the microparticles were collected from the bottom of the cooling chamber and stored in polyethylene closed bottles, at 25 °C. The batch size was 10 g for each formulation. The composition of the produced microparticles is reported in Table 2.

3.4. HPLC Analysis

HPLC analysis was performed for the quantification of both PZQ and the related impurities (or other detectable related products), in the activated samples, after the milling process and for the determination of the real drug content in the microparticles. HPLC analysis was also used to calculate the solubility and the dissolution properties of the different samples. The method was adapted from literature [26] and have been already validated and employed for PZQ quantification in previous studies [11–13]. Briefly, the HPLC system consisted of two mobile phase delivery pumps (LC-10ADvp, Shimadzu, Japan) and a UV–vis detector (SPD-10Avp, Shimadzu, Japan). An autosampler (SIL-20A, Shimadzu, Japan) was used to inject samples (20 µL) onto a Kinetex 5 µm C18 column (150 mm × 4.60 mm; Phenomenex, Bologna, Italy). The mobile phase was methanol and water at a ratio of 65:35 V/V, the flow rate and the wavelength of the UV detector were set at 1 mL/min and 220 nm, respectively. The linear calibration curve of the PZQ was obtained in the range of 0.4–40 mg/L (r^2 = 0.99985). The retention time of PZQ was about 5.5 min and the run time was set at 12 min. The PZQ standard solution was prepared by dissolving 10 mg of PZQ in 20 mL of methanol and diluting 1:200 in the mobile phase, in order to have a final standard PZQ concentration in solution of 2.5 mg/L. The calibration curve of the PZQ specified impurities (impurity A and B according to the PZQ monograph) [27], was obtained in the range of 0.01–1 mg/L (r^2 = 0.99927 and 0.99941, for impurity A and B, respectively). The retention time of the impurities were at 3.45 min and 11.2 min.

PZQ content was determined by dissolving a variable quantity of the specific sample (PZQ milled, PZQ:PVP cryo-comilled and loaded-microparticles) accurately weighed in 20 mL of methanol. In the case of the milled samples, the obtained solution was then diluted 1:200, in the mobile phase, corresponding to about 2.5 mg/L of PZQ, while for the microparticles, the solution was stirred for 24 h, protected from light, to assure a complete solubility. Finally, the solution was diluted in the mobile phase 1:100. For all samples prior to injection, the solutions were filtered through a 0.2 µm membrane filter and the drug content was assayed by HPLC. Each sample was analyzed in triplicates and the mean of the sum of the peak responses of PZQ was then calculated and have been reported along with the SD. Moreover, for the activated materials, the PZQ recovery was expressed as the percentage of PZQ, with respect to the sum of all peaks (PZQ and related impurities or other detectable related products). However, for the microparticles, the encapsulation efficiency (%) was then calculated by dividing the experimental drug content with the theoretical one and then multiplying it by 100.

3.5. Solubility and Dissolution Studies

Solubility and dissolution studies were carried out for both raw and activated materials and for the microparticles. For the solubility test, an excess amount of sample was added to the 10 mL of distilled water. The samples were magnetically stirred for 48 h, at 20 °C, and were protected from light by means of an aluminum foil, throughout the experiment. After equilibrium, the samples were centrifuged at 10,000 rpm, for 20 min, and the supernatant was filtered through a 0.20 mm membrane filter. After diluting the samples in a ratio of 1:200 in the mobile phase, they were finally analyzed by HPLC. The measurements were performed in triplicates, for each formulation, and the mean ± SD was reported. The statistical assessment of the obtained values was performed using one-way ANOVA, while comparison between means was performed using the t-Test. Differences were considered statistically significant for p values < 0.01.

In vitro dissolution studies of all samples were performed in 1000 mL of water, maintained at 37 ± 0.5 °C, and stirred at 100 rpm, using a paddle apparatus (Erweka DT800, Heusenstamm, Germany). Sink conditions were ensured by considering the PZQ water solubility at 37 °C of 215.0 ± 4.9 mg/L, as found by Trastullo et al. [7], and different amount of samples (corresponding to 16 mg of PZQ) were added to the vessel, according to their composition. Then aliquots of 2 mL were withdrawn at specified times, through a 8 µm filter, in order to only collect the dissolution media and leave the formulation in the vessel. At each sampling time, the PZQ content was assayed by the HPLC. Withdrawn samples were replaced with an equal volume of fresh medium. The dissolution tests were performed, at least in triplicates, and the mean ± SD was reported. Comparison between drug release profiles from the pellets were carried out, using the similarity factor (f_2).

$$f_2 = 50 * \log\left\{1 + \left[\frac{1}{n} * \sum_{t=1}^{n}(R_t - T_t)^2\right]^{-0.5} * 100\right\}$$

where n is the sampling number, R_t and T_t are the cumulative percentage drug dissolved of the reference and the test products at each time point t. For the f_2 values calculation, sampling number obtained within 20 min of the dissolution test were considered. The similarity factor fits the result between 0 and 100. The two drug release profiles were similar if the f_2 was greater than or equal to 50.

3.6. Wettability Studies

The measurements of contact angle were carried out according to the sessile drop method on compressed non-disintegrating disks, using deionized water as a wetting liquid, as previously reported for Gelucire-based matrices [28]. Disks were prepared by compressing the mixtures in a manual press Perkin Elmer, imparting a force of 1 tons for 1 min. The flat tablets produced were then analyzed with the Drop Shape Analysis System (Krüss DSA 30, Krüss GmbH, Germany), using a single drop of purified water (25 µL). The contact angle (between the disk and the drop) measurements, performed in triplicates, were taken after 10 s. The pure PZQ and milled PZQ were analyzed and the mean of at least three determinations, was calculated.

3.7. Viscosity Measurements

The viscosity of the molten mixtures, heated to the temperature set for the spray congealing process (60 °C), was measured with a Brookflied DVzT viscosimeter (Ametek GmbH, Lorch, Germany) using a spindle number RV04 and a rotating speed of 200 rpm.

3.8. Scanning Electron Microscopy (SEM)

Powder samples (pure PZQ, polymorphic form B and cryo-coground) were metallized with S150A Sputter Coater (Edwards High Vacuum, Crawley, West Sussex, UK) and then observed under a scanning electron microscope Leica Stereoscan 430i (Leica Cambridge Ltd., Cambridge, UK).

3.9. Environmental Scanning Electron Microscopy (ESEM)

The shape and surface characteristics of the microspheres were observed by environmental scanning electron microscopy (ESEM) (Quanta 200 FEI, FEI Company, Czech Republic).

3.10. Particle Size Analysis

Particle size measurements of the starting PZQ, Form B, cryo-coground and corresponding PM, were carried out, using a laser diffractometer (Malvern Mastersizer 2000, Malvern, UK). Before analysis, about 10 mg of each sample were dispersed by sonication in 100 mL of water containing 0.1% of polisorbate 80; sample containing PVP were dispersed in silicon oil (Silico DC 245 DOW Corning, Biesterfeld Spezialchemie GmbH, Hamburg, Germany).

Size distribution of microparticles was evaluated by sieve analysis, using a vibrating shaker (Octagon Digital, Endecotts, London UK), and a set of six sieves, ranging from 75 to 500 μm (Scientific Instrument, Milan, Italy).

3.11. Differential Scanning Calorimetry (DSC) Studies

Raw materials, activated materials, and MPs were analyzed by DSC, using a Perkin Elmer DSC 6 (Perkin Elmer, Beaconsfield, UK), with nitrogen as the purge gas (20 mL/min). The instrument was calibrated with indium and lead, for temperature, and with indium for the measurements of enthalpy. About 8–9 mg of the sample were placed in an aluminum pan and heated from 30 to 180 °C, at a scanning rate of 10 °C/min. Residual crystallinity (%) was then calculated by the measurements of the enthalpy of fusion (ΔH) of the PZQ, using the following equation: Residual crystallinity (%) = $(\Delta H_{sample} \times 100)/\Delta H_{PZQ}$.

3.12. Hot Stage Microscopy (HSM) Analysis

HSM was performed on raw PZQ, PZQ:PVP physical mixture, activated materials and MPs, using a hot stage apparatus (Mettler-Toledo S.p.A., Novate Milanese, Italy), under a Nikon Eclipse E400 optical microscope. Solubilization processes, phase transitions, or polymorphic changes occurred during the heating (from 25 to 200 °C, scanning rate 10 °C/min) were observed and images were captured by means of a Nikon Digital Net Camera DN100. The magnification was set at 10×.

3.13. X-Ray Powder Diffraction Studies (PXRD)

PXRD patterns were recorded using a Bruker AXS D5005 X-ray Diffractometer (Karlsruhe, Germany) with Cu-Kα radiation (1.5418 Å), monochromatized by a secondary flat graphite crystal. The analyses were performed in duplicates, using a current of 30 mA and the voltage was set at 40 kV. The powder samples were scanned in the range of 3°–35° of the 2θ angle, steps were of 0.05° of 2θ, and the counting time was of 5 sec/step. The samples subjected to the analysis were the following: Gelucire 50/13, raw PZQ, polymorphic form B, MPs B and MPs D, and the corresponding PMs.

3.14. Fourier Transform-Infrared Spectra (FT-IR) Analysis

Studies of infrared spectra of the excipients (PVP and Gelucire 50/13), raw PZQ, PZQ:PVP physical mixture, the activated materials, and the MPs were conducted with an IR spectrophotometer (Jasco FT-IR A-200, Pfungstadt, Germany), using the KBr disc method. The samples were mixed with KBr and compressed into a tablet (10 mm in diameter and 1 mm in thickness), using a hydraulic press

(Perkin Elmer, Beaconsfield, UK), at 3 tons, for 3 min. The scanning range was 650–4000 cm^{-1} and the resolution was 1 cm^{-1}.

3.15. Physical Stability During Storage

In order to check possible modifications of the solid state within time, PXRD analyses was done, and dissolution tests of the MPs were carried out after 1 year.

3.16. In Vitro Studies on S. Mansoni

Newly transformed Schistosomula (NTS) were obtained using the mechanical transformation method [29] and maintained at 37 °C, 5% CO_2, for 24 h before the experiments. One hundred NTS were placed in each well of a 96 well plate, containing Medium 199, supplemented with 5% iFCS and 1% penicillin/streptomycin. A concentration of 100 µg/mL PZQ and Form B and MPs B were used and serially diluted 1:3, up to 1.23 µg/mL. The plate was then incubated at 37 °C, 5% CO_2, and monitored at 24 h, 48 h, and 72 h. The assay was performed with biological triplicates, for each concentration. At each time point the NTS were assessed, microscopically, using a viability scale (3 = motile, no changes to morphology; 2 = reduced motility or some damage to the tegument noted; 1 = severe reduction to motility or damage to the tegument observed; 0 = dead). NTS incubated in the highest concentration of DMSO served as the negative controls.

Adult *S. mansoni* were collected from the hepatic portal and mesenteric veins of the infected mice. In a 24-well plate, 2–3 worm pairs were placed in the culture medium (RPMI supplemented with 5% iFCS and 1% penicillin/streptomycin) with 0.33, 0.11, and 0.037 µg/mL of the test compounds for 24 h, 48 h, and 72 h, at 37 °C, 5% CO2, in biological duplicates. Effects were assessed microscopically, as described above. Adult worms exposed to the maximum concentration of DMSO served as the negative control. IC_{50} values for both NTS and adult worms were calculated using the CompuSyn software (ComboSyn Inc., Paramus, NJ, USA).

4. Conclusions

Recent developments of praziquantel formulations that are useful for the preparation of a more appropriate pediatric dosage form, were demonstrated and discussed. The mechanochemical activation, whether in the presence or in the absence of povidone, had resulted in a remarkable increase of the solubility of the starting drug. The milling at room temperature and the cryo-comilling of PZQ, with the polymer, in a 1:1 weight ratio, enabled the improvement of the biopharmaceutical properties, by obtaining either a polymorph (Form B) or an amorphous/nanocrystalline form, with a reduction in crystallinity of about 70%, respectively. The results attested the great potentiality and effectiveness of the method, especially by neat grinding, where the absence of the polymer avoided the dilution of the drug and fulfilled the principle of minimizing the use of excipients in pediatric formulations. However, the unfavorable technological characteristics of the activated systems, such as poor wetting and flowability, caused by electrostatic and cohesive forces, might limit their use in the pharmaceutical production process. On the contrary, Gelucire® 50/13 microspheres, in which the activated systems were encapsulated by means of the spray congealing technology, showed a further increase in solubility and a marked improvement in the rate of dissolution of the drug. After the spray congealing process, the results evidenced that the cryo-coground and the milled PZQ formed either a solid dispersion (nanocrystalline and partial amorphous phase) or a solid solution (completely amorphous state), respectively. As a consequence, the MPs containing both activated systems showed a further increase of the biopharmaceutical properties, compared to the milled powders. The in vitro antischistosomal activity showed that MPs enabled the PZQ release, while maintaining its in vitro activity.

To conclude, the approach consisting in the association of the spray congealing with the mechanochemical activation grinding, in the absence of the polymer, was the most favorable, thus, it is a promising product for designing a new praziquantel formulation and a valid option for enhancing the performance of this antischistosomal drug, possibly permitting a significant reduction in the therapeutic dose.

Author Contributions: Conceptualization, B.A. and B.P.; Investigation and Analysis, S.B., D.Z., E.F., and F.L.; Data Curation, S.B., B.A., and B.P.; Writing – Original Draft Preparation, B.A., S.B., and B.P.; Writing – Review & Editing, J.K. and N.P.; Project Administration, B.A. and D.V.; Funding Acquisition, D.V. and N.P.

Funding: This research was funded by RFO 2017 from University of Bologna.

Acknowledgments: The authors thank Fatro s.r.l. for the gift of praziquantel E.P. grade and Gattefossé Italy for the supply of Gelucire 50/13.

Conflicts of Interest: The authors declare no conflict of interest.

References

1. Bergquist, R.; Utzinger, J.; Keiser, J. Controlling schistosomiasis with praziquantel: How much longer without a viable alternative? *Infect. Dis. Poverty* **2017**, *6*, 74–83. [CrossRef] [PubMed]
2. Siqueira, L.D.P.; Fontes, D.A.F.; Aguilera, C.S.B.; Timóteo, T.R.R.; Ângelos, M.A.; Silva, L.C.P.B.B.; de Melo, C.G.; Rolim, L.A.; da Silva, R.M.F.; Neto, P.J.R. Schistosomiasis: Drugs used and treatment strategies. *Acta Trop.* **2017**, *176*, 179–187. [CrossRef]
3. Utzinger, J.; N'Goran, E.K.; Caffrey, C.R.; Keiser, J. From innovation to application: Social-ecological context, diagnostics, drugs and integrated control of schistosomiasis. *Acta Trop.* **2011**, *120*, S121–S137. [CrossRef] [PubMed]
4. Colley, D.G.; Bustinduy, A.L.; Secor, W.E.; King, C.H. Human schistosomiasis. *Lancet* **2014**, *383*, 2253–2264. [CrossRef]
5. da Silva, V.B.R.; Campos, B.R.K.L.; de Oliveira, J.F.; Decout, J.L.; do Carmo Alves de Lima, M. Medicinal chemistry of antischistosomal drugs: Praziquantel and Oxamniquine. *Bioorg. Med. Chem.* **2017**, *25*, 3259–3277. [CrossRef] [PubMed]
6. WHO. *Model List of Essential Medicines for Children*, 6th ed. March 2017. Available online: http://www.who.int/medicines/publications/essentialmedicines/6th_EMLc2017_FINAL_amendedAug2017.pdf?ua=1 (accessed on 11 June 2018).
7. Trastullo, R.; Dolci, L.S.; Passerini, N.; Albertini, B. Development of flexible and dispersible oral formulations containing praziquantel for potential schistosomiasis treatment of pre-school age children. *Int. J. Pharm.* **2015**, *495*, 536–550. [CrossRef] [PubMed]
8. Olliaro, P.L.; Vaillant, M.; Hayes, D.J.; Montresor, A.; Chitsulo, L. Practical dosing of praziquantel for schistosomiasis in preschool-aged children. *TMIH* **2013**, *18*, 1085–1089. [CrossRef] [PubMed]
9. Yang, L.; Geng, Y.; Li, H.; Zhang, Y.; You, J.; Chang, Y. Enhancement the oral bioavailability of praziquantel by incorporation into solid lipid nanoparticles. *Pharmazie* **2009**, *64*, 86–89. [CrossRef] [PubMed]
10. Chaud, M.V.; Lima, A.C.; Vila, M.M.D.C.; Paganelli, M.O.; Paula, F.C.; Pedreiro, L.N.; Gremião, M.P.D. Development and evaluation of praziquantel solid dispersions in sodium starch glycolate. *Trop. J. Pharm. Res.* **2013**, *12*, 163–168. [CrossRef]
11. Perissutti, B.; Passerini, N.; Trastullo, R.; Keiser, J.; Zanolla, D.; Zingone, G.; Voinovich, D.; Albertini, B. An explorative analysis of process and formulation variables affecting comilling in a vibrational mill: The case of praziquantel. *Int. J. Pharm.* **2017**, *533*, 402–412. [CrossRef] [PubMed]
12. Zanolla, D.; Perissutti, B.; Passerini, N.; Chierotti, M.R.; Hasa, D.; Voinovich, D.; Gigli, L.; Demitri, N.; Geremia, S.; Keiser, J.; et al. A new soluble and bioactive polymorph of praziquantel. *Eur. J. Pharm. Biopharm.* **2018**, *127*, 19–28. [CrossRef]
13. Zanolla, D.; Perissutti, B.; Passerini, N.; Invernizzi, S.; Voinovich, D.; Bertoni, S.; Melegari, C.; Millotti, G.; Albertini, B. Milling and comilling Praziquantel at cryogenic and room temperatures: Assessment of the process-induced effects on drug properties. *J. Pharm. Biomed. Anal.* **2018**, *153*, 82–89. [CrossRef]
14. EMA Guideline on Pharmaceutical Development of Medicines for Paediatric Use. 2013. Available online: https://www.ema.europa.eu/en/documents/scientific-guideline/guideline-pharmaceutical-development-medicines-paediatric-use_en.pdf (accessed on 6 March 2019).

15. Passerini, N.; Perissutti, B.; Albertini, B.; Franceschinis, E.; Lenaz, D.; Hasa, D.; Locatelli, I.; Voinovich, D. A new approach to enhance oral bioavailability of Silybum Marianum dry extract: Association of mechanochemical activation and spray congealing. *Phytomed* **2012**, *19*, 160–168. [CrossRef]
16. Albertini, B.; Di Sabatino, M.; Melegari, C.; Passerini, N. Formulation of spray congealed microparticles with self-emulsifying ability for enhanced glibenclamide dissolution performance. *J. Microencapsul.* **2015**, *32*, 181–192. [CrossRef]
17. Perissutti, B.; Rubessa, F.; Princivalle, F. Solid dispersions of carbamazepine with Gelucire 44/14 and 50/13. *S.T.P. Pharma Sci.* **2000**, *10*, 479–484.
18. Albertini, B.; Di Sabatino, M.; Melegari, C.; Passerini, N. Formulating SLMs as oral pulsatile system for potential delivery of melatonin to pediatric population. *Int. J. Pharm.* **2014**, *469*, 67–79. [CrossRef]
19. Qi, S.; Marchaud, D.; Craig, D.Q.M. An investigation into the mechanism of dissolution rate enhancement of poorly water-soluble drugs from spray chilled gelucire 50/13 microspheres. *J. Pharm. Sci.* **2010**, *99*, 262–274. [CrossRef]
20. Šagud, I.; Zanolla, D.; Perissutti, B.; Passerini, N.; Škorić, I. Identification of degradation products of praziquantel during the mechanochemical activation. *J. Pharm. Biomed. Anal.* **2018**, *159*, 291–295. [CrossRef]
21. Hasa, D.; Voinovich, D.; Perissutti, B.; Grassi, G.; Fiorentino, S.; Farra, R.; Abrami, M.; Colombo, I.; Grassi, M. Reduction of melting temperature and enthalpy of drug crystals: Theoretical aspects. *Eur. J. Pharm. Sci.* **2013**, *50*, 17–28. [CrossRef]
22. Borrego-Sánchez, A.; Hernández-Laguna, A.; Sainz-Díaz, C.I. Molecular modeling and infrared and Raman spectroscopy of the crystal structure of the chiral antiparasitic drug Praziquantel. *J. Mol. Model.* **2017**, *23*, 106. [CrossRef]
23. Espinosa-Lara, J.C.; Guzman-Villanueva, D.; Arenas-García, I.J.; Herrera-Ruiz, D.; Rivera-Islas, J.; Román-Bravo, P.; Morales-Rojas, H.; Höpfl, H. Cocrystals of Active Pharmaceutical Ingredients—Praziquantel in Combination with Oxalic, Malonic, Succinic, Maleic, Fumaric, Glutaric, Adipic, And Pimelic Acids. *Cryst. Growth Des.* **2013**, *13*, 169–185. [CrossRef]
24. Costa, E.D.; Priotti, J.; Orlandi, S.; Leonardi, D.; Lamas, M.C.; Nunes, T.G.; Diogo, H.P.; Salomon, C.J.; Ferreira, M.J. Unexpected solvent impact in the crystallinity of praziquantel/poly (vinylpyrrolidone) formulations. A solubility, DSC and solid-state NMR study. *Int. J. Pharm.* **2016**, *511*, 983–993. [CrossRef]
25. Brubach, J.B.; Ollivon, M.; Jannin, V.; Mahler, B.; Bourgaux, C.; Lesieur, P.; Roy, P. Structural and thermal characterization of mono- and diacyl polyoxyethylene glycol by infrared spectroscopy and X-ray diffraction coupled to differential calorimetry. *J. Phys. Chem. B* **2004**, *108*, 17721–17729. [CrossRef]
26. Sun, Y.; Bu, S. Simple, cheap and effective high-performance liquid chromatographic method for determination of praziquantel in bovine muscle. *J. Chromatogr. B* **2012**, *899*, 160–162. [CrossRef]
27. Praziquantel. In *European Pharmacopoeia 8.0*; Directorate for the Quality of Medicines of the Council of Europe (EDQM): Strasbourg, France, 2014; pp. 3086–3087.
28. Kim, M.S.; Kim, J.S.; Hwang, S.J. Enhancement of wettability and dissolution properties of cilostazol using the supercritical antisolvent process: Effect of various additives. *Chem. Pharm. Bull.* **2010**, *58*, 230–233. [CrossRef]
29. Lombardo, F.C.; Pasche, V.; Panic, G.; Endriss, Y.; Keiser, J. Life cycle maintenance and drug-sensitivity assays for early drug discovery in Schistosoma mansoni. *Nat. Protoc.* **2019**, *14*, 461–481. [CrossRef]

International Journal of
Molecular Sciences

MDPI

Article

Tridimensional Retinoblastoma Cultures as Vitreous Seeds Models for Live-Cell Imaging of Chemotherapy Penetration

Ursula Winter [1], Rosario Aschero [2], Federico Fuentes [3], Fabian Buontempo [4], Santiago Zugbi [4], Mariana Sgroi [5], Claudia Sampor [6], David H. Abramson [7], Angel M. Carcaboso [8] and Paula Schaiquevich [1,4,*]

1 National Scientific and Technical Research Council (CONICET), Buenos Aires CP1425, Argentina; winter.u.a@gmail.com
2 Pathology Service, Hospital de Pediatría Prof. Dr. JP Garrahan, Buenos Aires CP1245, Argentina; rosarioaschero@gmail.com
3 Institute of Experimental Medicine (IMEX), National Academy of Medicine, Buenos Aires CP1425, Argentina; fedefuentes@gmail.com
4 Pharmacy, Hospital de Pediatría Prof. Dr. JP Garrahan, Buenos Aires CP1245, Argentina; fabuontempo@yahoo.com.ar (F.B.); santiagozugbi@gmail.com (S.Z.)
5 Ophthlamology Service, Hospital de Pediatría Prof. Dr. JP Garrahan, Buenos Aires CP1245, Argentina; marianasgroi@gmail.com
6 Hematolog-Oncology Service, Hospital de Pediatría Prof. Dr. JP Garrahan, Buenos Aires CP1245, Argentina; claudiasampor@hotmail.com
7 Ophthalmic Oncology Service, Memorial Sloan-Kettering Cancer Center, New York, NY 10065, USA; Abramsod@mskcc.org
8 Institut de Recerca Sant Joan de Deu, Barcelona, Spain and Department of Pediatric Hematology and Oncology, Hospital Sant Joan de Deu, 08950 Barcelona, Spain; amontero@fsjd.org
* Correspondence: paulas@conicet.gov.ar; Tel.: +54-11-4122-6000 (ext. 7138)

Received: 20 December 2018; Accepted: 30 January 2019; Published: 2 March 2019

Abstract: A preclinical model could aid in understanding retinoblastoma vitreous seeds behavior, drug penetration, and response to chemotherapy to optimize patient treatment. Our aim was to develop a tridimensional in vitro model of retinoblastoma vitreous seeds to assess chemotherapy penetration by means of live-cell imaging. Cell cultures from patients with retinoblastoma who underwent upfront enucleation were established and thoroughly characterized for authentication of human tumor origin. The correlation of the in vitro tridimensional structures resembling human spheres and dusts vitreous seeds was established. Confocal microscopy was used to quantify real-time fluorescence of topotecan as a measure of its penetration into different sizes of spheres. Cell viability was determined after chemotherapy penetration. The in vitro spheres and dusts models were able to recapitulate the morphology, phenotype, and genotype of patient vitreous seeds. The larger the size of the spheres, the longer the time required for the drug to fully penetrate into the core ($p < 0.05$). Importantly, topotecan penetration correlated with its cytotoxic activity. Therefore, the studied tridimensional cell model recapitulated several characteristics of vitreous seeds observed in patients with retinoblastoma and were successfully used to assess live-cell imaging of chemotherapy penetration for drug distribution studies.

Keywords: tumorspheres; retinoblastoma; topotecan; penetration; confocal microscopy

1. Introduction

Retinoblastoma is the most common intraocular tumor of childhood affecting 1 in 15,000 to 1 in 18,000 live births [1–3]. Retinoblastoma is highly curable if diagnosed in the early stages. For decades, surgical removal of the affected eye (or both eyes in cases of bilateral retinoblastoma) has been the first choice. Later on, the introduction of chemotherapy provided the basis for eye preservation. Over the last decade, retinoblastoma treatment has radically changed from using systemic chemotherapy infusion, with low bioavailability in the ocular tissues but high in plasma, and thus severe systemic adverse events, to highly-selective novel techniques of drug delivery, including direct ophthalmic artery chemotherapy and intravitreous injection [3–5]. Though a striking increase in ocular survival has been attained, eyes with tumors that grow from the retina to the vitreous humor, namely vitreous seeds, are more difficult to cure and may relapse. Therefore, vitreous seeds remain a challenge in the management of intraocular retinoblastoma and removal of the affected eye may be the only treatment option [5,6]. Recently, Munier et al. classified vitreous seeds into "dust", "spheres", and "clouds" based on their heterogeneous appearance at fundoscopy [6]. Dusts are composed of loose tumor cells in the vitreous, clouds are dense tumor fragments formed by translocation of the primary tumor content to the vitreous, and spheres are translucent solid tumors formed by further clonal growth of the dust or the cloud, or by sprouting of the primary retinal tumor [7]. Each class of seeds required a different cumulative dose and number of intravitreal injections of melphalan to achieve complete response to treatment [7–10]. Later on, each class of seeds is also correlated with histopathological features [11]. In general, clouds need the greatest number of melphalan injections (and cumulative doses) followed by spheres, and finally dusts [8–10]. Dusts might be more sensitive to treatment because they are more accessible to drugs, as they are composed of clusters of loose cells, while spheres or clouds grow as tight clusters with different layers of viable cells and may hamper homogeneous distribution of the drug after intravitreal injections [8,9]. Thus, assessment of the capacity of chemotherapy to penetrate into the tumor seeds, and thereby to become available to exert its cytotoxic effect would certainly add to the knowledge on and improvement of chemotherapy use and patient management.

Among the antineoplastic agents used for retinoblastoma, topotecan has been widely used based on its effect in preclinical models and clinical studies [12]. To quantify topotecan penetration into living tumorspheres, confocal microscopy could be used as topotecan in a fluorescent drug [13,14]. This powerful technique allows for the visualization of structures, including living cells and even thick living specimens, after noninvasive serial optical sectioning [15].

Over the last years, the development of three-dimensional (3D) in vitro tumor models allowed the establishment of structural complex cell-cell interactions in systems that resemble in vivo tumors [16,17]. Still, further developments using 3D tumor culture technology are needed for translational studies in retinoblastoma. Thus, we explored a scaffold-free culture method allowing self-aggregation of tumor cells into a 3D structure to closely resemble vitreous seeds. In serum-free, growth factor-supplemented culture conditions, retinoblastoma cells grow in suspension and form tumorspheres that may recapitulate vitreous seeds of the spheres class observed in vivo, providing a suitable in vitro tumor model to study drug penetration [18,19]. In contrast, commercial retinoblastoma cell lines grow as loose aggregates, resembling vitreous seeds classified as dust.

Thus, the aim of our study was to establish and characterize tumorpsheres derived from patients with intraocular retinoblastoma and use them to assess if differences in tridimensional conformations and sphere size affect topotecan penetration using confocal microscopy. The developed tridimensional cell model resembles several characteristics of vitreous seeds in pediatric patients with intraocular tumors. This model was useful to assess live-cell imaging of chemotherapy penetration for drug distribution studies and cytotoxicity assessment.

2. Results

2.1. Patient-Derived Tumorspheres Resemble the Original Tumor

Early-passage cell cultures derived from intraocular tumors of two patients who underwent upfront enucleation without receiving previous treatment were obtained and named after the codes

HPG-RBT-12L and HPG-RBT-26. They were considered established after three passages, each one performed during the log-phase growth.

Analysis of DNA showed that in both primary cells the *RB1* gene mutations were germline and single base substitutions. *RB1* mutation in HPG-RBT-12L cells, as well as in the parental tumor was identified as a point mutation in exon 15 (NM_000321.2(RB1):c.1421G>T) associated with altered splicing, while for the HPG-RBT-26 cells and tumor in exon 23 (NM_000321.2(RB1):c.2359C>T(p.Arg787*)), it was associated with a premature stop codon. Moreover, the short tandem profile (STR) for the DNA from the cell lines was identical to that obtained for the tumor DNA (Tumor samples: HPG-RBT-12T and HPG-RBT-26T), confirming the origin of the cell cultures (Table 1). There was no significant overlap between both primary cell lines and no cell line corresponded to the DNA profile of the present retinoblastoma cell lines in the STR database of the American Type Culture Collection.

Table 1. Short Tandem Repeat analysis of the cell lines and matched tumors.

Marker	Cell Line	Tumor	Cell Line	Tumor
	HPG-RBT-12L	**HPG-RBT-12T**	**HPG-RBT-26**	**HPG-RBT-26T**
Amelogenin	X	X	X	X
D3S1358	14–17	14–17	17–18	17–18
D13S317	12	12	14	14
Penta E	6–13	6–13	11–14	11–14
D16S539	10–13	10–13	12	12
D18S51	12–15	12–15	12–15	12–15
CSF1PO	12	12	12–14	12–14
Penta D	10–12	10–12	8–14	8–14
TH01	8	8	6	6
VWA	15–16	15–16	15–19	15–19
D21S11	29–31	29–31	29–33.2	29–33.2
D7S820	10–11	10–11	10–11	10–11
D5S818	11–13	11–13	11–13	11–13
TPOX	11–12	11–12	8	8
D8S1179	14	14	8	8
FGA	21–25	21–25	22–23	22–23

As shown in Figure 1, both primary cell cultures were positive for cone-rod homebox transcription factor (CRX) by RT-qPCR and presented similar expression levels with respect to the commercial cell line Y79.

Both small and large patient-derived tumorspheres, as well as Y79 cells, were positive stained for arrestin 3 and synaptophysin, confirming the retinal and neuroectodermic tumor cell origin (Figure 2A,B). Notwithstanding the size of the 3D model, all of them were histologically composed of viable and proliferative tumor cells, as shown in Figure 2C,D, respectively.

Taking into account the morphology, HPG-RBT-12L and HPG-RBT-26 cell cultures grew as freely floating tridimensional structures in serum-free medium, matching the classification of seeding for "spheres". These 3D structures formed spontaneously in the culture media resulting in different sizes. A representative image of small and large tumorspheres from both patients is depicted in Figure 3 showing the morphological features. Small tumorspheres from patient 1 and 2 (Figure 3A,B) showed a median (range) diameter of 91 μm (53–109) and 53 μm (41–79), respectively. In the case of large tumorspheres (Figure 3C,D), median diameters (range) were 353 μm (341–381) and 356 μm (334–385) for patients 1 and 2, respectively.

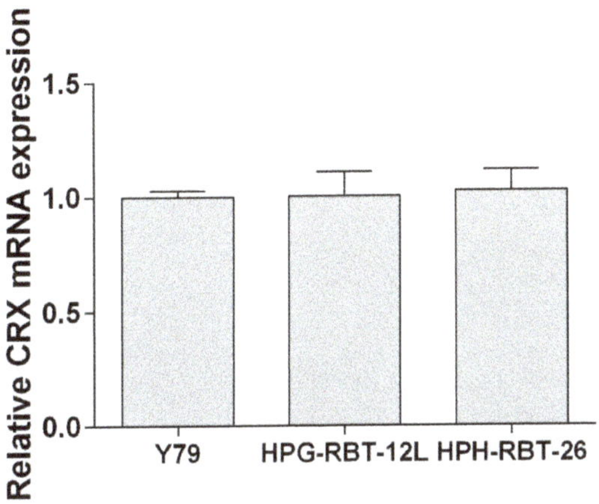

Figure 1. Expression of cone-rod homebox transcription factor in patient-derived cells.Expression of the cone-rod homebox transcription factor (CRX) was determined by RT-qPCR and transcript levels were quantified relative to the housekeeping gene and then normalized by the level of mRNA CRX detected in Y79 cells. Data is shown as mean (SEM).

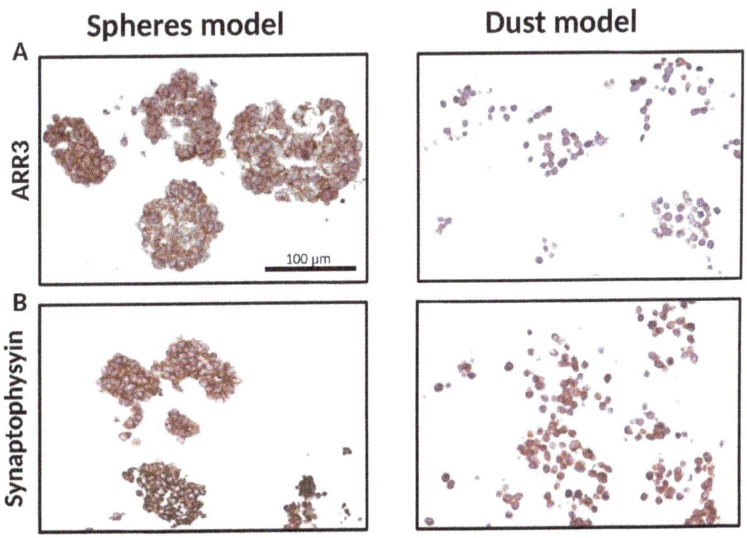

Figure 2. *Cont.*

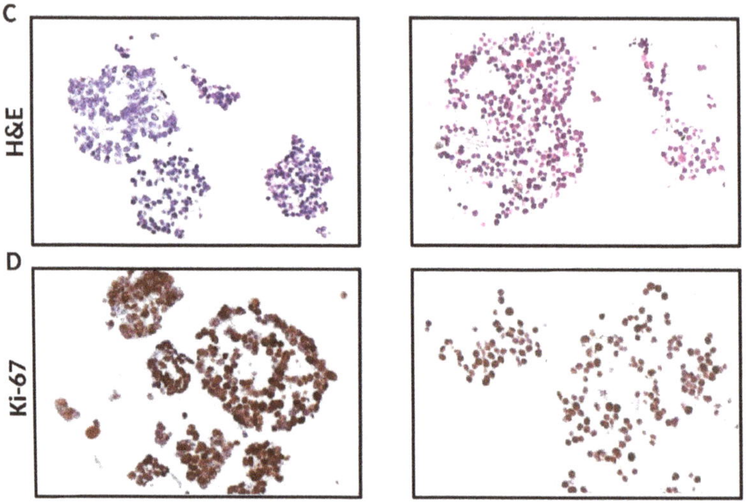

Figure 2. Characterization of the tridimensional retinoblastoma cell structures. Representative small and large 3D tumor cell cultures resembling both spheres (HPG-RBT-12L) and dusts (Y79) show: (**A**) Cone photoreceptor-specific staining (ARR3). (**B**) synaptophysin stain confirming neuronal characteristics of the cells. (**C**) strong nuclear basophilic staining demonstrating they are composed of viable tumor cells; and (**D**) Ki-67 expression showing positive brown-staining all through the 3D structure. Scale bar, 100 μm. Images taken at 20× magnification. Abbreviations: ARR3, arrestin3; H&E, hematoxylin and eosin.

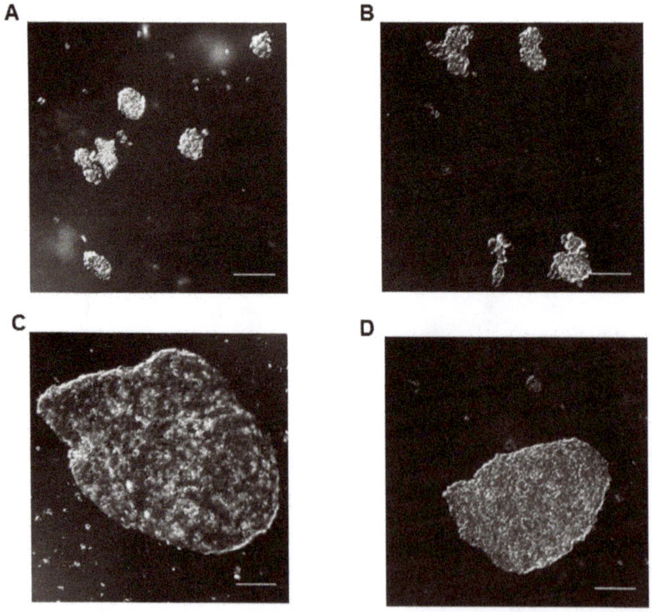

Figure 3. Representative confocal microscope images of in vitro tumorspheres derived from intraocular retinoblastoma tumors. Small (**A**,**B**) and large (**C**,**D**) spherical aggregates of primary tumor cells (HPG-RBT-12L and HPG-RBT-26 cells) obtained by culturing samples from intraocular tumors of patients. Images taken at 20× magnification. Scale bar, 100 μm.

2.2. Topotecan Penetration and Live-Cell Imaging

Confocal microscopy images were collected from 3D tumor models exposed to topotecan. Supplementary Video 1 shows sequentially acquired and recorded images from a representative experiment in a large HPG-RBT-12L tumorsphere before and 5 min after topotecan administration. The penetration of the green color corresponds to topotecan fluorescence from the bathing solution into the outer layers of the tumorsphere and thereafter into the core as a function of time. Different rates of topotecan penetration were observed in tridimensional structures formed by Y79 cells (fast penetration) and in patient-derived tumorspheres (slow penetration) (Figure 4). A uniform green signal corresponding to topotecan fluorescence was observed even in large clusters, as topotecan penetrated into the Y79 aggregates in less than 0.5 min due to the lack of strong cell-cell interactions, as shown in Figure 4A. Y79 cells grow as loose clusters in suspension, resembling dust vitreous seeds. This result was in contrast to the delayed penetration in patient-derived spheres. As shown in Figure 4B, after 0.5 min of topotecan exposure, the spheres are black due to the lack of topotecan penetration into the 3D structure. In large spheres, topotecan penetration into the tumorsphere was evidenced by the progressive acquisition of green fluorescence from the outer to the core of the sphere over 5 min of drug exposure (Supplementary Video 1).

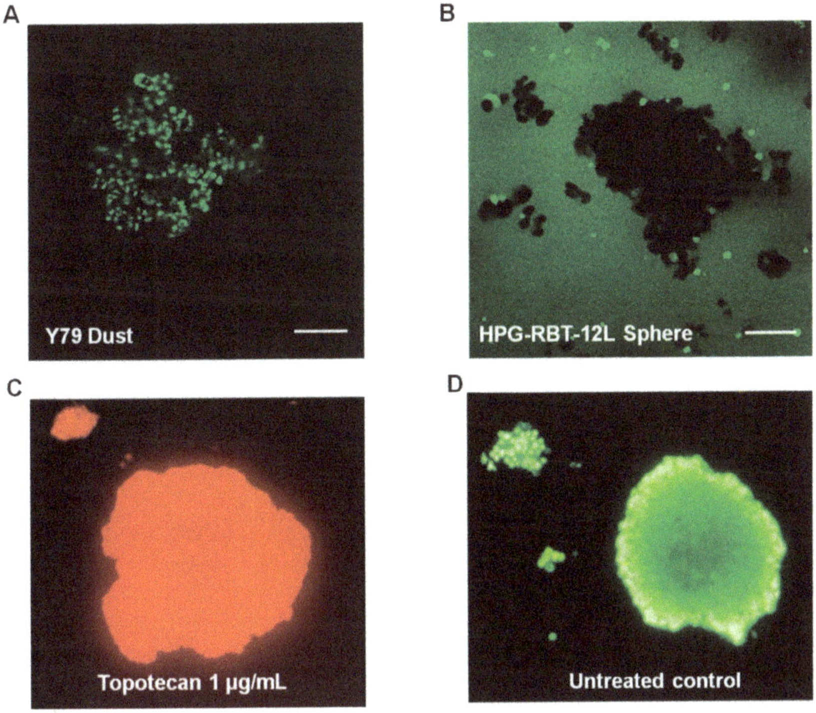

Figure 4. Topotecan penetration and activity in retinoblastoma 3D structures. (**A**) Topotecan fluorescence as green color after 0.5 min of incubation in Y79 cells (dust model) and (**B**) in small and large HPG-RBT-12L cells (sphere model). Note the absence of staining within small and large spheres after incubation with topotecan for 0.5 min. (**C**) Cytotoxic activity of topotecan (1 µg/mL) shown as accumulation of ethidium bromide in a tumorsphere treated for 10 min; dead nucleated cells are stained red. (**D**) Control tumorsphere (untreated control) stained green with acridine orange. Scale bar, 100 µm.

The time for maximum topotecan accumulation into tumorspheres (t_{max}) was defined as the time to achieve maximum fluorescence in the core of the sphere. A decrease in tumorsphere size correlated with a shorter time for topotecan accumulation from the bathing solution into the core of the tumorsphere for both cell models ($p < 0.05$; Table 2). However, no significant differences in t_{max} were found between cell models for a similar size range ($p > 0.05$).

Table 2. Time to achieve maximum fluorescence in the core of small and large retinoblastoma spheres from models HPG-RBT-12L and HPG-RBT-26.

	t_{max} (min) HPG-RBT-12L	t_{max} (min) HPG-RBT-26	P Value (t Test)
Small spheres	1.50 (0.29)	1.62 (0.12)	0.677
Large spheres	2.75 (0.14)	2.67 (0.17)	0.721
p-value (t test)	0.008	0.004	

Abbreviations: t_{max}, time to achieve maximum fluorescence in the core of the sphere. Data is showed as mean (SEM) of three independent experiments for each cell model.

2.3. Topotecan Cytotoxicity

In order to characterize the sensitivity of the cell lines to the chemotherapeutic agent, we determined the IC50 of topotecan for the three cell cultures. Mean (range) topotecan IC50 in Y79, HPG-RBT-12L, and HPG-RBT-26 cells was 28.6 nM (26.1–30.5), 12.3 nM (6.8–13.7), and 6.3 nM (5.4–8.6), respectively. Topotecan IC50 in Y79 was similar to the value previously reported and significantly higher than the values attained in HPG-RBT-12L and HPG-RBT-26 ($p < 0.05$). However, no significant difference could be detected in topotecan IC50 between primary cell cultures ($p > 0.05$, Figure S1).

Because topotecan needs at least 10 min to exert its cytotoxic activity in commercial cell cultures [20], we assessed the cytotoxic effect of topotecan penetration into the spheres by ethidium bromide staining after at least 10 min of topotecan exposure. As we used a concentration of topotecan that was clinically relevant and higher than the IC50 determined in the cell cultures, it was expected that topotecan would exert a cytotoxic effect in the tridimensional cell clusters. Complete cell damage was observed by ethidium bromide red staining as shown in Figure 4C, compared to a control (topotecan-free) tumorsphere stained as green with acridine orange (Figure 4D), supporting the hypothesis that once inside the cells that composed the spheres, topotecan exerted cytotoxic activity.

3. Discussion

In this study, we established patient-derived retinoblastoma tridimensional culture resembling the architecture of spheres, a class of vitreous seeds, and for the first time used live tumorsphere imaging to gain insight into the penetration process of a chemotherapeutic agent in such tumor models. After a thorough characterization of the cell lines, we took advantage of the fluorescence of topotecan in conditions compatible with live-cell monitoring to visualize and quantify its penetration into the core of tumorspheres derived from two patients with intraocular retinoblastoma and in cell clusters of Y79 resembling spheres and vitreous dust seeds, respectively [7,8,11]. The quantification of topotecan at different times based on the fluorescence intensity showed that drug penetration into the core of the cell aggregates was immediate in dust, and faster in small than in large spheres. This result is consistent with pathological observations stating that larger seeds consist of a greater number of cells arranged in multilayers around a central core. Therefore, spheres are not mere aggregates of cells but 3D structures that may hamper drug accessibility.

The property of topotecan fluorescence has been used to quantify the drug in several biological matrices for pharmacokinetic studies [12,14]. In addition, different researchers have exploited this property to study the release rates of topotecan from liposomes and nanoparticles as part of the development of new formulations, or to assess temporal variations in the disposition of topotecan in living cells or even whole animals using non-invasive fluorescence microscopy [21–24]. As a novel application, we previously reported the use of fluorescent visualization of topotecan to evaluate the safety of the injection technique and adequate drug delivery in retinoblastoma patients using a portable Wood's lamp [13].

Ours is the first study on the imaging of topotecan distribution in living retinoblastoma cells spontaneously clustered in a 3D structure, as occurs in vivo. Previous studies on topotecan penetration in tumorspheres were performed in other tumor cell lines forced to aggregate into a sphere by gravity using the hanging drop technique [24,25]. In brief, the hanging drop technique consists of plating droplets of cells in culture media on a lid, inverting the lid over a dish, and place it in an incubator to allow for the 3D cell structure to form. Other researchers have worked with bioengineered spheres consisting of commercial Y79 cells and gelatin microparticles as a scaffold to obtain the 3D structure [26]. Specifically in that study, Y79 cells showed genomic alterations, including changes in the expression of the collagen family when using microparticles as the matrix on which they were grown. Thus, it may be expected that an altered drug penetration into the Y79-microparticle induced by the culture conditions may not resemble the in vivo behavior of tumor seeds. In contrast, our patient-derived tumor cells spontaneously grow as tumorspheres without the need for external forces, or polymer support resembling floating vitreous seeds, as occurs in patients.

After upfront enucleation of two patients affected with retinoblastoma, we established two patient-derived cell models. Then, we first confirmed the origin of the cell lines by different genomic analysis and the identification of the expression of markers of retinal and neuronal tumor origin to sustain the basis of all further developments. Importantly, cell-growth conditions avoiding the use of serum for culturing the cells from the biopsy have been shown to dramatically improve tumor cell selection and to avoid differentiation of the cell culture to a different phenotype than the parental tumors [18,19]. Moreover, it is important to mention that cell-growth conditions and the cell line passage number may result in different pharmacological sensitivity of the cell cultures as previously reported for retinoblastoma and other tumors [27,28]. As shown in the present study, commercial retinoblastoma cells spontaneously grow in loose aggregates of cells that allow almost instantaneous drug penetration, likely resembling the dust vitreous seeds. Therefore, caution should be exercised if using commercial cell lines but also different cell growth conditions to establish primary cell cultures of retinoblastoma to use them as a basis of pharmacological screening and translation into the clinics.

The structural properties of the patient-derived tumorspheres were similar to vitreous seeds classified as "spheres", based on morphological and histopathological observations [7,8,11]. Patient sphere sizes were reported to be in the range of 15 to 300 μm or even larger. In line with this observation, our tumorspheres obtained from in vitro culture are also within that range (Figure 3). In addition, small patient spheres displayed a homogeneous positive staining for Ki-67 as a marker of cell proliferation widely used in tumor tissue biopsies. We also showed that large tumorspheres are composed of multilayers of cells and that all cells located in the core or the surroundings were proliferative with almost identical expression levels of Ki-67. Thus, our model was able to recapitulate the morphology and phenotype of dusts and spheres of human vitreous seeds.

Importantly, the 3D multilayer cell structures may hinder drug penetration, and thus, diffusion studies of active drugs in each in vitro model, and eventually the evaluation of drug sensitivity in 3D models should be performed [29,30]. We demonstrated, by means of confocal microscopy of topotecan-treated spheres, that the drug penetrated upon the core of the tridimensional structures in all cases. Also, we observed different rates of drug penetration, depending on the tridimensional structure, resembling dusts or spheres, and the size of the in vitro spheres. Topotecan penetration was almost immediate in clusters of Y79 cells as they gathered in a loose packing 3D structure. Thus, although we observed a lower sensitivity to topotecan (higher topotecan IC50) of Y79 compared with patient-derived tumor cells, the drug exerts its cytotoxic activity immediately after exposure of the loose aggregates of Y79 cells. On the contrary, as a result of the multicellular architecture of the 3D structures that allow cell-cell interactions, patient-derived spheres presented longer times of drug penetration into the core and also differences between sphere sizes. Nonetheless, topotecan penetrated through all the evaluated 3D models and exerted a cytotoxic effect. These results obtained in vitro show a strong cytotoxic activity of topotecan in Y79, and primary cell cultures may differ from the clinical observation in retinoblastoma patients. Topotecan is usually administered with carboplatin or melphalan based on the favorable ocular disposition, the synergistic cytotoxicity, and because of the lack of documented clinical efficacy if administered as a single-agent [31].

Finally, and based on previous reports on the classification of vitreous seeds and the present results, we may speculate that the longer it takes for topotecan to reach the inner layers of the tumorspheres, the higher the number of doses of chemotherapy required for tumor control, as topotecan penetration may be used as a surrogate for cytotoxicity. This assumption is based on the cytotoxic effect of topotecan revealed by ethidium bromide after its penetration into the spheres. Nonetheless, the relation between the number of doses and the actual dose of topotecan with the response in dusts, spheres, and clouds should be assessed in clinics.

There are some limitations of the present 3D cell culture model that could be acknowledged. Heterogeneity of patient samples may limit the generalization of drug response and penetration rates through spheres that could be addressed if a larger number of samples are obtained to estimate interindividual variability in drug response. However, the lack of fresh primary tumors, specifically from patients enucleated upfront, and the difficulties in establishing primary tumor cell cultures limit the availability of a wide range of 3D models for retinoblastoma. In addition, technical challenges of the 3D cell culture technique relates to the ability of generating the same morphology and size of the aggregates and to the mobility of free-floating spheroids in suspension, posing a challenge to temporal imaging of the same spheroid. Also, it should be taken into account that the present study was performed with topotecan and that other drugs used for retinoblastoma treatment, including melphalan, may show different penetration rates into the 3D structures depending on the physicochemical properties among other factors.

Altogether, we developed an in vitro tumor model that resembles retinoblastoma vitreous seeds based on a close phenotype-genotype correlation. Under the cell culture conditions, these cells grow as tumorspheres with the ability to interact in a 3D structure. Tumorspheres provide a valuable model to study in vitro drug penetration as a surrogate for drug exposure in vitreous seeds that may be useful to optimize drug therapy, and ultimately, to improve the efficacy of retinoblastoma treatment.

4. Materials and Methods

4.1. Ethics Statement

Patient retinoblastoma samples were obtained after protocol and informed consent approval by Hospital de Pediatria JP Garrahan Institutional Review Board (protocol number 904, date of approval: 26 February 2016). Written informed consent was obtained from parents or guardians before sample collection.

4.2. Retinoblastoma Cell Line

The commercial cell line Y79 was obtained from the American Type Culture Collection (HTB-18, Manassas, VA, USA). Cells were cultured at 37 °C with 5% CO2 in RPMI-1640 medium (Paisley, Scotland, UK) with 20% fetal bovine serum (FBS, Internegocios, Cordoba, Argentina), as previously reported elsewhere [19,27].

4.3. Establishment of Patient-Derived Cell Cultures

An ophthalmologist collected fresh primary tumor samples from patients with intraocular retinoblastoma who underwent upfront enucleation as primary treatment. Tumor samples were mechanically disaggregated and after centrifugation, and cells were cultured in serum-free neural stem-cell medium as previously described for other pediatric stem cells culture conditions [19,27].

Briefly, cells were grown in serum-free neurobasal medium (Thermo Fisher Scientific, Grand Island, NY, USA) containing DMEM/F12 (Thermo Fisher Scientific), supplemented with B-27 (Thermo Fisher Scientific), heparin (Sigma-Aldrich), EGF and FGF (epidermal and fibroblast growth factors, respectively, Thermo Fisher Scientific), and PDGF (Platelet derived growth factor, PeproTech, Rocky Hill, NJ, USA). Cultures were maintained at 37 °C in an incubator with a humidified atmosphere of 5% CO_2 and 95% air in T25 flasks at a density of approximately 10^6 cells/mL before starting the experiment.

4.4. Cell Authentication and Retinal Lineage Markers

Genomic DNA was isolated from cell lines, blood, and tumor samples using PureLink Genomic DNA mini kit (Invitrogen, Carlsband, CA, USA) following the manufacturer's instructions. After checking the quality and quantity of the DNA by Nanodrop 2000 (Thermo Scientific, Waltham, MA, USA), short tandem repeat (STR) profiling was performed in tumor samples and cell lines for authenticating the origin of the cell lines by means of the analysis of 15 autosomal STR loci and amelogenin. RB1 mutations were assessed by direct sequencing of the 27 exons and the promoter region of the RB1 gene, and mutations were described with reference to GenBank accession # L11910 [32]. Multiplex Ligation-dependent Probe Amplification assay (MLPA, MRC, Amsterdam, The Netherlands) was performed according to the manufacturer´s protocol to screen for deletions or duplications in the *RB1* gene.

The expression of the photoreceptor lineage marker cone-rod homeobox transcription factor (CRX) was analyzed to confirm the retinal tumor origin. Real-time quantitative PCR (RT-qPCR) was used to evaluate the expression of CRX mRNA in the patient-derived cell cultures using TaqMan technology in a 7500 Sequence Detection System (Applied Biosystems, Foster city, CA, USA) [33]. Briefly, total RNAs was isolated from tumor cells using PureLink RNA Mini Kit (Thermo Fisher Scientific) following the manufacturer´s instructions. After RNA quantitation using NanoDrop spectrophotometer, RNA was reverse-transcribed into cDNA using random primers and the SuperScript III kit (Invitrogen). The sequences of primers and probes used for RT-qPCR to analyze CRX mRNA expression were as follows: Forward: 5′-AGGTGGCTCTGAAGATCAATCTG-3′, Reverse: 5′-TTAGCCCTCCGGTTCTTGAA-3′, and probe 5′-FAM-CTGAGTCCAGGGTTC-3′-MGB. Relative expression of CRX mRNA was determined in cell cultures using the $2^{-\Delta\Delta Ct}$ method, where Ct is the threshold cycle of the target product and that of the housekeeping gene. Then, primary cell culture data was normalized against CRX mRNA expression levels obtained in Y79.

Additionally, immunohistochemistry was performed in paraffin embedded cells for synaptophysin (NCL-L-SYNAP-299, Leica BioSystems, Newcastle, UK) and arrestin 3 (ARR3, 11100-2-AP, Proteintech group, Chicago, IL, USA) expression to confirm the neuronal and cone-specific origin of the cells, and hematoxylin-eosin staining for morphological assessment [19]. We also evaluated the immunohistochemical staining for Ki-67 as a marker of proliferating but not quiescent (G0 phase) cells using Ki-67 antibody (Ki-67 anti-human clone, Dako, Denmark), and the percentage of Ki-67 positive cells was compared between tumorspheres of different sizes and origins. The percentage of Ki-67 positive cells was calculated using ImageJ software (NIH, Bethesda, MD, USA) [34].

4.5. Live-Cell Confocal Microscopy

To quantify the penetration of topotecan into retinoblastoma patient-derived spheres, two sizes of tumorspheres were considered based on visual inspection with the microscope. "Small" (< 110 μm) and "large" (110–400 μm) tumorspheres were classified according to the diameter calculated from an optical slice corresponding to approximately the middle of each sphere.

First, cell culture viability was evaluated by means of trypan blue exclusion assay. Then, ten microliters of each cell suspension of approximately 10^6 cells/mL were seeded on a slide covered by a clean 15 mm round coverslip, and an appropriate field with tumorspheres was located. Untreated tumorspheres were evaluated for natural fluorescence in a separate experiment. As shown in Supplemental Figure S2A, there was no acquisition of signal before topotecan exposure, while after

6 min of drug addition to the medium (Figure S2B), the sphere acquired a complete green fluorescence due to topotecan penetration. Then, 1 µL of a 10 µg/mL solution of topotecan (final concentration 1 µg/mL) was added. This concentration was selected based on the dose of topotecan commonly used for intravitreal injection (30 µg) and the volume of distribution in human vitreous humor (4 mL), resulting in an in vivo concentration similar to that used for in vitro evaluation [35,36]. Immediately after, and thereafter every 30 s, fluorescent images were obtained using an Olympus Fluoview FV1000 confocal laser scanning microscope (Olympus, Tokyio, Japan) with imaging software (Olympus Fluoview FV10-ASW v1.7c, Melville, NY, USA), and equipped with a UPlanSApo 20X/0.75 NA objective. Excitation was provided by a 458 nm line of a multiline argon laser at a 25% transmittance, and fluorescent images were collected with a 505–605 nm emission filter. Images were collected with a format of 1024 × 1024 pixels. Topotecan penetration into the tumorsphere was determined in terms of fluorescence intensity and images were processed using ImageJ software. At least three tumorspheres from each size were assessed for each patient-derived model. Autofluorescence contribution (natural fluorescence contribution of untreated tumorspheres) was evaluated in a different experiment. All optical images collected from each tumorsphere were sequentially acquired and recorded on video using ImageJ software.

Images were processed to quantify the penetration of topotecan into the sphere, as follows. First, anisotropic diffusion filtering was applied to smooth noise and defined region boundaries of each tumorsphere [37]. Then, the Otsu's thresholding segmentation method was used for segmenting the spheres from the background and also inside each tumorsphere to calculate the percentage of change of black and white pixels as a function of time as a surrogate of topotecan penetration [38,39]. The time to achieve maximum fluorescence in the core of the tumorsphere as a surrogate of complete penetration of topotecan was defined as the time needed for obtaining at least 90% of white pixels.

4.6. *Topotecan Cytotoxicity and Cell Viability Assay*

Topotecan cytotoxicity was evaluated in each cell culture to determine the concentration of drug that causes a 50% decrease in cell proliferation or IC50, as previously reported [20,27]. Briefly, cells were counted, seeded in 96-well plates, cultured for 24 hours, and thereafter exposed to different concentrations of topotecan (0.001–1.000 nM). After incubation for 72 h, cell proliferation was determined by 3-(4,5-dimethylthiazol-2-yl)-2,5-diphenyl tetrazolium bromide (MTT) colorimetric assay. The IC50 was calculated using GraphPad Prism v.6 (GraphPad, San Diego, CA, USA).

At the end of the study, tumorspheres were stained using ethidium bromide (100 µg/mL, Sigma Aldrich) and acridine orange (100 µg/mL) double-staining for the visualization of nucleic acids of membrane-damaged cells (necrotic or cells in late apoptosis induced by topotecan, observed as stained red) and live cells (stained green), respectively, and observed under fluorescence microscope. As both reagents were added simultaneously, non-specific signals to acridine orange of ethidium bromide could be assessed in topotecan-treated and untreated 3D structures, respectively.

4.7. *Statistical Analysis*

Comparison of the time needed for topotecan to fully penetrate into the core of the two sizes of tumorspheres derived from patients was performed by means of a t test with a significance p-value of 0.05.

Supplementary Materials: Supplementary materials can be found at http://www.mdpi.com/1422-0067/20/5/1077/s1.

Author Contributions: Conceptualization, U.W., D.H.A., A.M.C., and P.S.; methodology, U.W., F.F., F.B., S.Z., R.A., M.S., C.S., A.M.C., and P.S.; cell line establishment and analysis, U.W., S.Z., and R.A.; confocal microscopy, U.W. and F.F.; formal analysis, U.W., F.F., F.B., D.H.A., A.M.C., and P.S.; patient treatment, M.S. and C.S.; clinical data analysis, M.S. and C.S. All authors significantly contributed to the design of the study, to the interpretation of the data, and to the writing and editing.

Funding: This work was supported by Agencia Nacional de PromociónCientífica-FONCYT (PICT 2016- 1505), Consejo Nacional de InvestigacionesCientificas y Tecnológicas (CONICET), FondationNelia et Amadeo Barletta, Fundación Natali DafneFlexer, and Fund for Ophthalmic Knowledge, NY, USA.

Acknowledgments: The authors want to thank Guillermo Chantada for his thorough discussions in preclinical modeling and clinical therapeutics, and Debora Chan and Juliana Gambini for their technical support with image processing. We also acknowledge the technical support with genomic studies of Irene Szijan and Marcela Menna.

Conflicts of Interest: The authors declare no conflict of interest.

References

1. Abramson, D.H. Retinoblastoma: Saving life with vision. *Annu. Rev. Med.* **2014**, *65*, 171–184. [CrossRef] [PubMed]
2. PDQ®Pediatric Treatment Editorial Board. *PDQ Retinoblastoma Treatment*; National Cancer Institute: Bethesda, MD, USA, 2018. Available online: https://www.cancer.gov/types/retinoblastoma/hp/retinoblastoma-treatment-pdq (accessed on 23 November 2018).
3. Chantada, G.; Schaiquevich, P. Management of retinoblastoma in children: Current status. *Paediatr. Drugs* **2015**, *17*, 185–198. [CrossRef] [PubMed]
4. Francis, J.H.; Levin, A.M.; Zabor, E.C.; Gobin, Y.P.; Abramson, D.H. Ten-year experience with ophthalmic artery chemosurgery: Ocular and recurrence-free survival. *PLoS ONE* **2018**, *13*, e0197081. [CrossRef] [PubMed]
5. Abramson, D.H.; Shields, C.L.; Munier, F.L.; Chantada, G.L. Treatment of Retinoblastoma in 2015: Agreement and Disagreement. *JAMA Ophthalmol.* **2015**, *133*, 1341–1347. [CrossRef] [PubMed]
6. Munier, F.L.; Gaillard, M.C.; Soliman, S.; Balmer, A.; Moulin, A.; Podilsky, G. Intravitreal chemotherapy for vitreous disease in retinoblastoma revisited: From prohibition to conditional indications. *Br. J. Ophthalmol.* **2012**, *96*, 1078–1083. [CrossRef] [PubMed]
7. Munier, F.L. Classification and management of seeds in retinoblastoma. Ellsworth Lecture Ghent August 24th 2013. *Ophthalmic Genet.* **2014**, *35*, 193–207. [CrossRef] [PubMed]
8. Francis, J.H.; Abramson, D.H.; Gaillard, M.C.; Marr, B.P.; Beck-Popovic, M.; Munier, F.L. The classification of vitreous seeds in retinoblastoma and response to intravitreal melphalan. *Ophthalmology* **2015**, *122*, 1173–1179. [CrossRef] [PubMed]
9. Francis, J.H.; Marr, B.P.; Abramson, D.H. Classification of Vitreous Seeds in Retinoblastoma: Correlations with Patient, Tumor, and Treatment Characteristics. *Ophthalmology* **2016**, *123*, 1601–1605. [CrossRef] [PubMed]
10. Berry, J.L.; Bechtold, M.; Shah, S.; Zolfaghari, E.; Reid, M.; Jubra, R.; Kim, J.W. Not All Seeds Are Created Equal: Seed Classification Is Predictive of Outcomes in Retinoblastoma. *Ophthalmology* **2017**, *124*, 1817–1825. [CrossRef] [PubMed]
11. Amram, A.L.; Rico, G.; Kim, J.W.; Chintagumpala, M.; Herzog, C.E.; Gombos, D.S.; Chévez-Barrios, P. Vitreous Seeds in Retinoblastoma: Clinicopathologic Classification and Correlation. *Ophthalmology* **2017**, *124*, 1540–1547. [CrossRef] [PubMed]
12. Schaiquevich, P.; Carcaboso, A.M.; Buitrago, E.; Taich, P.; Opezzo, J.; Bramuglia, G.; Chantada, G.L. Ocular pharmacology of topotecan and its activity in retinoblastoma. *Retina* **2014**, *34*, 1719–1727. [CrossRef] [PubMed]
13. Francis, J.H.; Marr, B.P.; Schaiquevich, P.; Kellick, M.G.; Abramson, D.H. Properties and clinical utility of topotecan fluorescence: Uses for retinoblastoma. *Br. J. Ophthalmol.* **2015**, *99*, 1320–1322. [CrossRef] [PubMed]
14. Beijnen, J.H.; Rosing, H. Bioanalytical methods for anticancer drugs. In *Cancer Clinical Pharmacology*; Schellens, J.H., McLeod, H.L., Newell, D.R., Eds.; Oxford University Press: New York, NY, USA, 2005.
15. Pawley, J. (Ed.) *Handbook of Biological Confocal Microscopy*; Springer: New York, NY, USA, 2006.
16. Nath, S.; Devi, G.R. Three-dimensionalculturesystems in cancer research: Focus on tumorspheroidmodel. *Pharmacol. Ther.* **2016**, *163*, 94–108. [CrossRef] [PubMed]
17. Halfter, K.; Mayer, B. Bringing 3D tumor models to the clinic—Predictive value for personalized medicine. *Biotechnol. J.* **2017**, *12*. [CrossRef] [PubMed]
18. Bond, W.S.; Akinfenwa, P.Y.; Perlaky, L.; Hurwitz, M.Y.; Hurwitz, R.L.; Chévez-Barrios, P. Tumorspheres but not adherent cells derived from retinoblastoma tumors are of malignant origin. *PLoS ONE* **2013**, *8*, e63519. [CrossRef] [PubMed]

19. Pascual-Pasto, G.; Olaciregui, N.G.; Vila-Ubach, M.; Paco, S.; Monterrubio, C.; Rodriguez, E.; Winter, U.; Batalla-Vilacis, M.; Catala, J.; Salvador, H.; et al. Preclinical platform of retinoblastoma xenografts recapitulating human disease and molecular markers of dissemination. *Cancer Lett.* **2016**, *38*, 10–19. [CrossRef] [PubMed]
20. Laurie, N.A.; Gray, J.K.; Zhang, J.; Leggas, M.; Relling, M.; Egorin, M.; Stewart, C.; Dyer, M.A. Topotecan combination chemotherapy in two new rodent models of retinoblastoma. *Clin. Cancer Res.* **2005**, *11*, 7569–7578. [CrossRef] [PubMed]
21. Croce, A.C.; Bottiroli, G.; Supino, R.; Favini, E.; Zuco, V.; Zuino, F. Subcellularlocalization of the camptothecin analogues, topotecan and gimatecan. *Biochem. Pharmacol.* **2004**, *67*, 1035–1045. [CrossRef] [PubMed]
22. Fugit, K.D.; Jyoti, A.; Upreti, M.; Anderson, B.D. Insights into accelerated liposomal release of topotecan in plasma monitored by a non-invasive fluorescence spectroscopic method. *J. Control. Release* **2015**, *197*, 10–19. [CrossRef] [PubMed]
23. Centelles, M.N.; Wright, M.; So, P.W.; Amrahli, M.; Xu, X.Y.; Stebbing, J.; Miller, A.D.; Gedroyc, W.; Thanou, M. Image-guided thermosensitive liposomes for focused ultrasound drug delivery: Using NIRF-labelled lipids and topotecan to visualize the effects of hyperthermia in tumours. *J. Control. Release* **2018**, *280*, 87–98. [CrossRef] [PubMed]
24. Jyoti, A.; Fugit, K.D.; Sethi, P.; McGarry, R.C.; Anderson, B.D.; Upreti, M. An in vitro assessment of liposomal topotecan simulating metronomic chemotherapy in combination with radiation in tumor-endothelial spheroids. *Sci. Rep.* **2015**, *5*, 15236. [CrossRef] [PubMed]
25. Foty, R. A simple hanging drop cell culture protocol for generation of 3D spheroids. *J. Vis. Exp.* **2011**, 2720. [CrossRef] [PubMed]
26. Mitra, M.; Mohanty, C.; Harilal, A.; Maheswari, U.K.; Sahoo, S.K.; Krishnakumar, S. A novel in vitro three-dimensional retinoblastoma model for evaluating chemotherapeutic drugs. *Mol. Vis.* **2012**, *18*, 1361–1378. [PubMed]
27. Winter, U.; Mena, H.A.; Negrotto, S.; Arana, E.; Pascual-Pasto, G.; Laurent, V.; Suñol, M.; Chantada, G.L.; Carcaboso, A.M.; Schaiquevich, P. Schedule-Dependent Antiangiogenic and Cytotoxic Effects of Chemotherapy on Vascular Endothelial and Retinoblastoma Cells. *PLoS ONE* **2016**, *11*, e0160094. [CrossRef] [PubMed]
28. Ledur, P.F.; Onzi, G.R.; Zong, H.; Lenz, G. Culture conditions defining glioblastoma cells behavior: What is the impact for novel discoveries? *Oncotarget* **2017**, *8*, 69185–69197. [CrossRef] [PubMed]
29. Qureshi-Baig, K.; Ullmann, P.; Rodriguez, F.; Frasquilho, S.; Nazarov, P.V.; Haan, S.; Letellier, E. What Do We Learn from Spheroid Culture Systems? Insights from Tumorspheres Derived from Primary Colon Cancer Tissue. *PLoS ONE* **2016**, *8*, e0146052. [CrossRef] [PubMed]
30. Lee, J.; Shin, D.; Roh, J.L. Development of an in vitro cell-sheet cancer model for chemotherapeutic screening. *Theranostics* **2018**, *8*, 3964–3973. [CrossRef] [PubMed]
31. Schaiquevich, P.; Fabius, A.W.; Francis, J.H.; Chantada, G.L.; Abramson, D.H. Ocular pharmacology of chemotherapy for retinoblastoma. *Retina* **2017**, *37*, 1–10. [CrossRef] [PubMed]
32. Parma, D.; Ferrer, M.; Luce, L.; Giliberto, F.; Szijan, I. RB1 gene mutations in Argentine retinoblastoma patients. Implications for genetic counseling. *PLoS ONE* **2017**, *12*, e0189736. [CrossRef] [PubMed]
33. Torbidoni, A.V.; Laurent, V.E.; Sampor, C.; Ottaviani, D.; Vazquez, V.; Gabri, M.R.; Rossi, J.; de Dávila, M.T.; Alonso, C.; Alonso, D.F.; et al. Association of cone-rod homeobox transcription factor messenger RNA with pediatric metastatic retinoblastoma. *JAMA Ophthalmol.* **2015**, *133*, 805–812. [CrossRef] [PubMed]
34. Schneider, C.A.; Rasband, W.S.; Eliceiri, K.W. NIH Image to ImageJ: 25 years of image analysis. *Nat. Methods* **2012**, *9*, 671–675. [CrossRef] [PubMed]
35. Rao, R.; Honavar, S.G.; Sharma, V.; Reddy, V.A.P. Intravitreal topotecan in the management of refractory and recurrent vitreous seeds in retinoblastoma. *Br. J. Ophthalmol.* **2018**, *102*, 490–495. [CrossRef] [PubMed]
36. Francis, J.H.; Gobin, Y.P.; Dunkel, I.J.; Marr, B.P.; Brodie, S.E.; Jonna, G.; Abramson, D.H. Carboplatin +/− topotecan ophthalmic artery chemosurgery for intraocular retinoblastoma. *PLoS ONE* **2013**, *8*, e72441. [CrossRef] [PubMed]
37. Perona, J.; Malik, J. Scale-space and edge detection using anisotropic diffusion. *IEEE Trans. Pattern Anal. Mach. Intell.* **1990**, *12*, 629–639. [CrossRef]

38. Otsu, N. A threshold selection method from gray-level histogram. *IEEE Trans. Syst. Man Cybern.* **1979**, *19*, 62–66. [CrossRef]
39. Shi, Y.; Karl, W.C. Real-time Tracking Using Level Sets. *IEEE Trans. Image Process.* **2008**, *17*, 645–656. [PubMed]

International Journal of
Molecular Sciences

MDPI

Article

Dasatinib/HP-β-CD Inclusion Complex Based Aqueous Formulation as a Promising Tool for the Treatment of Paediatric Neuromuscular Disorders

Annalisa Cutrignelli [1,*,†], Francesca Sanarica [1,2,†], Antonio Lopalco [1], Angela Lopedota [1], Valentino Laquintana [1], Massimo Franco [1], Brigida Boccanegra [1,2], Paola Mantuano [1,2], Annamaria De Luca [1,2,‡] and Nunzio Denora [1,*,‡]

1 Department of Pharmacy-Drug Sciences, University of Bari "Aldo Moro", 70125 Bari, Italy; francesca.sanarica88@gmail.com (F.S.); antonio.lopalco@uniba.it (A.L.); angelaassunta.lopedota@uniba.it (A.L.); valentino.laquintana@uniba.it (V.L.); massimo.franco@uniba.it (M.F.); brigida.boccanegra@uniba.it (B.B.); paola.mantuano@uniba.it (P.M.); annamaria.deluca@uniba.it (A.D.L.)

2 Unity of Pharmacology, Department of Pharmacy-Drug Sciences, University of Bari "Aldo Moro", 70125 Bari, Italy

* Correspondence: annalisa.cutrignelli@uniba.it (A.C.); nunzio.denora@uniba.it (N.D.); Tel.: +39-080-544-2767/2766 (A.C.)

† These authors contributed equally to the work.

‡ Co-senior authors.

Received: 21 December 2018; Accepted: 27 January 2019; Published: 30 January 2019

Abstract: New scientific findings have recently shown that dasatinib (DAS), the first-choice oral drug in the treatment of chronic myeloid leukemia (CML) for adult patients who are resistant or intolerant to imatinib, is also potentially useful in the paediatric age. Moreover, recent preclinical evidences suggest that this drug could be useful for the treatment of Duchenne muscular dystrophy, since it targets cSrc tyrosin kinase. Based on these considerations, the purpose of this work was to use the strategy of complexation with hydroxypropyl-β-cyclodextrin (HP-β-CD) in order to obtain an aqueous preparation of DAS, which is characterized by a low water solubility (6.49×10^{-4} mg/mL). Complexation studies demonstrated that HP-β-CD is able to form a stable host-guest inclusion complex with DAS with a 1:1 apparent formation constant of 922.13 M^{-1}, as also demonstrated by the Job's plot, with an increase in DAS aqueous solubility of about 21 times in the presence of 6% *w*/*v* of HP-β-CD (0.014 mg/mL). The inclusion complex has been prepared in the solid state by lyophilization and characterized by Fourier Transform Infrared (FT-IR), Nuclear Magnetic Resonance (NMR), Differential Scanning Calorimetry (DSC) techniques, and its dissolution profile was studied at different pH values. Moreover, in view of potential use of DAS for Duchenne muscular dystrophy, the cytotoxic effect of the inclusion complex has been assessed on C2C12 cells, a murine muscle satellite cell line. In parallel, a one-week oral treatment was performed in wild type C57Bl/6J mice to test both palatability and the exposure levels of the new oral formulation of the compound. In conclusion, this new inclusion complex could allow the development of a liquid and solvent free formulation to be administered both orally and parenterally, especially in the case of an administration in paediatric age.

Keywords: dasatinib; Duchenne muscular dystrophy; cyclodextrin inclusion complex; phase solubility studies; paediatric age; liquid formulation

1. Introduction

The drug dasatinib (DAS), whose chemical name is N-(2-chloro-6-methylphenyl)-2-[[6-[4-(2-hydroxyethyl)piperazin-1-yl]-2-methylpyrimidin-4-yl]amino]-1,3-thiazole-5-carboxamide (IUPAC)

(Figure 1) is a double inhibitor of kinase proteins, including proto-oncogene tyrosine-protein Src (Src-TK) family kinases [1].

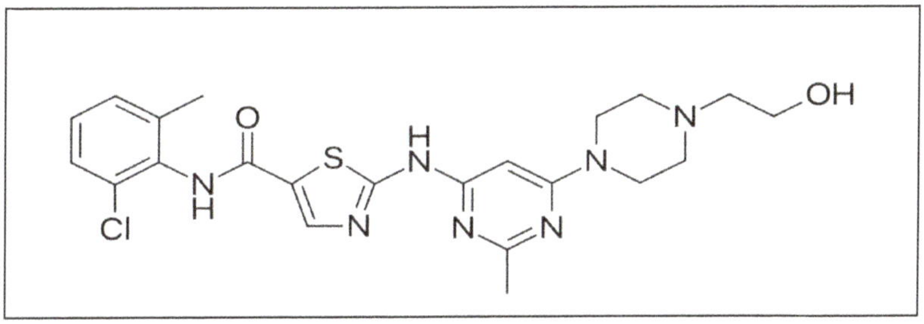

Figure 1. Chemical structure of dasatinib (DAS).

DAS is the first-choice oral drug in the treatment of chronic myeloid leukemia (CML) for those patients who are resistant or intolerant to imatinib. In fact, CML is a myeloprolifelative disorder that is caused by the BCR-ABL oncogene and DAS is a potent inhibitor of imatinib-resistant BCR-ABL mutants [2,3].

Until now, DAS was used exclusively for the treatment of adult patients, but new scientific findings have shown its potential in the treatment of CML in paediatric age, where its pharmacokinetic parameters, in particular, absorption and elimination time, were comparable with those in adult, with the same safety and efficacy profiles [4,5]. However, in these clinical trials, the drug was administered to children in the form of tablets or crushed tablets dispersed in fruit juice. In fact, DAS, formulated as monohydrate and marketed under the name of Sprycel®by Bristol Meyer Squibb, is presented in the form of coated tablets with a dosage ranging from 20 to 140 mg of the active ingredient. No liquid formulation is available on the market, and this may be a problem for paediatric patients who may not be able to swallow the tablets.

Moreover, recently, a study was published showing that DAS may be applied in the treatment of Duchenne muscular dystrophy (DMD), a genetic muscle-wasting disorder, whose symptoms occur around the age of four years in boys and get worse quickly. DMD is characterized by a progressive muscle degeneration and weakness and it is caused by the absence of the subsarcolemmal protein *dystrophin*. Dystrophin preserves sarcolemmal integrity by linking the cytoskeleton to the extracellular matrix via the interaction with the dystrophin-glycoprotein complex (DGC) and allowing for proper force transmission from contractile apparatus to extracellular matrix [6]. Thus, the primary structural defect causes an aberrant transmission of mechanical stimulus across the myofibers, leading to progressive muscle weakness and degeneration [7]. Similar defects occur in animals, such as the widely used C57Bl/10ScSn-Dmd^{mdx}/J (*mdx*) mouse model [7,8]. Recent studies in *mdx* mouse model have highlighted that, in dystrophin deficient muscles, Src-TK is both overactivated and overexpressed, due to the excessive ROS production, and contribute also to NOX activation, in an auto-reinforcing loop [9], then playing a key role in DMD pathogenesis. In addition, Src-TK is involved in phosphorylation and degradation of β-dystroglycan (β-DG), a member of DGC, contributing to the loss of this complex in dystrophic myofibers. Thus, either the pharmacological inhibition of Src-TK seems a feasible strategy to ameliorate the pathology [10,11]. Src-TK inhibitors are already clinically available as antitumor drugs, and DAS belongs to this class of drugs.

Based on what has just been outlined, it is evident that it would be useful to develop a new formulation of DAS, which is different from the one currently in use, possibly liquid, so that it could be readily used in paediatric patients either by oral or parenteral route [12].

Therefore, the purpose of the following work was to prepare an aqueous formulation of this drug, evaluating the possibility of using an inclusion complex with cyclodextrins (CDs), as it is a molecule that is characterized by a low water solubility [13]. Cyclodextrins, cyclic oligosaccharides consisting of glucose units joined by α 1,4-glycosidic bond have been widely used to improve the solubility and stability in water of different molecules due to their ability to form host-guest inclusion complexes [14–20]. Thus, in this work, we present an inclusion complex of DAS with the hydroxy-β-cyclodextrin (HP-β-CD), a semisynthetic cyclodextrin that is approved by FDA also for the parenteral administration. The DAS/HP-β-CD inclusion complex was first studied in solution by building the phase solubility diagram according to Higuchi-Connors [21] and a two-dimensional-NMR (2D-NMR) Heteronuclear Multiple Bond Correlation HMBC evaluation was carried out in order to investigate the portion of the molecule actually contained in the HP-β-CD cavity. Subsequently, this complex was prepared in solid state by lyophilization and characterized by Fourier Transform Infrared (FT-IR), Differential Scanning Calorimetry (DSC), evaluation of the incorporation degree, and study of dissolution profiles at different pH values. Finally, in view of potential use of DAS for DMD, we first assessed its cytotoxic action on C2C12 cells, a muscle satellite cell line; secondly, we conducted an in vivo study in wild type C57Bl/6J (WT) mice by administering the inclusion complex in drinking water for one week to test both palatability and the exposure levels of the complex.

2. Results and Discussion

2.1. Evaluation of the Inclusion Complex in Solution

First of all, the solubility of DAS was determined at 25 °C both in ultra-pure water and in buffered aqueous solutions at pH 1.2 (HCl 0.05 M, for oral administration) and at pH 7.4 (phosphate buffer 0.05 M, for parenteral administration). The results are shown in Table 1. DAS is a strong base with a pK_a value of 10.28 [11], so it is more soluble in acid environments where the protonation of the NH groups occurs.

The lowest solubility value was recorded in ultrapure water with a pH value of about 6.0, so the phase solubility diagram relating to the complexation of DAS with HP-β-CD has been studied at 25 °C in water. According to the Higuchi and Connors classification [20], it shows an A_P-type profile, as reported in Figure 2a, and this result clearly show that DAS solubility in water is linearly influenced by the presence of HP-β-CD until a percentage of cyclodextrin equal to about 6%, with the formation of an inclusion complex with 1:1 host:guest stoichiometry, while in the presence of major cyclodextrin percentages the formation of complexes with different stoichiometry occurs. From the analysis of the first linear portion of the Higuchi-Connors diagram (Figure 2b), it is possible to obtain the complexation constant for the complex with 1: 1 host:guest stoichiometry and it was found to be equal to 922.13 M^{-1}, with an increase of about 21 times of the DAS solubility in the presence of 6% *w/v* of cyclodextrin (0.014 mg/mL, 2.9×10^{-5} M) as compared to the solubility value of the drug in the absence of the complexant, which results to be 6.49×10^{-4} mg/mL (1.33×10^{-6} M).

Table 1. DAS solubility at 25 °C in presence of different environments.

Environment	DAS Solubility (mg/mL)	DAS Solubility (M)
HCl 0.05 M pH 1.2	4.31×10^{-2}	8.84×10^{-5}
Phosphate Buffer 0.05 M pH 7.4	7.65×10^{-4}	1.56×10^{-6}
Water	6.49×10^{-4}	1.33×10^{-6}

In order to determine the exact stoichiometric ratio between DAS and HP-β-CD in the formation of the inclusion complex, the Job's plot (Figure 2C) was constructed, as described in the experimental section. In detail, this study was conducted via ^{1}H-NMR observing the variation of chemical shifts of methyl hydrogens (CH_3) on the pyrimidine ring of DAS.

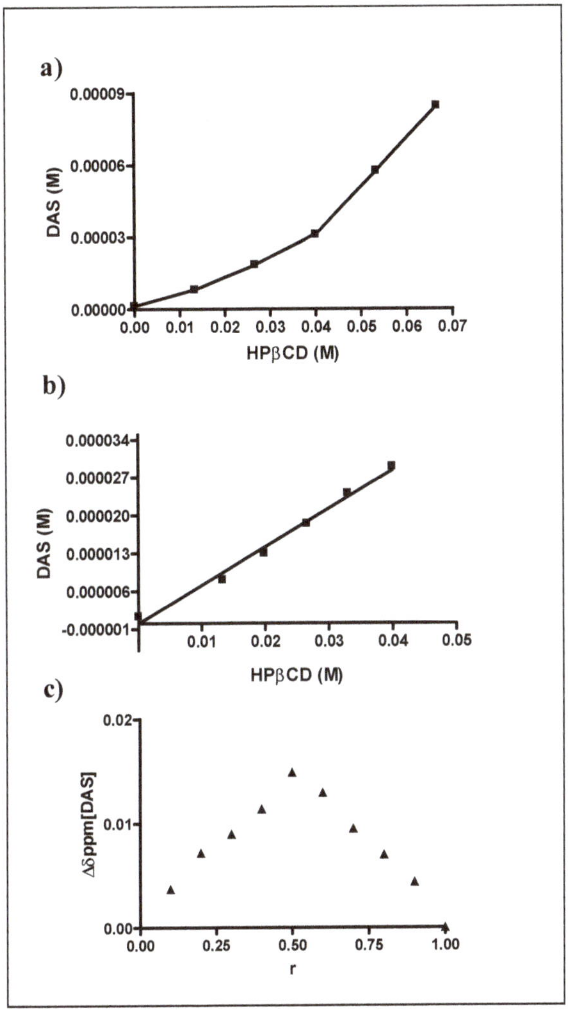

Figure 2. Phase solubility and Job's plot diagrams of DAS and hydroxypropyl-β-cyclodextrin (HP-β-CD) in water at 25 °C. (**a**) Phase solubility diagram in the HP-β-CD concentration range 0–10%; (**b**) Phase solubility diagram in the HP-β-CD concentration range 0–6%; (**c**) Job's plot diagram.

As shown in the graph, a highly symmetrical trend with a maximum value being recorded at $r = 0.5$ is observed, and this finding highlights the formation of a 1:1 inclusion complex. This result is quite in agreement with the phase solubility diagram because for the construction of Job's plot very low concentration of HP-β-CD are used and at low concentration of cyclodextrin the formation of an inclusion complex with a 1:1 host:guest stoichiometry occurs. This behavior has already been widely described in the literature [15,22]. It is in fact known that the balance of complexation between drug and cyclodextrin is strongly influenced by the concentrations in solution of the two components and that in the presence of high concentration of cyclodextrin different solubilization phenomena take place that modify the stoichiometry of the inclusion complex, leading to higher-order complexes.

Furthermore, the construction of the Job's diagram has been carried out on the basis of the displacement, in terms of chemical shift, of the methyl group protons on the DAS pyrimidine ring. It would therefore seem that this methyl group is directly involved in the formation of the inclusion complex with the cavity of the cyclodextrin and in order to obtain more information about the interactions of the drug with the cyclodextrin in solution, a ^{1}H- and 2D-NMR (HMBC) study was conducted, keeping the DAS concentration constant and varying the molar ratio DAS: HP-β-CD.

Figure 3 shows the 2D ^{1}H- ^{13}C-NMR spectrum of DAS in DMSO-d_6, which was used to make the correct assignment of DAS protons while in Figure 4a–c are reported the ^{1}H-NMR spectra of methyl resonances of CH_3 on pyrimidine and benzene rings at different DAS: HP-β-CD molar ratios.

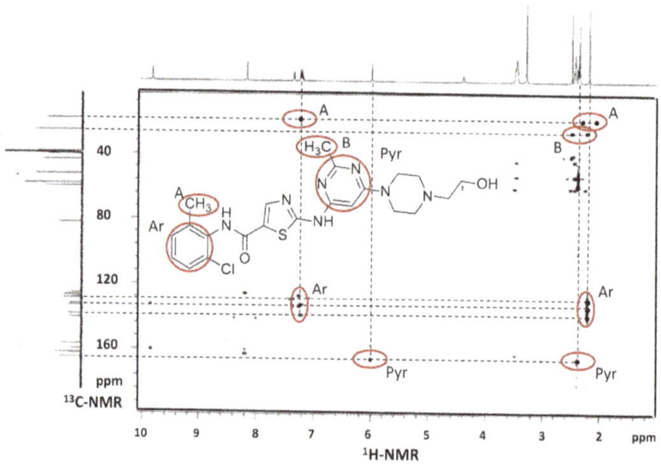

Figure 3. 2D ^{1}H- ^{13}C-NMR spectrum of DAS in DMSO-d_6. The cross-peaks displayed by HMBC were used to identify the structure of the drug, including the correlation of the δ of hydrogens and carbons separated from each other with two and three chemical bond.

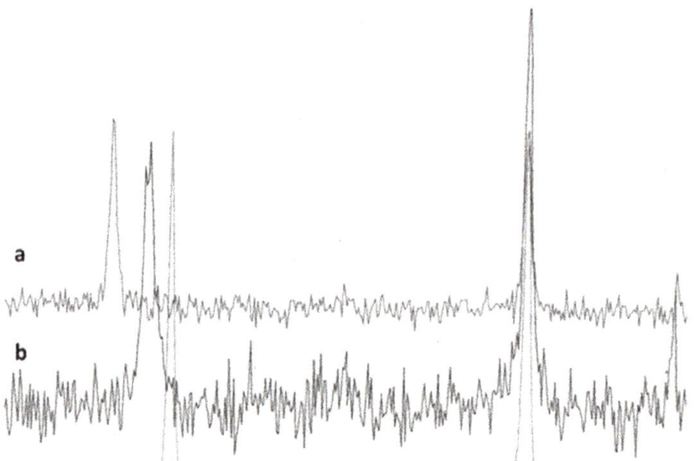

Figure 4. ^{1}H-NMR spectra of methyl resonances of CH_3 on pyrimidine (δ ~2.5) and benzene (δ ~2.3) rings in the presence of cyclodextrin at different DAS:HP-β-CD molar ratios. (**a**) DAS:HP-β-CD molar ratios 1:0; (**b**) DAS:HP-β-CD molar ratios 1:1; and (**c**) DAS:HP-β-CD molar ratios 1:10.

The relative positions of the peaks were in agreement with the assignment. ^{1}H- and 2D-NMR (HMBC) investigations confirmed the structure of the molecule and elucidated the interactions with the cyclodextrin in solution. Our results give a direct evidence of the formation of an inclusion complex between the drug and the cyclodextrin. In Table 2, we reported the variation of chemical shifts of methyl hydrogens (CH_3) on the pyrimidine ring of DAS in the presence of different concentrations of cyclodextrin (i.e., molar ratio drug:cyclodextrin 1:1, 1:2, and 1:3). As one can see from Table 2, the chemical shifts of this CH_3 are affected during complexation, showing changes in the ppm values. In particular, as shown in Figure 4a–c increasing the concentration of cyclodextrin in solution, we observed that the chemical shift of the CH_3 hydrogens on the pyrimidine ring shifted downfield (higher ppm). These findings suggest that the methyl hydrogens on the pyrimidine ring were directly involved in the complexation with cyclodextrin. In detail, the hydrogen nuclei of the drug included in the cyclodextrin cavity established hydrophobic interactions with cyclodextrin hydrogens, resulting in a their deshielding. No significant variation of the chemical shifts of the methyl hdrogens on the aromatic ring benzene was observed. This would suggest that this portion of the molecule is not interested in the complex formation with cyclodextrin.

Table 2. Shifts of CH_3 hydrogens in the presence of cyclodextrin at different DAS: HP-β-CD molar ratios.

Molar Ratio DAS: HP-β-CD	δ CH_3 (Pyrimidine)	δ CH_3 (Benzene)
1:0	2.498	2.292
1:1	2.513	2.293
1:10	2.534	2.294

2.2. Characterization of the Inclusion Complex in the Solid State

In order to exploit the complexation with HP-β-CD in the preparation of a powder formulation of the drug that instantly dissolves when placed in water to be administered orally or parenterally, the solid-state complex was prepared by lyophilization. The freeze-dried complex was characterized by the assessment of the degree of incorporation, expressed as g of DAS per 100 g of product and it was found to be 4.23 ± 0.42 g of DAS per 100 g of lyophilized powder. This solid inclusion complex has been characterized by FT-IR, DSC, and dissolution profile. Figure 5 shows the IR spectra of DAS, HP-β-CD, and HP-β-CD solid inclusion complex.

The IR spectrum of DAS shows an absorption band at 1609 cm^{-1} due to stretching of the carbonyl group in the amidic bound, and two absorption bands at 2945 and 2930 cm^{-1} due to C-H stretching of methylenic and alchilic groups. In addition, the bands at 1583, 1498, and 1417 cm^{-1} corresponding to the C-C strain of the aromatic ring, and the bands at 3461 and 3225 cm^{-1}, corresponding to the stretching of the N-H and O-H are highlighted, respectively.

The spectra of CDs inclusion complex appear to be very similar to those of cyclodextrin, since the cyclodextrins exhibit a high number of polar groups (OH, CO) that give rise to very broad absorption bands, which in some regions often overlap with those of DAS, also because a large excess of cyclodextrin is present in the complexes. This finding is confirmed by the DSC study reported in Figure 6.

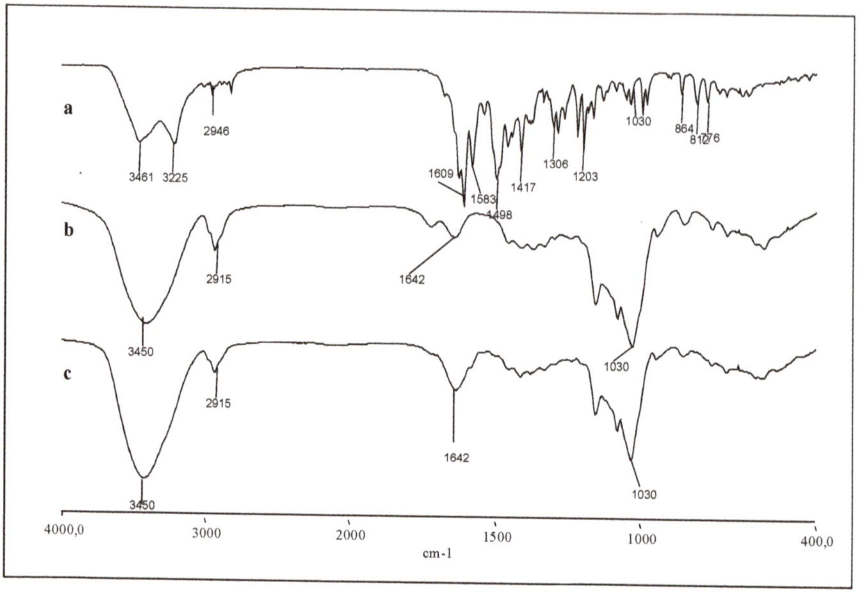

Figure 5. Fourier Transform Infrared (FT-IR) spectra (**a**) DAS, (**b**) HP-β-CD, and (**c**) DAS/HP-β-CD complex.

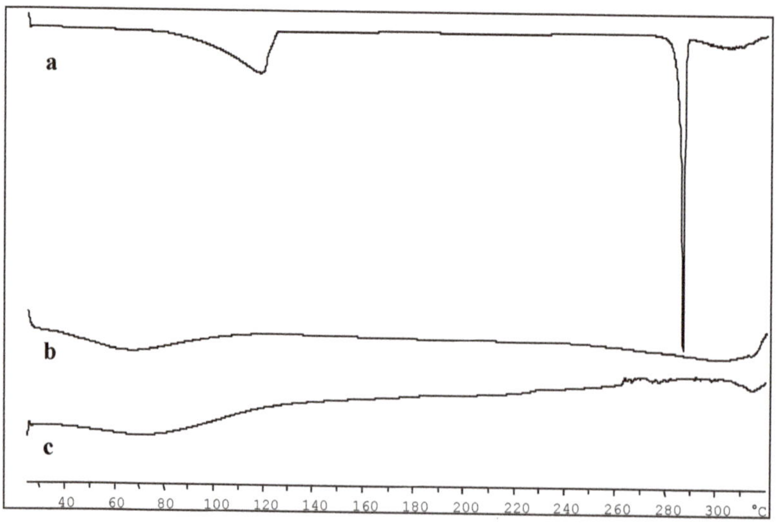

Figure 6. DSC thermograms: (**a**) DAS, (**b**) HP-β-CD, and (**c**) DAS/HP-β-CD complex.

In the DAS thermogram (Figure 6a), it is evident the crystalline nature and the high degree of purity of this compound that shows an endothermic spike at 285 °C, according with data reported in literature [11].

The HP-β-CD thermogram (Figure 6b) highlights the amorphous nature of the same, which does not exhibit an endothermic melting peak but only a sloping peak between 80 and 100 °C due to the loss of the water present in the sample. The thermogram of the DAS/HP-β-CD complex (Figure 6c) has a trend that is comparable to that of cyclodextrin alone and this indicates the drug's inclusion within the cavity of the complexing agent with its amorphization.

2.3. Dissolution Studies

Moreover, dissolution studies have been performed at 37 °C in two different media: phosphate buffer 0.05M pH = 7.4 and HCl 0.05M pH = 1.2. Figure 7 shows obtained dissolution profiles.

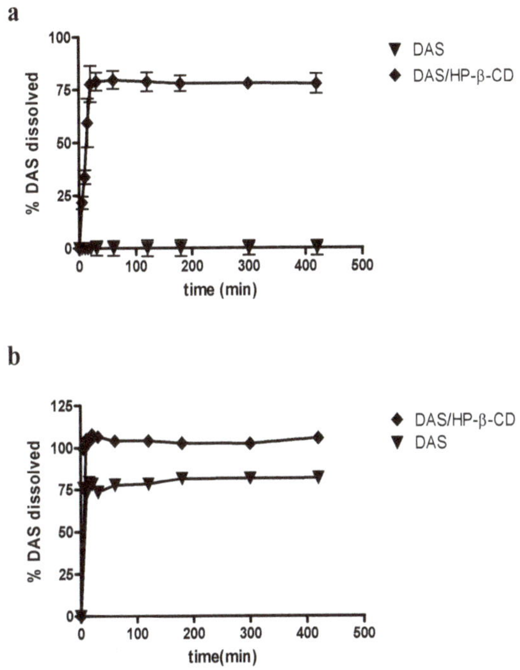

Figure 7. Dissolution profiles at 37 °C: (**a**) pH 7.4 and (**b**) pH 1.2 of DAS alone (▼) and DAS/HP-β-CD complex (♦). All values are mean ± SD, n = 3.

It is clear that the DAS/HP-β-CD lyophilized complex exhibits a better dissolution profile than the drug alone. In particular, this is especially evident at pH 7.4 where it was not possible to obtain the dissolution profile of DAS alone due to its very low solubility at this pH value, which prevents the quantitative determination of the drug via HPLC in the dissolution medium. The hydrophobic nature of the drug prevented its contact with the dissolution medium, causing it to float on the surface and hindering its dissolution. Instead, in the same dissolution medium the presence of HP-β-CD allows for the achievement of a quantity of dissolved drug equal to about 77% after 420 min. At pH 1.2, DAS appears to be more soluble, as demonstrated by the previously described solubility analysis. In this case, hence, it was possible to obtain the solubility profile of the drug alone at this pH value and it is evident that after approximately 420 min the quantity of drug dissolved is approximately 81%, as compared to the 100% that is reached from the complex with the HP-β-CD.

Therefore, the complexation with the selected cyclodextrin certainly represents an effective strategy for improving the solubility characteristics and the dissolution profile of DAS, also allowing an improvement of these characteristics with respect to those of monohydrate and polymorphic forms that are patented [23,24], and enabling it to be administered parenterally, in addition to oral administration, which is currently the only possible DAS route of administration.

2.4. Cytotoxicity Studies

The cell viability study was performed to compare the cytotoxicity, and then the pharmacological activity, of the DAS/HP-β-CD inclusion complex with that of the free drug both solubilized in DMEM.

In view of potential use for DMD, this effect has been assessed on C2C12 myoblasts. For DAS alone, due to its very low water solubility, a DMSO solution was prepared and this solution was subsequently diluted in DMEM so that the final DMSO concentration in each well was less than 0.15% in order to ensure cellular vitality. The two vehicles were also tested, i.e., HP-β-CD and DMSO, both diluted in DMEM and both at the highest concentration tested in the presence of DAS. The results that were obtained in terms of cellular viability are shown in Figure 8.

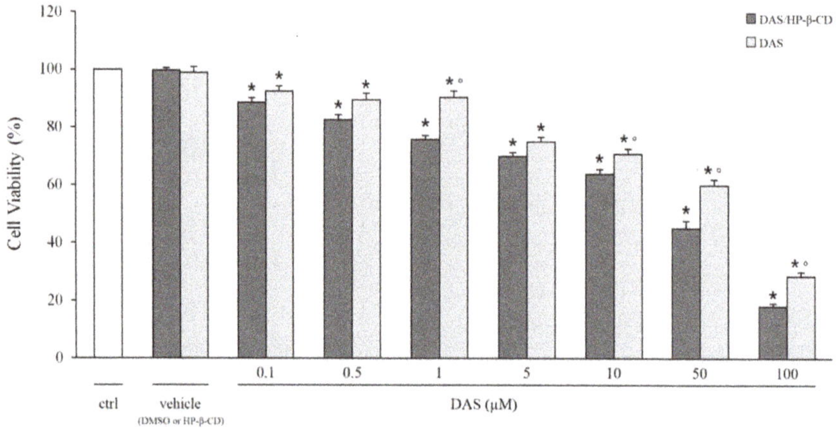

Figure 8. Effect of DAS on cell viability. The figure shows the cytotoxic effect on cell viability of increasing concentration of DAS (0.1–100 μM) alone or complexed with HP-β-CD. The results are expressed as the percentage of the control (ctrl) and presented as the mean ± S.E.M. Each data is from 24–48 replicates (wells) and 6–9 different culture dishes. The statistical significance between groups was evaluated by Student's *t*-test, as follows: significantly different with respect to * the control value ($0.001 < p < 0.05$); ° DAS/HP-β-CD at the same concentration ($0.001 < p < 0.05$).

It is clear that all of the compounds tested show, as expected, a cytotoxicity that is concentration dependent. In particular, the DAS/HP-β-CD complex has a relatively higher effect on cell viability than free DAS, with significantly different statistical results ($0.001 < p < 0.005$ and $0.025 < p < 0.001$). Furthermore, since both vehicles are not cytotoxic, because they guarantee 100% cellular viability, the cytotoxicity recorded in the test is attributable exclusively to the effect of the drug. The obtained result suggests that DAS complexation with HP-β-CD increases the cytotoxicity of the drug, and this effect is probably a consequence of the increased solubility of DAS in water-like phase. Anyway, it is important to underline that the concentration at which DAS exerted cytotoxic actions on C2C12 cells, both free than complexed with HP-β-CD, is higher that the IC50 values known to inhibit cancer cell growth, which are in the nM range. Therefore, this in vitro experiment underlines that DAS is relatively safe on satellite muscle precursors being cytotoxic only at high concentrations. In fact, the concentrations that are used in the cell viability test are above the therapeutic plasma levels of DAS, which range in the low μM values.

2.5. Pharmacokinetic Results

HPLC analyses were carried out to evaluate the DAS traceability in main target tissues (quadriceps and liver) of treated mice. Appreciable drugs' levels were found in quadriceps and livers of treated animals (Figure 9). These results are in line with the finding that DAS is rapidly distributes in tissues [25]. Also, the level reached in skeletal muscle allows for predicting sufficient exposure for the action of DAS to take place, considering that the inhibition of Src-TK occurs in the nanomolar range [26].

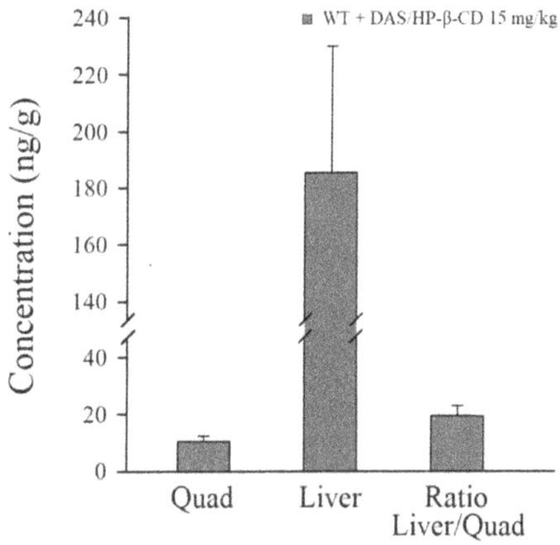

Figure 9. Pharmacokinetic analysis in quadriceps and livers of DAS/HP-β-CD inclusion complex administered at 15 mg/kg in drinking water for 1 week. All values are mean ± S.E.M. from 7–8 mice for each group. No significant difference was found by Student *t*-test analysis.

3. Materials and Methods

3.1. Materials

DAS (MW = 488 g/mol), was purchased from Sigma Aldrich (Milan, Italy). HP-β-CD (hydroxypropil-β-cyclodextrin, MW = 1396 Dalton, substitution degree 0.65) was kindly provided by Roquette (FR). HCl and phosphate salts for the preparation of buffers were purchased from Fluka (Sigma Aldrich, Milan, Italy). Bidistilled water was bought from Carlo Erba (Milan, Italy). The cell counting Kit-8 (CCK-8) used for cytotoxicity studies was purchased from Sigma Aldrich (Milan, Italy). All other products and reagents used in this work were of analytical grade.

3.2. Quantitative Analysis of DAS

The quantitative analysis of DAS was performed by High-performance liquid chromatography (HPLC). In detail, a HPLC station composed of a Agilent 1260 LCVL quaternary pump, a variable wavelength UV-visible detector and a fixed 20 μL loop manual injector was used. The analytical data were processed with the Agilent OpenLab LC software. For the analysis, a C_{18} Zorbax SB – Aq (4.6 × 150 mm) column was eluted in isocratic mode with a methanol/ammonium acetate pH = 3 60/40 v/v mixture, continuously monitoring the eluent at 280 nm. In these conditions, the retention time of the drug was about 2.8 min.

Standard calibration curves were prepared at a wavelength of 280 nm using the same analysis conditions and they resulted in a linear plot (r^2 = 0.999) in the range of tested concentrations (from 2.17×10^{-4} M and 6.78×10^{-6} M).

3.3. Solubility and Phase-Solubility Studies

The phase solubility study was conducted in accordance with Higuchi and Connors [21]. In detail, DAS was added in excess to an aqueous solution containing HP-β-CD in the appropriate concentration (0–10% w/v) until saturation and the suspensions thus obtained were placed in 4 mL vials with screw cap to avoid changes that are caused by evaporation. The obtained mixtures were vortexed for about 5 min and then placed in a thermostat bath at 25 °C for three days.

Subsequently, an aliquot of the aqueous phase of each mixture was transferred into a 5 mL glass syringe and filtered through a 0.22 μm cellulose acetate membrane filter (Millipore®, Milan, Italy).

The obtained filtrate was suitably diluted and subjected to subsequent HPLC analysis for the quantification of the drug. All of the determinations were conducted in triplicate.

The obtained data were used to determine the apparent stability 1:1 constant ($K_{1:1}$) of the DAS/HP-β-CD inclusion complex, using the slope of the phase solubility diagrams straight line, as reported by Higuchi and Connors in the following equation:

$$K_{1:1} = \frac{slope}{S_0(1 - slope)} \tag{1}$$

where S_0 represents DAS solubility in absence of cyclodextrin determined in the same way.

3.4. Preparation of Solid DAS /HP-β-CD Inclusion Complex

The DAS/HP-β-CD inclusion complex was prepared in the solid state by freeze drying [18]. The lyophilized product was prepared by adding DAS and HP-β-CD in water in equimolar amounts. The obtained suspension was vigorously vortexed for about five minutes, left under stirring for two days, filtered through 0.22 μm cellulose acetate filters (Millipore), then frozen, and lyophilized (Lio 5P, Milan, Italy). The obtained product was characterized by DSC and FT-IR.

3.5. Determination of DAS Incorporation Degree in the Solid Cyclodextrin Inclusion Complex

The amount of DAS that is present in the DAS/HP-β-CD solid complex was determined by solubilizing about 5 mg of sample in 5 mL of deionized water. Samples were injected in HPLC after filtration with 0.22 μm cellulose acetate filters (Millipore®). The incorporation degree of DAS into the inclusion complex was determined from the peak areas obtained and expressed as g of DAS per 100 g of complex.

3.6. Job's Plot Method

The stoichiometry of the inclusion complex DAS/HP-β-CD in aqueous solution was determined by the continuous variation method or Job's method [15]. Briefly, equimolar (1.02×10^{-3} M) CD_3OD/D_2O (50/50, v/v) solutions of DAS and HP-β-CD were mixed to a fixed volume by varying the molar ratio from 0 to 1, keeping the total molar concentration of the species constant. After stirring for 1 h, for each solution the ^{1}H-NMR spectra were registered and the chemical shifts of the host's protons were calculated and expressed as ppm. The Δchemical shift was determined as the difference between chemical shifts with and without HP-β-CD. Subsequently, Δppm × [DAS] was plotted versus r, where:

$$r = \frac{[DAS]}{[DAS] + [HP - \beta - CD]} \tag{2}$$

3.7. ^{1}H-NMR and Heteronuclear Multiple Bond Correlation (HMBC) Spectroscopic Studies

^{1}H nuclear magnetic resonance (^{1}H-NMR) spectra were recorded using a NMR Agilent Technologies 500/54 Premium Shielded instrument and ^{1}H chemical shifts were referred to DHO as internal standard. For the Heteronuclear Multiple Bond Correlation (HMBC) experiment, an Agilent 500 mHz spectrometer was used. The concentration of the drug was 10 mg/mL in a 5-mm NMR tube. Sample temperature was set to 25 °C. The following parameters were used for 2D ^{1}H- ^{13}C-heteronuclear multiple bond correlation (HMBC) experiment: number of scans, 2, number of complex data points (experiments) in F1, 128; number of complex data points in F2, 2048; sweep width in F1 and F2, 222 and 13ppm, respectively; spectrometer offset for ^{1}H and ^{13}C, 6 and 100 ppm, respectively; interscan delay, 1.5 s. Data were processed with the software Topspin. For 2D, the spectrum was zero filled to 512 data apodization function in both dimensions prior to Fourier transform and phase correction. Chemical shifts were expressed in parts per million (ppm) with respect to the DMSO-d_6 signal for carbon and H_2O_2 for proton [27–31].

3.8. Fourier Transform Infrared (FT-IR) Spectroscopy

The FT-IR spectra of DAS, HP-β-CD, and DAS/HP-β-CD solid complex were recorded with a Perkin-Elmer 1600 FTIR spectrophotometer dispersing each sample in KBr for spectroscopy (2 mg of sample in 200 mg of KBr) [32]. The scan range used was 400–4000 cm^{-1}, with a resolution of 1 cm^{-1}. The instrument was periodically calibrated.

3.9. Differential Scanning Calorimetry (DSC) Analysis

The thermal analysis of DAS and DAS/HP-β-CD solid complex were performed using a Mettler Toledo DSC 822e Star^{e} 202 system (Mettler Toledo, Switzerland) equipped with a thermal analysis automatic program, as described in a previous work [33]. The instrumentation was calibrated periodically, using indium as reference.

3.10. Dissolution Studies

Dissolution experiments were performed at 37 °C using a BIODIS USP III apparatus (Varian Inc., Cary North Carolina, CA, USA), equipped with a rod stirrer maintaining a rotational speed of 100 rpm during the test. Samples of DAS or DAS/HP-β-CD solid complex, equivalent to about 2 mg of DAS, were suspended in the dissolution medium (80 mL of 0.05 M phosphate buffer at pH 7.4 or HCl 0.05 M pH = 1.2). The volume of 80 mL was chosen taking into account the HPLC quantification limit for the determination of DAS.

At predetermined time intervals, 600 μL of suspension were collected and, in order to keep constant the initial volume, 600 μL of the same dissolution medium previously thermostated at the same temperature were added. Samples were subsequently filtered using a 0.22 μm membrane filter (Millipore®cellulose acetate), and the filtrates thus obtained were subjected to HPLC analysis after appropriate dilution. For quantitative analysis the calibration curve previously constructed was used and the dissolution profiles shown correspond to the average of three determinations.

3.11. Cytotoxicity Studies

C2C12 myocytes were cultured in DMEM that was supplemented with 10% fetal bovine serum, 1% penicillin, 1% streptomycin and 1% glutamine and were maintained at 37 °C in 5% CO_2/95% air. Cell viability was evaluated by measuring the succinic dehydrogenases activity in the cell suspension using the cell counting Kit-8 (CCK-8) (Sigma Aldrich), which utilizes a highly water-soluble tetrazolium salt and whose detection sensitivity is higher than other tetrazolium salts [34].

Cells were seeded in 96-well cultures at a density of approximately 4.5×10^3 cells per well and then cultured for 16 h. Afterwards, the cells were treated for 5 h with free DAS (in DMSO <0.15% in order to ensure cellular vitality) or complexed with HP-β-CD, but at the same concentration calculated on the basis of the incorporation degree, both being dissolved in DMEM. Following exposure, 10 μL of CCK-8 were added into each well and then the plate was incubated for additional two hours. The absorbance at 450 nm was measured using a spectrophotometer (microplate reader Victor V31420–40; PerkinElmer, Wellesley, Massachusetts). Cell viability (%) is expressed according to the following formula:

$$\text{cell viability (\%)} = [(\text{test value} - \text{blank})/(\text{control value} - \text{blank}) \times 100] \quad (3)$$

where the blank value represents that of a cell-free wells; the control value represents that of wells of cells do not treated with DAS and the test value represents that of wells of cells treated with DAS. The results are expressed as the percentage of the control and presented as the mean ± SD. Each data is from 24–48 replicates (wells) and 6–9 different culture dishes.

3.12. In Vivo Study

A total of 10 sedentary WT male mice C57Bl/6J (Charles River, Italy for Jackson Laboratories), homogeneous for age and body weight (BW) were divided into 2 groups as follows: 4 WT mice vehicle-treated (HP-β-CD 10%) and 6 WT mice treated with DAS/HP-β-CD inclusion complex at the dose of 15 mg/Kg. Drug and vehicle were administered in drinking water for 1 week. The dose was chosen based on data in literature, in fact, the human dose commonly administered in clinical practice, converted in the appropriate animal equivalent, resulted to be approximately 20 mg/kg per day [35]. Care in animal handling and environment conditions was used to avoid any animal discomfort and stress during the study period. Food intake was monitored, and composition was maintained constant [36]. No abnormal gross findings in animal well-being and no animal deaths were observed during the study period.

3.13. Ex vivo Study: Pharmacokinetic Analysis

The pharmacokinetic (PK) analysis were commissioned to the CRO "XenoGesis Ltd.—Preclinical DMPK & Bioanalysis services, Nottingham, UK. In detail, analysis was performed in quadriceps (Quad) and livers of treated animals. Tissues were individually weighed into a "FastPrep" tube and PBS was added (3:1 ratio). Each tube was placed in the fast prep homogenizer on a predetermined 1min cycle to ensure complete homogenization. 40 μL of each homogenate was aliquoted to a fresh tube and 50 μL of MeOH plus 150 μL of Methanol-containing Internal standard (25 ng/mL Imipramine HCl) was added. Each sample was mixed on a Bioshake for 1 min and then transferred to the freezer at −20 °C for at least two hours prior centrifugation at 2500 × *g* for 20 min. The supernatants were then transferred to a 96-well plate for sampling by the LC-MS/MS. A Thermo TSQ Quantiva with Thermo Vanquish UHPLC system was used (Thermo Fisher Scientific Inc, Milan, Italy). Separation was achieved on a ACE-AR C18 (50 × 2.1 mm, 1.7 μm) column, with MilliQwater 0.1% formic acid (solvent A) and methanl-0.1% formic acid (solvent B) at 65 °C and at a flow rate of 0.8 mL/min. Positive ion spray voltage and vaporizer temperature were set at 3500 V and 450 °C, respectively, while the ion transfer tube temperature was set at 365 °C. Finally, sheath gas and auxiliary gas pressures were fixed at 54 and 17 bar, respectively. Detection was performed using a multiple reaction monitoring (MRM) via a positive ESI source spray voltage. Quantitative analysis was conducted by MRM at 232.06 to 401.11 m/z for DAS inclusion complex and at 86.10 to 193.04 m/z for the internal standard Imipramine. Mass transitions were combined for each compound to maximize sensitivity.

3.14. Statistic

In the elaboration of results in the cytotoxicity studies, the statistical significance between groups was evaluated by Student's t-test, as follows: * significantly different with respect to control value ($0.001 < p < 0.005$); significantly different with respect to DAS complexed with HPβCD ($0.025 < p < 0.001$).

4. Conclusions

From the results that were obtained in this study, it is possible to state that the complexation of DAS with HP-β-CD is successful both in solution and in the solid state. In particular, the presence of the cyclodextrin allows for obtaining an increase in the drug water solubility and a favorable dissolution profile especially at pH 7.4 as compared to the non-complexed drug Moreover, cytotoxicity studies highlight that DAS complexation with HP-β-CD increases the cytotoxicity of the drug and PK results of a one-week pilot study with DAS/HP-β-CD inclusion complex in WT mice provided the basis for further long-term in vivo treatment with this new oral formulation of DAS in treadmill-exercised *mdx* mice.

In conclusion, this new inclusion complex could allow the development of a liquid formulation to be administered orally, which could be a valid alternative to the one currently present on the market that is solid, especially in the case of an administration in paediatric age. Moreover, considering that HP-β-CD is FDA approved for parenteral formulations, the DAS/HP-β-CD inclusion complex could also be an interesting tool for the administration of DAS by this route.

Author Contributions: Conceptualization, A.D.L. and N.D.; methodology, A.C. and F.S.; software, V.L.; validation, N.D., A.L. (Angela Lopedota) and M.F.; formal analysis, A.L. (Antonio Lopalco); investigation, V.L., B.B., P.M. and A.L. (Antonio Lopalco); resources, A.L. (Angela Lopedota) and M.F.; data curation, A.C. and F.S.; writing—original draft preparation, A.C. and F.S.; writing—review and editing, A.C. and F.S.; visualization, A.C., F.S and A.L. (Antonio Lopalco); supervision, A.D.L. and N.D.; project administration, N.D.; funding acquisition, A.D.L.

Funding: The University of Bari (Italy), and the Inter-University Consortium for Research on the Chemistry of Metal Ions in Biological Systems (C.I.R.C.M.S.B.) are gratefully acknowledged. This research has been supported by PRIN-MIUR (Research Project of National Interest—Ministry of Education, University and Research) project no. 20108YB5W3_004. This work has been supported also by a grant to ADL from Dutch Duchenne Parent Project (NL_DPP) 2015, entitled "Preclinical studies to validate cSrc tyrosine kinase as therapeutic target in Duchenne muscular dystrophy".

Acknowledgments: We thank Antonio Palermo for his skillful technical assistance.

Conflicts of Interest: The authors declare no conflict of interest.

References

1. Lombardo, L.J.; Lee, F.Y.; Chen, P.; Norris, D.; Barrish, J.C.; Behnia, K.; Castaneda, S.; Cornelius, L.A.; Das, J.; Doweyko, A.M.; et al. Discovery of *N*-(2-chloro-6-methylphenyl)-2-[[6-[4-(2-hydroxyethyl)–piperazin-1-yl]-2–methylpyrimidin-4-ylamino] thiazole 5–carboxamide (BMS-354825), a dual Src/Abl kinase inhibitor with potent antitumor activity in preclinical assays. *J. Med. Chem.* **2004**, *47*, 6658–6661. [CrossRef] [PubMed]
2. Tokarski, J.S.; Newitt, J.A.; Chang, C.Y.J.; Cheng, J.D.; Wittekind, M.; Kiefer, S.E.; Kish, K.; Lee, F.Y.; Borzillerri, R.; Lombardo, L.J.; et al. The structure of dasatinib (BMS-354825) bound to activated ABL kinase domain elucidates its inhibitory activity against imatinib-resistant ABL mutants. *Cancer Res.* **2006**, *66*, 5790–5797. [CrossRef]
3. Talpaz, M.; Shah, N.P.; Kantarjian, H.; Donato, N.; Nicoll, J.; Paquette, R.; Cortes, J.; O'Brien, S.; Nicaise, C.; Bleickardt, E.; et al. Dasatinib in imatinib-resistant Philadelphia chromosome-positive leukemias. *N. Engl. J. Med.* **2006**, *354*, 2531–2541. [CrossRef] [PubMed]
4. Melville, N.A. Dasatinib in Children: An Effective Alternative to Imatinib. In Proceedings of the European Hematology Association (EHA) 2017 Congress, Madrid, Spain, 22–25 June 2017.
5. Zwaan, C.M.; Rizzari, C.; Mechinaud, F.; Lancaster, D.I.; Lehrnbecher, T.; Van der Velden, V.H.J.; Beverloo, B.B.; den Boer, M.L.; Pieters, R.; Reinhardt, D.; et al. Dasatinib in Children and Adolescent With Relapsed or Refractory Leukemia: Results of the CA180-018 Phase I Dose-Escalation Study of the innovative Therapies for Children With Cancer Consortium. *J. Clin. Oncol.* **2013**, *31*, 2460–2469. [CrossRef] [PubMed]
6. Prins, K.W.; Humston, J.L.; Mehta, A.; Tate, V.; Ralston, E.; Ervasti, J.M. Dystrophin is a microtubule-associated protein. *J. Cell Biol.* **2009**, *186*, 363–369. [CrossRef] [PubMed]
7. Hoffman, E.P.; Dressman, D. Molecular pathophysiology and targeted therapeutics for muscular dystrophy. *Trends Pharm. Sci.* **2001**, *22*, 465–470. [CrossRef]
8. De Luca, A. Pre-clinical drug tests in the mdx mouse as a model of dystrophinopathies: An overview. *Acta Myol. Myopathies Cardiomyopathies Off. J. Mediterr. Soc. Myol.* **2012**, *31*, 40–47.
9. Camerino, G.M.; Cannone, M.; Giustino, A.; Massari, A.M.; Capogrosso, R.F.; Cozzoli, A.; De Luca, A. Gene expression in mdx mouse muscle in relation to age and exercise: Aberrant mechanical-metabolic coupling and implications for pre-clinical studies in Duchenne muscular dystrophy. *Hum. Mol. Genet.* **2014**, *23*, 5720–5732. [CrossRef]
10. Paletta-Silva, R.; Rocco-Machado, N.; Meyer-Fernandes, J.R. NADPH oxidase biology and the regulation of tyrosine kinase receptor signaling and cancer drug cytotoxicity. *Int. J. Mol. Sci.* **2013**, *14*, 3683–3704. [CrossRef]
11. Lipscomb, L.; Piggot, R.W.; Emmerson, T.; Winder, S.J. Dasatinib as a treatment for Duchenne muscular dystrophy. *Hum. Mol. Genet.* **2016**, *25*, 266–274. [CrossRef]

12. Lopalco, A.; Curci, A.; Lopedota, A.; Cutrignelli, A.; Laquintana, V.; Franco, M.; Denora, N. Pharmaceutical preformulation studies and paediatric oral formulations of sodium dichloroacetate. *Eur. J. Pharm. Sci.* **2019**, *127*, 339–350. [CrossRef] [PubMed]
13. Korashy, H.M.; Rahman, A.F.M.M.; Kassem, G.M. *«Dasatinib» in Profiles of Drug Substances, Excipients, and Related Methodology*; Academic Press: New York, NY, USA, 2012.
14. Loftsson, T.; Duchene, D. Cyclodextrins and their pharmaceutical applications. *Int. J. Pharm.* **2007**, *329*, 1–11. [CrossRef] [PubMed]
15. Cutrignelli, A.; Lopedota, A.; Denora, N.; Iacobazzi, R.M.; Fanizza, E.; Laquintana, V.; Perrone, M.; Maggi, V.; Franco, M. A new complex of curcumin with sulfobutylether-β-cyclodextrin: Characterization studies and in vitro evaluation of cytotoxic and antioxidant activity on HepG-2 cells. *J. Pharm. Sci.* **2014**, *103*, 3932–3940. [CrossRef] [PubMed]
16. Cutrignelli, A.; Lopedota, A.; Denora, N.; Laquintana, V.; Tongiani, S.; Franco, M. Characterization and release studies of liposomal gels containing glutathione/cyclodextrins complexes potentially useful for cutaneous administration. *J. Pharm. Sci.* **2014**, *103*, 1246–1254. [CrossRef]
17. Lopedota, A.; Trapani, A.; Cutrignelli, A.; Laquintana, V.; Denora, N.; Franco, M.; Trapani, G.; Liso, G. Effect of cyclodextrins on physico-chemical and release properties of Eudragit RS 100 microparticles containing glutathione. *J. Incl. Phenom. Macrocycl. Chem.* **2007**, *57*, 425–432. [CrossRef]
18. Tricarico, D.; Maqoud, F.; Curci, A.; Camerino, G.; Zizzo, N.; Denora, N.; Cutrignelli, A.; Laquintana, V.; Lopalco, A.; la Forgia, F.; et al. Characterization of minoxidil/hydroxypropyl-β-cyclodextrin inclusion complex in aqueous alginate gel useful for alopecia management: Efficacy evaluation in male rat. *Eur. J. Pharm. Biopharm.* **2018**, *122*, 146–157. [CrossRef]
19. Lopedota, A.; Denora, N.; Laquintana, V.; Cutrignelli, A.; Lopalco, A.; Tricarico, D.; Maqoud, F.; Curci, A.; Mastrodonato, M.; la Forgia, F.; et al. Alginate-Based Hydrogel Containing Minoxidil/Hydroxypropyl-β-Cyclodextrin Inclusion Complex for Topical Alopecia Treatment. *J. Pharm. Sci.* **2018**, *107*, 1046–1054. [CrossRef]
20. Lopedota, A.; Cutrignelli, A.; Denora, N.; Laquintana, V.; Lopalco, A.; Selva, S.; Ragni, L.; Tongiani, S.; Franco, M. New ethanol and propylene glycol free gel formulations containing a minoxidil-methyl-β-cyclodextrin complex as promising tools for alopecia treatment. *Drug Dev. Ind. Pharm.* **2015**, *41*, 728–736. [CrossRef]
21. Higuchi, T.; Connors, K.A. Phase solubility techniques. *Adv. Anal. Chem. Instrum.* **1965**, *4*, 117–212.
22. Ventura, C.A.; Tommasini, S.; Falcone, A.; Giannone, I.; Paolino, D.; Sdrafkakis, V.; Mondello, M.R.; Puglisi, G. Influence of modified cyclodextrins on solubility and percutaneous absorption of celecoxib through human skin. *Int. J. Pharm.* **2006**, *314*, 37–45. [CrossRef]
23. Purohit, P.; Rampalli, S.; Murali, M.S.V.; Upalla, L.; Pothana, P. Process for Preparing Dasatinib Monohydrate. U.S. Patent No 9,145.406 B2, 29 September 2015.
24. Yan, R.; Yang, H.; Xu, Y. Polymorphs of Dasatinib, Preparation Methods and Pharmaceutical Composition Thereof. U.S. Patent No 2012/0309968 A1, 6 December 2012.
25. Duckett, D.R.; Cameron, M.D. Metabolism considerations for kinase inhibitors in cancer treatment. *Expert Opin. Drug Metab. Toxicol.* **2010**, *6*, 1175–1193. [CrossRef]
26. Rao, S.; Larroque-Lombard, A.L.; Peyrard, L.; Thauvin, C.; Rachid, Z.; Williams, C.; Jean-Claude, B.J. Target modulation by a kinase inhibitor engineered to induce a tandem blockade of the epidermal growth factor receptor (EGFR) and c-Src: The concept of type III combi-targeting. *PLoS ONE* **2015**, *10*, e0117215. [CrossRef]
27. Caruso, G.; Fresta, C.G.; Martinez-Becerra, F.; Antonio, L.; Johnson, R.T.; de Campos, R.P.S.; Siegel, J.M.; Wijesinghe, M.B.; Lazzarino, G.; Lunte, S.M. Carnosine modulates nitric oxide in stimulated murine RAW 264.7 macrophages. *Mol. Cell. Biochem.* **2017**, *431*, 197–210. [CrossRef] [PubMed]
28. Lopalco, A.; Stella, V.J. Effect of Molecular Structure on the Relative Hydrogen Peroxide Scavenging Ability of Some α-Keto Carboxylic Acids. *J. Pharm. Sci.* **2016**, *105*, 2879–2885. [CrossRef]
29. Lopalco, A.; Dalwadi, G.; Niu, S.; Schowen, R.L.; Douglas, J.; Stella, V.J. Mechanism of Decarboxylation of Pyruvic Acid in the Presence of Hydrogen Peroxide. *J. Pharm. Sci.* **2016**, *105*, 705–713. [CrossRef] [PubMed]
30. Lopalco, A.; Douglas, J.; Denora, N.; Stella, V.J. Determination of pKa and Hydration Constants for a Series of α-Keto-Carboxylic Acids Using Nuclear Magnetic Resonance Spectrometry. *J. Pharm. Sci.* **2016**, *105*, 664–672. [CrossRef] [PubMed]

31. Denora, N.; Margiotta, N.; Laquintana, V.; Lopedota, A.; Cutrignelli, A.; Losacco, M.; Franco, M.; Natile, G. Synthesis, characterization, and in vitro evaluation of a new TSPO-selective bifunctional chelate ligand. *ACS Med. Chem. Lett.* **2014**, *5*, 685–689.
32. Cutrignelli, A.; Lopedota, A.; Trapani, A.; Boghetich, G.; Franco, M.; Denora, N.; Laquintana, V.; Trapani, G. Relationship between dissolution efficiency of Oxazepam/carrier blends and drug and carrier molecular descriptors using multivariate regression analysis. *Int. J. Pharm.* **2008**, *358*, 60–68. [CrossRef] [PubMed]
33. Lopedota, A.; Cutrignelli, A.; Laquintana, V.; Denora, N.; Iacobazzi, R.M.; Perrone, M.; Fanizza, E.; Mastrodonato, M.; Mentino, D.; Lopalco, A.; et al. Spray Dried Chitosan Microparticles for Intravesical Delivery of Celecoxib: Preparation and Characterization. *Pharm. Res.* **2016**, *33*, 2195–2208. [CrossRef] [PubMed]
34. Berridge, M.V.; Herst, P.M.; Tan, A.S. Tetrazolium dyes as tools in cell biology: New insights into their cellular reduction. *Biotechnol. Annu. Rev.* **2005**, *11*, 127–152.
35. Reagan-Shaw, S.; Nihal, M.; Ahmad, N. Dose translation from animal to human studies revisited. *FASEB J. Off. Publ. Fed. Am. Soc. Exp. Biol.* **2008**, *22*, 659–661. [CrossRef] [PubMed]
36. Capogrosso, R.F.; Mantuano, P.; Uaesoontrachoon, K.; Cozzoli, A.; Giustino, A.; Dow, T.; Srinivassane, S.; Filipovic, M.; Bell, C.; Vandermeulen, J.; et al. Ryanodine channel complex stabilizer compound S48168/ARM210 as a disease modifier in dystrophin-deficient mdx mice: Proof-of-concept study and independent validation of efficacy. *FASEB J. Off. Publ. Fed. Am. Soc. Exp. Biol.* **2018**, *32*, 1025–1043. [CrossRef] [PubMed]

International Journal of
Molecular Sciences

MDPI

Review

Bcr-Abl Tyrosine Kinase Inhibitors in the Treatment of Pediatric CML

Francesca Carofiglio, Antonio Lopalco, Angela Lopedota, Annalisa Cutrignelli, Orazio Nicolotti, Nunzio Denora, Angela Stefanachi * and Francesco Leonetti *

Dipartimento di Farmacia Scienze del Farmaco Università degli Studi di Bari "Aldo Moro", via Orabona 4, 70125 Bari, Italy; francescacarofiglio94@gmail.com (F.C.); antonio.lopalco@uniba.it (A.L.); angelaassunta.lopedota@uniba.it (A.L.); annalisa.cutrignelli@uniba.it (A.C.); orazio.nicolotti@uniba.it (O.N.); nunzio.denora@uniba.it (N.D.)

* Correspondence: angela.stefanachi@uniba.it (A.S.); francesco.leonetti@uniba.it (F.L.); Tel.: +39-08-0544-2783 (A.S.); +39-08-0544-2784 (F.L.)

Received: 14 May 2020; Accepted: 18 June 2020; Published: 23 June 2020

Abstract: The therapeutic approach to Chronic Myeloid Leukemia (CML) has changed since the advent of the tyrosine kinase inhibitor (TKI) imatinib, which was then followed by the second generation TKIs dasatinib, nilotinib, and, finally, by ponatinib, a third-generation drug. At present, these therapeutic options represent the first-line treatment for adults. Based on clinical experience, imatinb, dasatinib, and nilotinib have been approved for children even though the studies that were concerned with efficacy and safety toward pediatric patients are still awaiting more specific and high-quality data. In this scenario, it is of utmost importance to prospectively validate data extrapolated from adult studies to set a standard therapeutic management for pediatric CML by employing appropriate formulations on the basis of pediatric clinical trials, which allow a careful monitoring of TKI-induced adverse effects especially in growing children exposed to long-term therapy.

Keywords: chronic myeloid leukemia; tyrosine kinase inhibitors; pediatric age; imatinib; dasatinib; nilotinb; ponatinib; formulation

---

1. Introduction

Protein kinases (PK) enable the phosphorylation of hydroxyl groups of tyrosine, serine, and threonine residues. Based on this signaling, a cascade of molecular events is activated to promote a number of biochemical actions responsible for cells proliferation, survival, and functioning. There are two main classes of tyrosine kinases. The first class is made up by receptors tyrosine kinases (RTK) linked to transmembrane receptors. The second class is known as cytoplasmic non-receptor tyrosine kinases (NRTK) [1]. Receptors activated by a vascular endothelial growth factor (VEGF) and a platelet-derived growth factor (PDGF) and overexpressed in sarcoma, Breakpoin cluster region-Abelson Kinase (BCR-ABL) in myeloid leukemia, and JAK (Janus Kinases) are of oncology relevance as far as RTK and NRTK are concerned [1].

The overexpression or mutation of tyrosine kinases (TK) is a hallmark of cell cycle dysregulation often anticipating the tumor onset. For these reasons, selective TK inhibition is considered a target therapy in cancer [2].

Chronic myeloid leukemia (CML) is a myeloproliferative disorder, characterized by an abnormal granulocyte cells proliferation determining a high increase of white blood cell count in addition to a spleen enlargement (splenomegaly). The CML pathogenesis stems from the Philadelphia chromosome, discovered by Peter Noweel in 1960. It arises from a translocation of the TK ABL (Abelson) gene from chromosome 9 to chromosome 22, on the BCR gene (breakpoint cluster region). The translocation generates the oncogenic BCR-ABL1 fusion gene in hematopoietic stem cells, which encodes an abnormal protein, with constitutive TK activity, responsible for proliferative and anti-apoptotic signals. The occurrence of BCR-ABL protein kinases was observed in more than 90% of CML patients [3].

The BCR-ABL targets are downstream pathways including RAS, PI3K/AKT, and JAK/STAT that address the transformation of healthy cells toward neoplastic cells responsible for CML pathogenesis.

CML has three clinical phases. The first is the chronic phase (CP) without subjective symptoms after 3–5 years from diagnosis but a high white blood cell and platelet count. The second is the accelerated phase (AP) with an incremented differentiation of abnormal granulocytes. The third phase consists in the blast crisis (BC) with an increase of undifferentiated blasts. Patients in the chronic phase can be treated with tyrosine kinases inhibitors (TKIs). Unfortunately, accelerated and blast phases are not responsive to TKIs likely because their progression is not affected by BCR-ABL.

The 3D structure of BCR-ABL kinase, like other tyrosine kinases, is characterized by two lobes, including the N-terminal and the C-terminal one, connected by a short peptide strand, known as a hinge region. ATP (adenosine triphosphate) binds to the fissure between the lobes through two hydrogen bonds with a residue of Glu316 and Met318.

Moreover, in the rear portion of the ATP binding pocket, there is a residue of Thr315 with a key role for selectivity. This residue is called "gatekeeper" because of a resistance phenomenon associated with its related single point mutations.

The Bcr-Abl activation is determined by a flexible protein segment in the N-terminal lobe, called the activation loop. In its active state, it adopts an open conformation with the amino acid triad Asp-Phe-Gly (DFG, corresponding to 381–383 residues in Abl) directed toward the ATP binding site. This is the DFG IN conformation, whichh allows ATP to approach the binding site, where the Phe residue of the triad is located.

When the activation loop is closed as it is in the inactive state, the Asp-Phe-Gly amino acid triad adopts the DFG out conformation (Figure 1) [4].

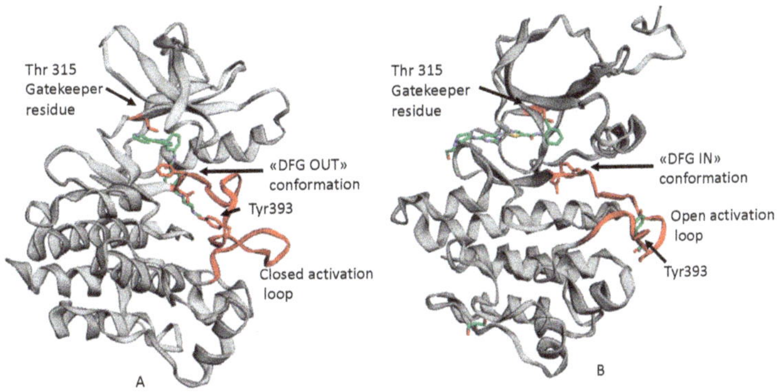

Figure 1. An example of ATP competitive inhibitors: (**A**) X-ray solved structure of Abl kinase domain in complex with Imatinib (PDB code: 1IEP) with DFG outconformation and closed activation loop. (**B**) X-ray solved structure of Abl Kinase domain with Dasatinib (PDB code: 2GQG) with DFG in conformation, open activation loop [5].

2. CML Therapeutic Approach

In the past decades, the first therapeutic approach to CML involved the use of ordinary chemotherapeutical drugs (such as a busulfan, hydroxyurea, cyclophosphamide, and vincristine), which is followed by allogeneic hematopoietic stem cells transplantation (allo-HSCT). This is considered a potentially curative procedure for a variety of hematological malignancies in order to reconstitute hematopoiesis [6].

However, allo-HSCT is limited by the availability of donors, transplant-related mortality, and early morbidity, so that it is considered a third-line treatment for most pediatric cases. Therefore, Bcr-Abl inhibitors may be accounted as a targeted therapy toward CML, avoiding all the side effects associated with canonical chemotherapy, and graft-versus-host disease, which is one of the major complications of allogeneic hemapoietic cell transplantation (HCT).

Currently, Imatinib is not the only TK inhibitor approved for patients with CML-CP (CML-Chronic Phase) including pediatric patients, but it remains the first-line therapy for patients with CML-AP (CML-Accellerate Phase) and CML-BC (CML-Blast Crisis).

Imatinib (Figure 2A) was discovered in 1992 as a selective inhibitor of Bcr-Abl TK. It is able to link to the kinase in its inactive form preventing the ATP binding (Figure 1A). It was approved in 2001 by FDA as first-line therapy for CML patients by obtaining optimal results compared to those who had been treated with hydroxyurea and IFN-α in terms of CHR (Complete Hematological Response), CCR (Complete Cytogenetic Response), and MMR (Major Molecular Response). Unfortunately, the appearance of resistance phenomena required the development of a second generation of TKI [6].

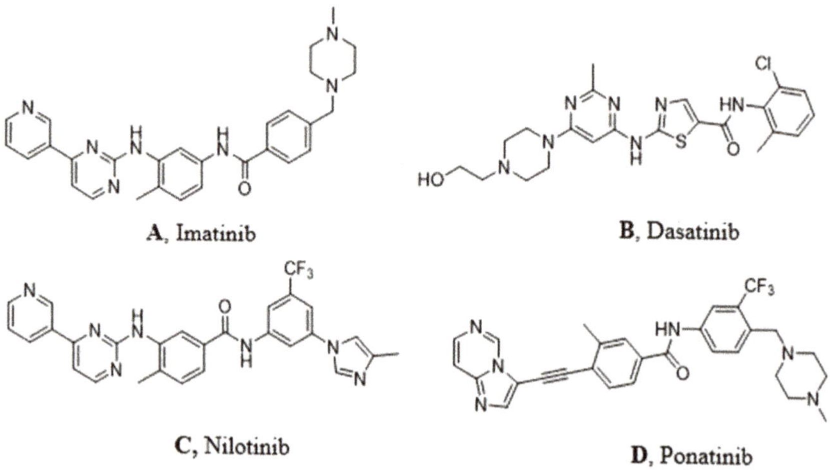

Figure 2. Molecular structures of the principal BCR-ABL TKIs used for treating pediatric chronic myeloid leukemia (CML): (**A**) Imatinib; (**B**) Dasatinib; (**C**) Nilotinib; (**D**) Ponatinib.

2.1. Resistance and Intolerance to Imatinib

The urgent need to develop a second-generation of TK inhibitors was principally due to the resistance and the intolerance of some patients to imatinib. A patient can be defined resistant to imatinib when presenting an increasing white blood cell or platelets counts that prove a primary resistance or a hematologic relapse. Moreover, resistance to imatinib consists of suboptimal cytogenetic or molecular response or failure, progression to AP/BC, reappearance of Ph+ bone marrow cells following achievement of complete cytogenetic response (CCR) and, at least, increase of more than 30% in Ph+ cells in peripheral blood or bone marrow, or loss of a molecular response. Frequently, resistance is associated with emerging mutations in some BCR-ABL amino acid residues. We can count 12 residues with risk of mutations that can be classified in resistant (T315I), less sensitive (Y235H, F359V, E255K), and sensitive (all other mutations except those mentioned before) [7].

Other resistance phenomena such as P-gp (Glycoprotein P) efflux pump overexpression, OCT1 (Organic Cation transporter) reduced expression, and the activation of alternating oncogenic pathways (Src (proto oncogene Sarcome tyrosin kinase), PI3K (Phosphoinositide 3-kinases), PDGFR, KRAS (Kirsten rat sarcoma Kinase), and JAK2) must be reminded [8,9].

Furthermore, intolerance is related to imatinib side effects due to its off-target activity. Among all, we can remember the increase of blood bilirubin, headache, rash, vomiting, cardiotoxicity, neutropenia, obesity, and abdominal pain [1,10].

A patient can be defined intolerant to imatinib only when he develops serious adverse events (AEs) such as renal failure or liver injury (hepatic transaminase and bilirubin elevations), which require discontinuation or change of therapy [11].

2.2. Second Generation TKI

Dasatinib (Figure 2B) is the first second generation TKI approved by FDA in 2006. It is able to inhibit BCR-ABL both in the DGF in (Figure 1B) and in the DGF out conformation and also other TK such as PDGFR and Src [12]. Based on X-ray analysis, it was observed that dasatinib binds the ABL protein in a less strict conformation compared to imatinib. Despite the lower number of binding interactions observed for dasatinib vs. ABL with respect to that occurring between imatinib vs. ABL. The former is provided with inhibiting activity 325-fold higher and it is able to overcome many BCR-ABL mutations resistant to imatinib with the exception of the T315I one [13]. Furthermore, dasatinb is not P-gp substrate and, thus, can be used in patients resistant to imatinb and nilotinib.

Nilotinib (Figure 2C) is another second-generation TKI. It is a BCR-ABL, c-kit, and PDGFR inhibitor from 10-fold to 30-fold more powerful than imatinib on the BCR-ABL oncoprotein. It inhibits the proliferation of cells exhibiting the mutated TK. Nilotinib binds the protein in its inactive conformation (DFG OUT) through interactions similar to those observed for imatinib, such as four hydrogen bonds and many Van der Waals contacts. The successful introduction of the trifluoromethyl group on the phenyl ring and the substitution of the methyl-piperazine with imidazole are responsible for its greater activity. It can also overcome some Bcr-Abl mutations (L248V, G250E, Q252H) apart from T315I [14]. According to Redaelli et al. [15], by comparing the nilotinib activity vs. imatinib and dasatinib, the former showed a mild resistance to 13 out of 18 mutations of BCR-ABL with high resistance to T315I and E255V changes. Based on some clinical trials of phase III in 2011 (ENESTnd Study), [16] a major complete cytogenetic remission (CCR) was observed and a reduction in the percentage of patients proceeded to the accelerated phase of CML after an annual treatment with nilotinib in comparison with imatinib. For these reasons, nilotinib was approved as first-line therapy for newly diagnosed patients with CML in 2011.

Ponatinib (Figure 2D) is a very powerful third-generation pan-inhibitor of TKs. It was synthetized as a ligand of ATP binding domain of BCR-ABL in closed conformation. Ponatinib binds the DFG OUT conformation of the protein. The peculiarity of this molecule lies in the occurrence of a triple bond, which allows us to mitigate the steric hindrance due T315I mutation. The triple bond between the phenyl and the heterocycle ring in ponatinib can hold the isoleucine side chain. Having said that, ponatinib can be considered as a TK inhibitor unaffected from T315I mutation. Its binding is ensured by five hydrogen bonds. Moreover, there are several Van der Waals interactions supporting affinity in the case of multiple mutations [17]. Notably, ponatinib kept its activity in other CML mutations, such as M244V, G250E, and Q252H. However, its use drew attention to the appearance of an important side effect, which is the inhibition of parental cells Ba/F3 proliferation [18].

3. CML in Pediatric Age

Horibe et al. combined and analyzed data from more than 3856 Japanese children and 1803 young adults up to the age of 29 years with leukemia. Among pediatric leukemic patients under 15 years of age, the proportion of patients with CML was about 1.8% [19]. The relative incidence of CML starts to increase in the last teen years, between the ages of 15 and 19, up to 8.3%. Specifically, the average annual incidence of CML in children younger than 15 years is 0.6–1.0 cases per million and, for patients between 15 and 19 years of age, it is 2.1 per million [20]. Several published reports on pediatric CML have shown a male predominance of the disease with a male:female ratio of 1.3:1.7 [21,22], as reported in studies conducted on adults [23]. The incidence of CML among males and females, comparing statistics by age in the UK, is reported in Figure 3 [24]. As shown, CML mainly affects old people with a median age of CML diagnosis from 60 to 65 years, especially in western countries, [25] while it is considered a rare condition in adolescents and an ultra-rare condition in children. Only 2% of all leukemias in children younger than 15 years of age and 9% of all leukemias in adolescents between 15 and 19 years, with an annual incidence of 1 and 2.2 cases per million in these two ages groups, respectively, can be ascribed to CML [26]. As shown in Figure 3, 23% of new cases occurred in the United Kingdom (UK) in the period from 2014 to 2016, which were relative to people aged more than 75 years old. Incidence rates are significantly lower in females than in males [24] (Cancer research UK).

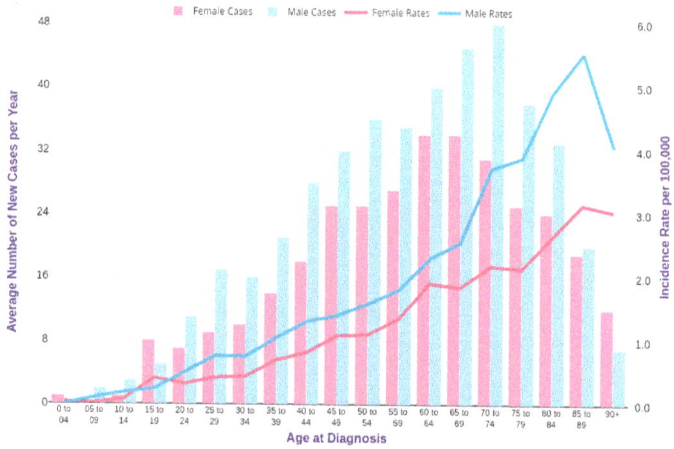

Figure 3. CML incidence rates in the United Kingdom (UK) [24].

The graphs in Figure 4 show the NAACCR (North American Association of central cancer registries, 2019) age-adjusted incidence rates of CML in male and female patients under 20 years of age [27]. It goes without saying that, in spite of the rareness of pediatric CML, its incidence from 2012 and 2016 is, to some extent, steadily increased in young males while a more fluctuating trend is observed in young females during the same lapse of time.

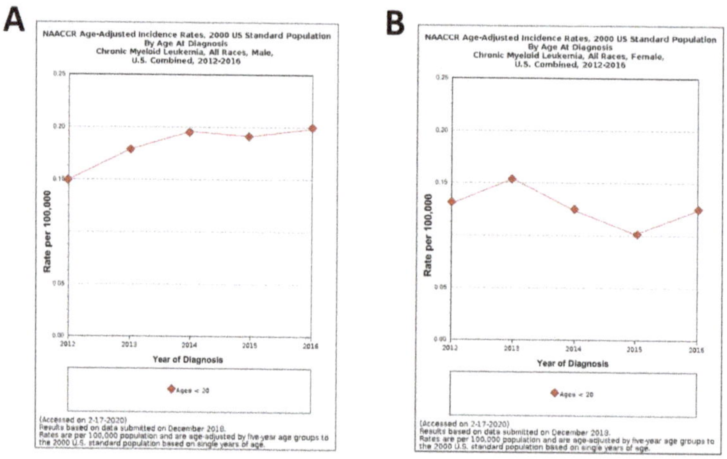

Figure 4. North American Association of central cancer registries (NAACCR) Fast Stats: age-adjusted incidence rates of CML in male and female patients under 20 years of ages (**A**) and (**B**), respectively [27].

First of all, it must be considered that CML in children and adolescents is different from CML in adults. Pediatric CML shows more aggressive features, such as larger spleen size in proportion to body weight, higher white blood cell and platelet counts, and more frequent diagnosis of advanced phases [28]. There are also some genetic differences between pediatric and adult CML such as a higher number of mutations prompting cancer progression. For example, about 60% of pediatric patients have ASXL1 mutation compared to only 15% of adults [29]. These differences between pediatric and adult CML are related to host factors and CML cell biology, which leads to different clinical presentation and disease spread [28]. However, in the absence of a representative sample of cases of pediatric CML, there are currently not enough clinical studies to establish practice standards for treating this serious disorder. For this reason, many pediatric oncologists can only follow the ELN (European Leukemia Net) [28] and NCCN (National Comprehensive Cancer Network) [30] guidelines, designed for adult patients.

The scoring system commonly used to prognosticate and manage CML in adults such as SOKAL (Sokal relative risk score), Hasford (Hasford relative risk score), and EUTOS (European treatment and outcome study score), are not applicable to pediatric patients [31,32]. The International Registry for Chronic Myeloid Leukemia (I-CML-Peds study) in Children and Adolescents, after evaluating and comparing the risk group allocations and outcome between the prognostic scores in a population of 350 pediatric patients, proved that the ELTS (EUTOS Long Term Survival) score was the best metric in terms of differentiation of progression-free survival [31]. However, more data are necessary to extend and confirm ELTS applicability to children and adolescents.

3.1. TKI Pediatric Therapy: Issues and Concerns

Since the development of TKIs, which have replaced what is now the third line CML treatment, the allo-HSTC, life expectancy for adult patients with CML has grown significantly. However, TKI therapy for pediatric patients can be considered a thorny matter because children are actively growing during the therapy, so they can face a growth disturbance, which is an adverse event never seen in adults [33]. Moreover, they have a much longer life expectancy than adults and, consequently, should experience a longer exposure to TKI (some children must follow TKI therapy for all life). However, long-term effects of TKIs beyond 15 years of age are still missing [34]. It was, thus, wise to undertake an accurate monitoring of the length of therapy to identify the appearance of resistance or intolerance phenomena. To minimize TKI-related side effects, a discontinuation approach has been challenged in pediatric patients as done in the case of adults with a deep and sustained molecular response [28,35,36]. Unfortunately, the only discontinuation reported for children and adolescents are due to poor adherence of the therapy. In general, only a few pediatric patients discontinued TKI successfully [37,38]. As a result, there is an urgent need for perspective studies based on a much larger number of pediatric CML cases.

The goals of CML therapy are the same for adults and children, which include disease remission, reduced risk of progression, and survival. However, the treatment of pediatric CML must take into account the challenge of achieving these goals while minimizing toxicities for six or seven decades [39]. Although cure is the ideal goal for all patients regardless of age, for older adults, it may be sufficient to approach CML as a chronic disease with the aim of maintaining patients in CML-CP (Chronic Phase) for a few decades with TKIs. In the GIMEMA (Gruppo Italiano Malattie ematologiche) study [40], the percentage of young adults who were treated with TKI and had cumulative probability of progression to AP (Accelerated Phase) and BP (Blast Phase) at 8 years was 16%, which was higher than in adults or elderly equal to 5% and 7%, respectively [28,36].

As discussed before, prolonged treatment with TKIs has potential long-term effects and, thus, different undesired effects in growing children when compared to adults. These can be cumulatively higher in the pediatric CML population in case of lifetime exposure to TKIs. Furthermore, reports from pediatric oncologists observed poor adherence more frequently in adolescents than older adults or younger children. For these reasons, extended use of TKIs is a less viable option in these patients, which makes the issue even more complicated. When selecting a TKI, it is important to consider the adherence of patients to the therapy. Twice-daily dosing of nilotinib may be more challenging for pediatric patients than once-daily dosing of imatinib or dasatinib. Formulations of TKIs with better palatability should be developed because they may improve adherence among pediatric patients. Other methods to improve adherence such as direct clinical supervision or various reminder systems may be needed in patients with suboptimal responses to TKIs. Another factor that needs to be considered is the cumulative cost of TKI therapy in children who may need decades of treatment, even though this may become a minor issue in the near future with the introduction of generic products that are much cheaper than the proprietary medicinal products.

3.2. HSCT vs. TKI in the Therapy of Pediatric CML

As already mentioned, HSCT (Hemapoietic Stem Cell Transplantation) is recommended only in CP-CML patients resistant or intolerant to TKIs, [41] even if it is the sole treatment option able to completely eliminate leukemic stem cells [42]. Clearly, children have longer life expectancies, and the perspective to receive TKIs lifelong could increase their risk of morbidities and, unfortunately, decrease their life quality [33]. If an appropriate donor is available, HSCT could be considered resolutive in pediatric CML by avoiding the need for prolonged TKIs and reducing the therapy cost.

Recently, Chaudhury et al. evaluated outcomes, reported in the CIBMTR (Center for International Blood and Marrow Transplant Research) database in 449 CML pediatric patients who received myeloablative HCT [43].

They analyzed several parameters influencing the outcomes, such as patient age and pre-HCT TKI therapy. They evaluated a probability of 5 years of overall survival (OS) and leukemia-free survival (LFS) post-HCT of 75% and 59%, respectively, without consequences due to a difference of age or due to pre-HCT TKI therapy. Very favorable effects were obtained if HCT was performed in more recent years and if it was a Matching Sibling Donor (MSD) HTC.

The estimated LFS in pediatric CP-CML patients treated with imatinib is about 98% at 3 years with a complete hematologic response of about 98% [44]. However, a comparison among patients treated with TKIs and others receiving HCT cannot be performed because HCT could not always be a therapy option.

An important parameter to take into account for the efficacy of CML therapy is evaluating Health-Related Quality of Life (HRQOL), recently investigated in CML patients receiving imatinib or HSTC [45,46].

In 2014, Mo et al. demonstrated that HRQOL of patients treated with Identical Sibling Donor (ISD)-HSCT is comparable to that of imatinib-treated patients [47].

Based on the evidence that young patients are able to obtain better HRQOL after HSCT, which is the only lasting cure for CML, as mentioned above. It could be considered a valuable therapy option for pediatric CP-CML.

Very recently, Athale et al. of the Children's Oncology Group CML Working Group, in the absence of pediatric CML treatment guidelines, analyzed the issues and the concerns derived from the comparison between TKI and Allo-HSCT in pediatric age and made several recommendations about the HSCT therapy [48]. Taking into account that, in recent years, HSCT in children determined fewer complications than in adults, they recommend HSCT in pediatric age when CML progresses from CP to AP or BP, when the therapy with two different TKIs fails or when the patients show intolerance or resistance to TKIs. HSCT may also be considered an option for children and young adults in case of compliance problems, but only after a deep risk/benefit analysis.

3.3. TKIs Discontinuation: Treatment-Free Remission as an Additional Goal in CMLTKI

The optimal responses of CML patients to TKI treatment resulted in a large increase of the life expectancy, which is a figure now approaching that of the general population [49]. However, as mentioned before, long-term use of TKIs can be responsible for more or less severe adverse events. Among these, the most remarkable and common side effects, such as fatigue and musculoskeletal pain, may affect patient quality of life (QoL) [45]. Such a similar concern, in addition to the high up-front cost of TKI treatment, is responsible for nonadherence to therapy. For all these reasons, TKI discontinuation in order to achieve treatment-free remission (TFR) seems to be the most encouraging option for CML patients showing a sustained deep molecular response (DMR). Needless to say, such considerations need to be done prior to stopping TKI treatment. First of all, it is necessary to define the stage and the course of the disease at the time of the diagnosis. Precisely, the Sokal risk score at the time of diagnosis has been identified as an important prognostic factor for successful TFR with TKI, especially with imatinib. The Sokal risk score is an index used to predict prognosis at the time of CML diagnosis before starting treatment. It can be useful to determine risk and decide on therapy guided by the NCCN guidelines for CML. According to the TWISTER (The Australasian Leukaemia & Lymphoma Group (ALLG) CML8 study, ACTRN 12606000118505) study, a high-risk Sokal score at diagnosis was associated with molecular relapse [50]. In contrast, STIM (STOP Imatinib) study reported that patients with low-risk Sokal score had an estimated survival without relapse at 18 months of about 54%, compared with about 35% and 13% in those with an intermediate and high score, respectively [51]. Another important factor to consider is the molecular response, which is a measure of treatment success in CML. To better understand what MR (Molecular Response) is, it is necessary to remind that Bcr-Abl is the final genetic marker for Ph+.

If TKI therapy is going well, first, white blood cells return to a normal level (complete hematologic response), then Ph+ cells cannot be found in bone marrow (complete cytogenetic response) with a very small amount of Bcr-Abl (major molecular response), as long as it totally disappears from the bone marrow, which achieves the deep molecular response as a sign of disease remission. Specialists may also refer to molecular response 4.5 or MR4.5. This means a 4.5-log reduction or a decrease of 10,500 times in BCR-ABL1 transcripts, which is lower than they were before treatment started. In particular, several discontinuation trials established sustained MR4 for at least two years as the fundamental criterion for considering treatment discontinuation, [52] even though the specific eligibility criteria vary across the TRF (Treatment Free Remission) trial, which leads to choosing MR4.5 as a benchmark. Furthermore, it is necessary to clarify that the duration of DMR (Deep Molecular Response) may be more important that the duration of therapy in order to choose TKI discontinuation, as shown in the European Stop Tyrosine Kinase Inhibitor Trial (EURO-SKI), which points out that the duration of response was the most relevant factor [53]. Once the necessary criteria for TKI discontinuation have been met, why should patients choose to cease TKI therapy? Above all, they may be motivated by prospects for a better life. In particular, the mitigation or the disappearance of adverse events experienced during the treatment may improve patient quality of life (QoL). This is a fundamental concern for both children and female patients of childbearing potential, so they could particularly benefit from TFR as a treatment goal due to the adverse outcome they may experience [54–56]. On the other hand, discontinuing TKI can be responsible for a withdrawal syndrome, characterized by musculoskeletal pain that frequently improves spontaneously or with anti-inflammatory agents, but occasionally (albeit rarely) may require resumption of TKI therapy [57,58]. Then, there is not only an economic consideration to do in choosing the option of TKI discontinuation because it is a question of such importance as the emotional and cognitive components since patient wishes and fears must be considered and respected. All these factors help to motivate adherence to therapy discontinuation when circumstances permit it. Many trials in the past years have investigated outcomes for TFR as a treatment goal in CML. Worthy of mention is the STOP-2G-TKI study, which enrolled patients who achieved MR4.5 after receiving 2G-TKIs (dasatinib or nilotinib) for at least two years [59]. During the treatment-free phase, there was no progression to AP and all relapsing patients regained MMR (Major Molecular Response) and MR4.5 after restarting therapy. Lots of trials are currently ongoing in order to verify if it is possible to lower the standards for TFR (responses less than MR4.5 sustained for less than two years) in order to increase the pool of eligible patients. In addition, before deciding on TFR, it is necessary to accurately monitor levels of BCR-ABL1 transcripts and continue to check once treatment has stopped. The molecular monitoring could give information about a potential relapse so that patients can restart TKI immediately. The most sensitive and accurate methods for quantifying transcripts are Real-time

Quantitative Polymerase Chain Reaction (RQ-PCR) or digital PCR technology. European Society for Medical Oncology (ESMO) [60] guidelines highlight the need for a rapid turnaround of PCR test results within four weeks and the capacity to provide PCR tests every four–six weeks when required, which are both feasible only in a suitably equipped laboratory [60]. Current NCCN guidelines suggest monthly monitoring for the first 12 months after discontinuation of TKI, every six weeks during months 13–24, and every 12 weeks thereafter [52]. Considering the low cost and the minimal discomfort of continued molecular monitoring, if the laboratories are capable of returning results rapidly, ideally in less than four weeks, even the patient compliance will be guaranteed. Even if lots of ongoing trials seem to prove that TKI discontinuation is a great strategy for CML management, it became evident that, while some patients show a steady increase of transcript levels after discontinuation, in many instances, low transcript levels remained. According to NCCN and ESMO guidelines, this has led to the notion that relapse, and, thus, reinitiating therapy, should be considered when MMR is lost in most instances. Understandably, further studies on TKI discontinuation are needed to help define the best predictive factors for identifying the most appropriate patients, especially for children and adolescents, because the limited available data are mainly based on case reports of noncompliant patients [61]. Therefore, current adult guidelines for stopping TKI cannot be applied to pediatric patients without proper prospective clinical trials.

3.4. Imatinib in Pediatric CML

Glivec (that is imatinib mesylate) is a tyrosine kinase inhibitor previously approved by the FDA for treating CML and for treating patients with Kit (CD117) positive metastatic and/or unresectable malignant gastrointestinal stromal tumors (GIST). Glivec was also approved in 2003 for use in children with Ph+ chronic phase CML, which was recurrent after stem cell transplantation or resistant to INF-α therapy. On January 2013, Glivec was also approved by FDA to treat children newly diagnosed with Philadelphia chromosome positive (Ph+) acute lymphoblastic leukemia (ALL) [62].

Since its introduction as a selective BCR-ABL1 TKI, imatinib replaced the CML first-line curative treatment, the allo-HSTC [63]. Similar to the strategy in adults, imatinib soon became the recommended initial standard of care in pediatric patients [64].

The results of clinical trials with imatinib in the adult patient population have been transferred to children, and, thus, imatinib is now the front-line treatment for childhood CML. Since the approval by the US FDA in 2003, several reports have been published on the effects and toxicity of imatinib in children. With regard to its dosage, pediatric doses of 260 mg/m^2 and 340 mg/m^2 have been found to be on par with 400 mg and 600 mg, respectively, in adults. Therefore, the recommended starting dose in children is 300 mg/m^2 once daily (maximum absolute dose, 400 mg) and 400–500 mg/m^2 for advanced-stage disease [64,65]. The lack of any specific guideline for children prompts the consideration of such adult-based criteria to measure the response and decide upon clinical status such as imatinib resistance, intolerance, noncompliance, or disease progression. In patients with an optimal response, imatinib may be continued until allo-HSCT is undertaken. In case of failure, second-generation TKIs and allo-HSCT need to be considered [66].

However, even if clinical experience with imatinib in the pediatric population is limited, a phase I trial conducted by the Children's Oncology Groupp [65] showed that a daily oral dose of imatinib (260–570 mg/m^2) was well tolerated in children [67]. A European group reported a multi-center phase II study of imatinib in 30 children with CML showing comparable results with those achieved in adult patients [68]. In addition to these studies, a phase III trial was conducted with 156 patients (91 male, 65 female, median age 13.2 years, range 1.2–18.0) with a newly diagnosed CML, recruited from March 2004 until December 2015 [37]. On the basis of this phase III trial, it was concluded that imatinib at the recommended dose results in both an excellent response and tolerable side effect rates in children and adolescents with newly diagnosed CML. First-line imatinib treatment allows patients to avoid major consequences of allo-HSCT, such as transplant-related mortality and chronic graft-versus-host disease. However, only careful follow-up on long-term administration of imatinib will provide information on the sustainability of responses, rates of resistance development, and treatment-related complications, especially when transition from a pediatric setting into adult hematology becomes mandatory [69,70]. Worth of mention is also a French national phase IV trial [22]. The results of this study pointed out a high rate of progression-free survival (98% at 36 months) and the achieving of CHR (Complete Hematological Respnse) for 43 patients (98%). Moreover, during the follow-up, 25 patients (75%) achieved an MMR. No treatment-related death occurred and toxicities were generally reversible with temporary treatment discontinuation or dose reduction. The proportion of patients discontinuing imatinib (30%) is very similar to that reported for adults (from 25% to 28%) [71,72]. In conclusion, this study focused on a median follow-up of 30 months that shows a satisfactory rate of response with acceptable adverse effects of Imatinib as initial therapy in children and adolescents with newly diagnosed CML in CP (Chronic Phase), which confirms its effectiveness in children and adolescents with CML in CP with response rates similar to those reported in adults. However, longer follow-up studies are necessary to have a more comprehensive view about long-term exposure [22].

As mentioned above, several studies have indicated that imatinib has a negative impact on growth and development in children with CML. In this respect, growth retardation, [33] dysregulation of bone remodeling [73], and alterations in bone metabolism [74] have been associated with imatinib treatment. A retrospective study reported significant growth deceleration after 12 months of first-line Imatinib therapy in pediatric patients [75].

Moreover, TKI response rates vary among different individuals in which pharmacokinetics is a key factor for successful CML treatments. Adherence to imatinib intake may be the most prominent factor influencing treatment outcome in teenagers, which points towards the potential benefits of regular drug monitoring [76].

3.5. Dasatinib in Pediatric CML

From 25% to 29% of patients that discontinued imatinib due to a poor response or toxicity [22,77] were left without approved therapies to treat children with Imatinib resistant/intolerant CML-CP.

On November 2017, the US FDA granted regular approval to the second generation TKI Dasatinib (SPRYCEL, Bristol-Myers Squibb Co.) for treating pediatric patients with Ph+ CML in the chronic phase. It is now available in tablet formulation [78].

In order to evaluate the safety and the efficacy of dasatinib in pediatric patients, many trials are still needed. Among the studies already done, a phase I trial aimed at determining suitable dosing for children with Ph+ leukemias. CA180–226/NCT00777036 is a phase II, open-label, non-randomized prospective trial of patients, 18 years of age receiving dasatinib [79]. Major cytogenetic response of >30% was reached by 3 months in the imatinib resistant/intolerant group and CCyR > 55% was reached by 6 months in the newly diagnosed CML-CP group. CCyR and major molecular response by 12 months, respectively, were 76% and 41% in the imatinib resistant/intolerant group and 92% and 52% in the newly diagnosed CML-CP group. Progression-free survival by 48 months was 78% and 93% in the imatinib resistant/intolerant and newly diagnosed CML-CP groups, respectively. No dasatinib-related pleural or pericardial effusion, pulmonary edema, or pulmonary arterial hypertension was reported. Bone growth and development events were reported in 4% of patients. Based on this trial, it was demonstrated that dasatinib is effective for treating pediatric CML-CP with a safety comparable in both children and adults [79].

3.6. Nilotinib in Pediatric CML

After imatinib and dasatinib, on March 2018, the FDA approved nilotinib (TASIGNA, Novartis Pharmaceuticals Corporation) for pediatric patients one year of age or older with newly diagnosed Ph+ CML-CP or Ph+ CML-CP resistant or intolerant to prior tyrosine-kinase inhibitor (TKI) [80]. Afterward, the European Medicine Agency (EMA) approved nilotinib for pediatric patients as well.

A phase II trial was conducted when imatinib was the only TKI indicated for pediatric patients with Ph+ CML-CP, with this implying alternative treatment options particularly for patients developing resistance/intolerance (R/I) to imatinib. This phase II study enrolled pediatric patients with either Ph+ CML-CP R/I to imatinib/dasatinib or newly diagnosed Ph+ CML-CP. The trial confirmed the clinical activity of nilotinib 230 mg/m^2 twice per day in pediatric patients with newly diagnosed Ph+ CML-CP or R/I to imatinib or dasatinib. Nilotinib was associated again with a manageable safety profile. In addition, cardiovascular events were not observed in this study and no new safety signals were identified. The most frequently reported drug-related AEs included the increase in bilirubin, alanine aminotransferase (ALT), and aspartate aminotransferase (AST) as well as headache and rash [81].

Considering that, as mentioned before, the imatinib therapy could be the cause of a slow growth and a delayed puberty in pediatric CML patients, it is still necessary to verify the nilotinib safety with regard to this kind of issue.

Despite the limited size due to the rarity of CML in children, a five-year update of the randomized ENESTnd trial [16] demonstrates the efficacy of nilotinib in pediatric patients at the recommended 230 mg/m^2 twice per day dose as well as a manageable safety profile comparable to that of Nilotinib in adults with Ph+ CML-CP [82,83]. Based on these results, nilotinib 230 mg/m^2 twice per day has been approved in Europe [84] for treating pediatric patients with R/I or newly diagnosed CML-CP. Nilotinib can be seen as a valuable additional therapeutic option for treating pediatric CML, according to its efficacy and manageable safety profile in children.

The safety profile of nilotinib in pediatric patients was generally similar to that observed for adults. As such, no new safety signals were identified.

Therefore, a nilotinib phase II study was conducted in 2019. Nilotinib demonstrated efficacy and a manageable safety profile in pediatric patients with Ph+ CML-CP [81]. A grade 1 SAE (Severe Adverse Event) of hormone deficiency that was suspected to be related to nilotinib treatment was reported in the R/I cohort, even though slowing of growth had already been observed when the patient was receiving first-line treatment with imatinib. However, because of the limited number of patients and short follow-up period for this analysis, few conclusions can be drawn regarding the impact of nilotinib of these parameters [81].

3.7. Ponatinib in Pediatric CML

An important mutation in CML, the threonine-to-isoleucine mutation at position 315 (T315I), has resulted in the development of ponatinib, which is currently the only TKI effective against T315I CML. However, it has not been studied sufficiently in children [85]. As a matter of fact, ponatinib, which is the TKI most recently approved, has demonstrated efficacy in adult patients with refractory CML but comes with an increased risk of arterial hypertension as well as serious arterial occlusive and venous thromboembolic events in adults [86]. As already mentioned, mutations of the BCR-ABL1 kinase domain may cause TKI resistance and mutation screening is recommended in patients with a poor response (primary resistance) as well as those who lose the initial response (secondary resistance) [87]. A sharp and sudden increase in the BCR-ABL1 transcript ratio should always raise suspicion of poor adherence [61], as TKI resistance due to mutations in the ABL1 kinase domain are characterized by a slower expansion of the mutated clone. Each TKI has different patterns of inactivity against defined mutations. Therefore, the specific mutation needs to be considered when selecting a TKI [65]. Due to the fact that, only a few case reports are available regarding the use of this un-licensed drug in children and adolescents, it is necessary to mention the experience of the international registry of childhood chronic myeloid leukemia, according to which data from 11 children with CML registered to the I-CML-Ped-Study and from three children with Ph+ acute lymphoblastic leukemia (Ph+ ALL) treated with ponatinib were retrospectively collected [88]. According to the results of the trial and with the limitation of the retrospective nature of this study, ponatinib may be a reasonable additional treatment option for children with Ph+ leukemias who failed several lines of therapy. Another study reported the treatment of a young adolescent with chronic phase CML treated with ponatinib [85]. The patient was a 10-year-old male found to have a white blood cell count of 96,000 mL with 1% blasts and the (9;22) (q34,q11.2) translocation. In particular, the patient was started on Imatinib and, after 6 months, achieved a CMR (Complete Molecular Response). However, at 10 months, BCR-ABL was again detected by PCR and transcript levels increased over time. At 18 months, gene sequencing identified the T315I mutation and the patient had no HLA-matched donors [85]. While CML with the T315I mutation has been considered an indication for allogeneic stem cell transplant, this case demonstrates that Ponatinib may be a reasonable alternative treatment in children and adolescents.

Likewise, Imatinib and Ponatinib may also impair growth given the patient growth curve. As a matter of fact, during the therapy, the child has had a significant decline in height velocity. However, it is encouraging that it did not prevent pubertal development [85].

4. TKI Dosage Forms for Children: Formulation Considerations

Currently, no oral liquid formulation for TKIs is available on the market allowing exact dosing in children, according to a patient body weight or surface, age, and physio-pathological conditions. Oncology pharmacists face a constant challenge with young patients unable to swallow oral anti-cancer drugs. Thus, this makes extemporaneous oral liquid preparation a necessary requirement. Inappropriate extemporaneous preparations of TKIs may increase the risk of overdosing or underdosing. Based on a review of the literature, a few compounding recipes are available for these drugs, as shown in Table 1 [89]. Currently, imatinib (Glivec®), dasatinib (Sprycel®), nilotinib (Tasigna®), and ponatinib (Iclusig®) formulations as film-coated tablets or capsules are available on the market [90–93].

Table 1. Instructions to prepare extemporaneous oral liquid suspensions of tyrosine kinase inhibitors (TKIs) [94].

TKI	Instructions
Imatinib	Tablets may be dispersed in water or apple juice using 50 mL for 100 mg tablet or 200 mL for 400 mg tablet. The contents must be stirred until dissolved and used immediately. For children < 3 years old, it is recommended that at least 120 mL of water or food be taken to avoid esophageal irritation.
Dasatinib	Tablets can be allowed to dissolve over 20 min at room temperature in 30 mL of lemonade, preservative-free apple juice, or preservative-free orange juice. After ingestion, rinse the residue off glass with 15 mL of the juice and administer.
Nilotinib	Capsules may be dispersed in 5 mL of applesauce and ingested immediately on an empty stomach and abstain from eating for at least 1 h.

These oral solid dosage forms are limited by their rigid dose content and the ability of the children to swallow them. Tablets can be scored to allow splitting to reduce their size. However, this can result in inaccurate dosages within the fragmented tablets. Capsules can be opened and the contents taken to improve acceptability in children. However, the capsule contents may taste unpleasant and the bioavailability of the opened capsule may differ from that of the intact product [94].

The recommended daily doses of imatinib for the pediatric population with CML in chronic phase are 260 mg/ m^2 and 340 mg/m^2 [76]. Imatinib tablets are not manufactured in a dosage form suitable for children, and often need to be fractioned to obtain 50 mg parts with even lower doses required in young children or infants. The drug should be taken usually in the morning after breakfast or at school, at the evening before going to sleep to reduce nausea as a side effect [95]. Since imatinib acts as a local irritant, it is recommended to take the tablets with at least 120 mL of water for children < 3 years old. For young patients who cannot swallow whole tablets, the tablet should be grounded and the resulting powder should be dispersed in water or apple juice or mixed with apple puree or yogurt. The acidic pH of apple juice or puree (pH 3.5) is advantageous as the solubility of imatinib strongly increases at pH values below 6.5. However, the drug is unstable in orange juice, cola, or milk under these acidic conditions [96].

Dasatinib (Sprycel®), which is a second generation TKI, is also formulated as coated tablets with a dosage ranging from 20 to 140 mg [97]. Until today, this drug has been used exclusively for treating adult patients, but clinical trials have shown its potential use in the treatment of CML in children, where its pharmacokinetic parameters of absorption and elimination time were comparable with those in adults with the same safety and efficacy profiles [22,98,99]. However, in these studies, dasatinib was administered to children in the form of tablets or crushed tablets dispersed in fruit juice.

There is a paucity of data in terms of stability, bioequivalence, and safety of these extemporaneously compounded oral formulations. Before an oral liquid formulation is prepared extemporaneously, it is essential to have an adequate understanding of the pharmacokinetic characteristics of the drug, its stability, compatibility with excipients, and palatability of the final preparation, ease of administration, and safety concerns [100–103].

TKI oral liquid formulations must be appropriate for the children in terms of dose, convenience, and acceptability to ensure compliance with the medication and, at the same time, safety and efficacy of the therapy. The design of a pediatric formulation needs to take into account the differences in pediatric anatomy and physiology (especially in newborns and infants) to ensure that the pharmacokinetic profile of the drug is not compromised.

Formulation can lead to differences in pharmacokinetic profiles for a drug by highlighting the risks associated with manipulating to enable administration to children. Prescribers need to be aware of the consequences of manipulating medicinal formulations, particularly for anti-cancer drugs with a narrow therapeutic index, even in extemporaneous compounding by an oncology pharmacist, where there is insufficient evidence on product quality and safety.

The choice of excipients represents another major factor involved in extemporaneous oral liquid preparation for children. The excipients used in pediatric formulations need to be appropriate for the age to minimize excipient toxicity. Usually, the major barrier in development of oral liquid formulations is taste-masking of the active ingredient. Drug taste and its palatability are the greatest barriers for completing a treatment. The excipients used in the development of a pediatric liquid formulation need to be safe and acceptable. Excipients are typically used to improve palatability, shelf-life, and/or manufacturing processes [104–107].

5. Conclusions

The therapeutic approach to CML has changed since the introduction of the TKI imatinib, followed by dasatinib, nilotinib, which was approved for use in children, and ponatinib, which became the first-line treatment in adults and expanded the therapeutic options, pushing allogeneic stem cell transplantation to a third-line treatment for most pediatric cases. Unfortunately, the selection of a TKI continues to rely on clinical experience in adults [108] without sufficient data on efficacy and safety specific to pediatric patients.

The availability of three TKIs is challenging to choose a first-line option, but it also gives clinicians additional treatment chances in case of a suboptimal response [108].

More experience has been gained about efficacy, toxicity profiles, and comorbidities of Imatinib compared to the other TKIs [22,37,109]. The appearance of resistance or intolerance phenomena has required the development of a second generation of TKI. Drug availability, ease of administration, and financial issues should also be considered.

Evidence-based recommendations have been established for treating CML in adults treated with TKIs. Appropriate guidelines are, however, difficult to extend to pediatric patients given the very rare occurrence in children and adolescents [105]. As a matter of fact, the CML incidence increases with age, from 0.09/100,000 at ≤ 15 years of age to 7.88/100,000 at ≥ 75 years of age. Needless to say, here are several biological and clinical differences between pediatric and adult CML. Markedly increased leukocyte count and a higher incidence of splenomegaly are characteristic features at diagnosis of pediatric patients.

TKIs are designed to inhibit BCR-ABL1 kinase, but they have unfavorable effects, so-called "off-target" complications, such as growth impairment, especially important in children, because they are constantly and actively growing. Long-term morbidity due to TKIs is unknown. Furthermore, the adverse effects on growing children have not been clearly elucidated, even though the exposure period to Imatinib is relatively short. To establish the standard therapeutic management for pediatric

CML, it is important to prospectively confirm the attractive outcomes obtained in adult studies via pediatric clinical trials with a careful monitoring system for TKI-induced adverse effects, especially in growing children [110].

Limited experience with very young children, the transition of teenagers to adult medicine, and the goal of achieving treatment-free remission for this rare leukemia are more significant obstacles that require further clinical investigations [48,108].

Lastly, in order to carry out a possible and viable therapy with TKIs in the pediatric age, a key role is played by developing appropriate formulations specifically customized toward this kind of patients, since these drugs are often available in solid dosage forms, which are difficult to administer in children.

Funding: Authors acknowledge Project POR Puglia FESR-FSE 2014–2020 Azione 1.6. Innonetwork progetto D.I.V.A., cod. JD6EDJ7 and the University of Bari "Aldo Moro" (Italy) for their financial support.

Conflicts of Interest: The authors declare no conflict of interest.

Abbreviations

CML	Chronic Myeloid Leukemia
PK	Protein Kinase
TK	Tyrosine Kinase
TKI	Tyrosine Kinase Inhibitor
RTK	Receptor Tyrosine Kinase
NRTK	Non-Receptor Tyrosine Kinase
VEGFR	Vascular Endothelial Growth Factor
PDGF	Platelet Derived Growth Factor
Src	Sarcoma
JAK	Janus Kinase
ABL	Abelson (gene)
BCR	Breakpoint Cluster Region (gene)
PI3K	Phosphoinositide 3-Kinase
STAT	Signal Transducer and Activator of Transcription (protein)
CP	Chronic Phase
AP	Accelerated Phase
BC	Blast Crisis
Asp	Aspartate
Phe	Phenylalanine
Gly	Glycine
ATP	Adenosine Triphosphate
DFG	Aspartate-Phenilalanine-Glycine
Thr	Threonine
HSCT	Hematopoietic Stem Cells Transplantation
HCT	Hematopoietic Cell Transplantation
INF	Interferon
CHR	Complete Hematological Response
MMR	Major Molecular Response
CCR	Complete Cytogenetic Remission
Ph	Philadelphia (chromosome)
Ph+	Philadelphia positive
P-gp	Glycoprotein P
OCT	Organic Cation Transporter
PDGFR	Platelet Derived Growth Factor Receptor
UK	United Kingdom
NAACCR	North American Association of Central Cancer Registries

ELN	European Leukemia Net
NCCN	National Comprehensive Cancer Network
CI	Confidence Interval
STIM	Stop Imatinib
LSC	Leukemic Stem Cells
ALL	Acute Lymphoblastic Leukemia
CMR	Complete Molecular Response
FDA	Food and Drug Administration
EMA	European Medicinal Agency
R/I	Resistance/Intolerance
SAE	Severe Adverse Event
HLA	Human Leukocyte Antigen
CIBMTR	Center for International Blood and Marrow Transplant Research
OS	Overall Survival
LFS	Leukemia-Free Survival
MSD	Matching Sibling Donor
HCT	Hematopoietic Cell Transplantation
HRQOL	Health-Related Quality of Life
ISD	Identical Sibling Donor
TFR	Treatment-free Remission
QoL	Quality of Life
DMR	Deep Molecular Response
EURO-SKI	European Stop Tyrosine Kinase Inhibitor
ESMO	European Society for Medical Oncology
RQ-PCR	Real-time Quantitative Polymerase Chain Reaction

References

1. Chen, M.H.; Kerkela, R.; Force, T. Mechanisms of cardiomyopathy associated with tyrosine kinase inhibitor cancer therapeutics. *Circulation* **2008**, *118*, 84–95. [CrossRef] [PubMed]
2. Tyrosine-kinase-inhibitor. Available online: https://www.cancer.gov/about-cancer/treatment/types/targeted-therapies/targeted-therapies-fact-sheet (accessed on 20 January 2020).
3. Cortes, J.E.; Talpaz, M.; Beran, M.; O'Brien, S.M.; Rios, M.B.; Stass, S.; Kantarjian, H.M. Philadelphia chromosome-negative chronic myelogenous leukemia with rearrangement of the breakpoint cluster region. Long-term follow-up results. *Cancer* **1995**, *75*, 464–470. [CrossRef]
4. Reddy, E.P.; Aggarwal, A.K. The Ins and Outs of Bcr-Abl Inhibition. *Genes Cancer* **2012**, *3*, 447–454. [CrossRef]
5. Reynolds, C.R.; Islam, S.A.; Sternberg, M.J.E. EzMol: A web server wizard for the rapid visualisation and image production of protein and nucleic acid structures. *J. Mol. Biol.* **2018**, *430*, 1–5. [CrossRef]
6. Gyurkocza, B.; Rezvani, A.; Storb, R.F. Allogeneic hematopoietic cell transplantation: The state of the art. *Expert Rev. Hematol.* **2010**, *3*, 285–299. [CrossRef] [PubMed]
7. Nardi, V.; Azam, M.; Daley, G.Q. Mechanisms and implications of imatinib resistance mutations in BCR-ABL. *Curr. Opin. Hematol.* **2004**, *11*, 35–43. [CrossRef]
8. Widmer, N.; Colombo, S.; Buclin, T.; Decosterd, L.A. Functional consequence of *MDR1* expression on imatinib intracellular concentrations. *Blood* **2003**, *102*, 1142. [CrossRef] [PubMed]
9. Watkins, D.; Hughes, T.; White, D. OCT1 and imatinib transport in CML: Is it clinically relevant? *Leukemia* **2015**, *29*, 1960–1969. [CrossRef] [PubMed]
10. Cavalluzzi, M.M.; Imbrici, P.; Gualdani, R.; Stefanachi, A.; Mangiatordi, G.F.; Lentini, G.; Nicolotti, O. Human ether-à-go-go-related potassium channel: Exploring SAR to improve drug design. *Drug Discov. Today* **2020**, *25*, 344–366. [CrossRef] [PubMed]
11. Ángeles-Velázquez, J.L.; Hurtado-Monroy, R.; Vargas-Viveros, P.; Carrillo-Muñoz, S.; Candelaria-Hernández, M. Imatinib Intolerance Is Associated with Blastic Phase Development in Philadelphia Chromosome-Positive Chronic Myeloid Leukemia. *Clin. Lymphoma Myeloma Leuk.* **2016**, *16*, S82–S85. [CrossRef] [PubMed]

12. Talpaz, M.; Shah, N.P.; Kantarjian, H.; Donato, N.; Nicoll, J.; Paquette, R.; Cortes, J.; O'Brien, S.; Nicaise, C.; Bleickardt, E.; et al. Dasatinib in Imatinib-Resistant Philadelphia Chromosome–Positive Leukemias. *N. Engl. J. Med.* **2006**, *354*, 2531–2541. [CrossRef] [PubMed]
13. Tokarski, J.S.; Newitt, J.A.; Chang, C.Y.; Cheng, J.D.; Wittekind, M.; Kiefer, S.E.; Kish, K.; Lee, F.Y.; Borzillerri, R.; Lombardo, L.J.; et al. The structure of Dasatinib (BMS-354825) bound to activated ABL kinase domain elucidates its inhibitory activity against imatinib-resistant ABL mutants. *Cancer Res.* **2006**, *66*, 5790–5797. [CrossRef] [PubMed]
14. Bose, P.; Park, H.; Al-Khafaji, J.; Grant, S. Strategies to circumvent the T315I gatekeeper mutation in the Bcr-Abl tyrosine kinase. *Leuk. Res. Rep.* **2013**, *2*, 18–20. [CrossRef] [PubMed]
15. Redaelli, S.; Piazza, R.; Rostagno, R.; Magistroni, V.; Perini, P.; Marega, M.; Gambacorti-Passerini, C.; Boschelli, F. Activity of Bosutinib, Dasatinib, and Nilotinib Against 18 Imatinib Resistant BCR/ABL Mutants. *J. Clin. Oncol.* **2009**, *27*, 469–471. [CrossRef] [PubMed]
16. Saglio, G.; Kim, D.-W.; Issaragrisil, S.; le Coutre, P.; Etienne, G.; Lobo, C.; Pasquini, R.; Clark, R.E.; Hochhaus, A.; Hughes, T.P.; et al. Nilotinib versus Imatinib for Newly Diagnosed Chronic Myeloid Leukemia. *N. Engl. J. Med.* **2010**, *362*, 2251–2259. [CrossRef]
17. O'Hare, T.; Shakespeare, W.C.; Zhu, X.; Eide, C.A.; Rivera, V.M.; Wang, F.; Adrian, L.T.; Zhou, T.; Huang, W.-S.; Xu, Q.; et al. AP24534, a Pan-BCR-ABL Inhibitor for Chronic Myeloid Leukemia, Potently Inhibits the T315I Mutant and Overcomes Mutation-Based Resistance. *Cancer Cell* **2009**, *16*, 401–412. [CrossRef]
18. Azam, M.; Nardi, V.; Shakespeare, W.C.; Metcalf III, C.A.; Bohacek, R.S.; Wang, Y.; Sundaramoorthi, R.; Sliz, P.; Veach, D.R.; Bornmann, W.G.; et al. Activity of dual SRC-ABL inhibitors highlights the role of BCR/ABL kinase dynamics in drug resistance. *Proc. Natl. Acad. Sci. USA* **2006**, *103*, 9244–9249. [CrossRef]
19. Horibe, K.; Tsukimoto, I.; Ohno, R. Clinicopathologic characteristics of leukemia in Japanese children and young adults. *Leukemia* **2001**, *15*, 1256–1261. [CrossRef]
20. Mattano, L.; Nachman, J.; Ross, J.; Stock, W. Leukemias. In *Cancer Epidemiology in Older Adolescents and Young Adults 15 to 29 Years of Age, Including SEER Incidence and Survival: 1975–2000*; Bleyer, A., O'Leary, M., Barr, R., Ries, L.A.G., Eds.; National Cancer Institute, NIH: Bethesda, MD, USA, 2006; pp. 39–51.
21. Millot, F.; Traore, P.; Guilhot, J.; Nelken, B.; Leblanc, T.; Leverger, G.; Plantaz, D.; Bertrand, Y.; Bordigoni, P.; Guilhot, F. Clinical and biological features at diagnosis in 40 children with chronic myeloid leukemia. *Pediatrics* **2005**, *116*, 140–143. [CrossRef] [PubMed]
22. Millot, F.; Baruchel, A.; Guilhot, J.; Petit, A.; Leblanc, T.; Bertrand, Y.; Mazingue, F.; Lutz, P.; Verite, C.; Berthou, C.; et al. Imatinib is effective in children with previously untreated chronic myelogenous leukemia in early chronic phase: Results of the French national phase IV trial. *J. Clin. Oncol.* **2011**, *29*, 2827–2832. [CrossRef] [PubMed]
23. Chen, Y.; Wang, H.; Kantarjian, H.; Cortes, J. Trends in chronic myeloid leukemia incidence and survival in the United States from 1975 to 2009. *Leuk. Lymph.* **2013**, *54*, 1411–1417. [CrossRef] [PubMed]
24. Cancer Research UK. Available online: https://www.cancerresearchuk.org/health-professional/cancer-statistics/statistics-by-cancer-type/leukaemia-cml/incidence#heading-One (accessed on 20 January 2020).
25. Gugliotta, G.; Castagnetti, F.; Apolinari, M.; Pirondi, S.; Cavo, M.; Baccarani, M.; Rosti, G. First-Line Treatment of Newly Diagnosed Elderly Patients with Chronic Myeloid Leukemia: Current and Emerging Strategies. *Drugs* **2014**, *74*, 627–643. [CrossRef] [PubMed]
26. Ries, L.A.G.; Smith, M.A.; Gurney, J.G.; Linet, M.; Tamra, T.; Young, J.L.; Bunin, G.R. *Cancer Incidence and Survival among Children and Adolescents: United States SEER Program 1975–1995*; National Cancer Institute: Bethesda, MD, USA, 1999; NIH Pub. No. 99–4649. Available online: https://seer.cancer.gov/archive/publications/childhood/ (accessed on 22 June 2020).
27. North American Association of Central Cancer Registries. NAACCR Fast Stats: An Interactive Tool for Quick Access to Key NAACCR Cancer Statistics. Available online: http://www.naaccr.org/ (accessed on 17 February 2020).
28. Hijiya, N.; Schultz, K.R.; Metzler, M.; Millot, F.; Suttorp, M. Pediatric chronic myeloid leukemia is a unique disease that requires a different approach. *Blood* **2016**, *127*, 392–399. [CrossRef]
29. Ernst, T.; Busch, M.; Rinke, J.; Ernst, J.; Haferlach, C.; Beck, J.F.; Hochhaus, A.; Gruhn, B. Frequent ASXL1 mutations in children and young adults with chronic myeloid leukemia. *Leukemia* **2018**, *32*, 2046–2049. [CrossRef]

30. Radich, J.P.; Deininger, M.; Abboud, C.N.; Altman, J.K.; Berman, E.; Bhatia, R.; Bhatnagar, B.; Curtin, P.; DeAngelo, D.J.; Gotlib, J.; et al. Chronic Myeloid Leukemia, Version 1.2019 NCCN clinical practice guidelines in oncology. *JNCCN* **2018**, *16*, 1108–1135. [CrossRef]
31. Millot, F.; Guilhot, J.; Suttorp, M.; Güneş, A.M.; Sedlacek, P.; De Bont, E.; Li, C.K.; Kalwak, K.; Lausen, B.; Culic, S.; et al. Prognostic discrimination based on the EUTOS long-term survival score within the International Registry for Chronic Myeloid Leukemia in children and adolescents. *Haematologica* **2017**, *102*, 1704–1708. [CrossRef]
32. Salas, D.G.; Glauche, I.; Tauer, J.T.; Thiede, C.; Suttorp, M. Can prognostic scoring system for chronic myeloid leukemia as established in adults be applied to pediatric patients? *Ann. Hematol.* **2015**, *94*, 1363–1371. [CrossRef] [PubMed]
33. Shima, H.; Tokuyama, M.; Tanizawa, A.; Tono, C.; Hamamoto, K.; Muramatsu, H.; Watanabe, A.; Hotta, N.; Ito, M.; Kurosawa, H.; et al. Distinct Impact of Imatinib on Growth at Prepubertal and Pubertal Ages of Children with Chronic Myeloid Leukemia. *J. Pediatr.* **2011**, *159*, 676–681. [CrossRef]
34. Jeyaraman, P.; Naithani, R. Discontinuation of Imatinib in a Child with Chronic Myeloid Leukemia. *J. Pediatr. Hematol. Oncol.* **2020**, *42*, e64–e65. [CrossRef]
35. Saussele, S.; Richter, J.; Guilhot, J.; Gruber, F.X.; Hjorth-Hansen, H.; Almeida, A.; Janssen, J.J.W.M.; Mayer, J.; Koskenvesa, P.; Panayiotidis, P.; et al. Discontinuation of tyrosine kinase inhibitor therapy in chronic myeloid leukaemia (EURO-SKI): A prespecified interim analysis of a prospective, multicentre, non-randomised, trial. *Lancet Oncol.* **2018**, *19*, 747–757. [CrossRef]
36. Hijiya, N.; Millot, F.; Suttorp, M. Chronic myeloid leukemia in children: Clinical findings, management, and unanswered questions. *Pediatr. Clin. N. Am.* **2015**, *62*, 107–119. [CrossRef] [PubMed]
37. Suttorp, M.; Schulze, P.; Glauche, I.; Göhring, G.; von Neuhoff, N.; Metzler, M.; Sedlacek, P.; de Bont, E.S.J.M.; Balduzzi, A.; Lausen, B.; et al. Front-line imatinib treatment in children and adolescents with chronic myeloid leukemia: Results from a phase III trial. *Leukemia* **2018**, *32*, 1657–1669. [CrossRef] [PubMed]
38. Giona, F.; Saglio, G.; Moleti, M.L.; Piciocchi, A.; Rea, M.; Nanni, M.; Marzella, D.; Testi, A.M.; Mariani, S.; Laurino, M.; et al. Treatment-free remission after imatinib discontinuation is possible in paediatric patients with chronic myeloid leukaemia. *Br. J. Haematol.* **2015**, *168*, 305–308. [CrossRef]
39. Mangiatordi, G.F.; Alberga, D.; Altomare, C.D.; Carotti, A.; Catto, M.; Cellamare, S.; Gadaleta, D.; Lattanzi, G.; Leonetti, F.; Pisani, L.; et al. Mind the gap! A journey towards computational toxicology. *Mol. Inform.* **2016**, *35*, 294–308. [CrossRef]
40. Castagnetti, F.; Gugliotta, G.; Baccarani, M.; Breccia, M.; Specchia, G.; Levato, L.; Abruzzese, E.; Rossi, G.; Iurlo, A.; Martino, B.; et al. A GIMEMA CML Working Party. Differences among young adults, adults and elderly chronic myeloid leukemia patients. *Ann. Oncol.* **2015**, *26*, 185–192. [CrossRef]
41. Passweg, J.R.; Baldomero, H.; Peters, C.; Gaspar, H.B.; Cesaro, S.; Dreger, P.; Duarte, R.F.; Falkenburg, J.H.F.; Farge-Bancel, D.; Gennery, A.; et al. Hematopoietic SCT in Europe: Data and trends in 2012 with special consideration of pediatric trans-plantation. *Bone Marrow Trans.* **2014**, *49*, 744–750. [CrossRef]
42. Baccarani, M.; Deininger, M.W.; Rosti, G.; Hochhaus, A.; Soverini, S.; Apperley, J.F.; Cervantes, F.; Clark, R.E.; Cortes, J.E.; Guilhot, F.; et al. European LeukemiaNet recommendations for the management of chronic myeloid leukemia. *Blood* **2013**, *122*, 872–884. [CrossRef]
43. Chaudhury, S.; Sparapani, R.; Hu, Z.-H.; Nishihori, T.; Abdel-Azim, H.; Malone, A.; Olsson, R.; Hamadani, M.; Daly, A.; Bacher, U.; et al. Outcomes of Allogeneic Hematopoietic Cell Transplantation in Children and Young Adults with Chronic Myeloid Leukemia: A CIBMTR Cohort Analysis. *Biol. Blood Marrow Trans.* **2016**, *22*, 1056–1064.
44. Millot, F.; Suttorp, M.; Guilhot, J.; Sedlacek, P.; De Bont, E.S.; Li, C.K.; Kalwak, K.; Lausen, B.; Srdjana, C.; Dresse, M.-F.; et al. The International Registry for Chronic Myeloid Leukemia (CML) in Children and Adolescents (I- CML-Ped-Study): Objectives and preliminary results. *Blood* **2012**, *120*, 3741. [CrossRef]
45. Efficace, F.; Baccarani, M.; Breccia, M.; Alimena, G.; Rosti, G.; Cottone, F. Health related quality of life in chronic myeloid leukemia patients receiving long-term therapy with imatinib compared with the general population. *Blood* **2011**, *118*, 4554–4560. [CrossRef] [PubMed]
46. Chang, G.; Orav, E.J.; McNamara, T.K.; Tong, M.Y.; Antin, J.H. Psychosocial function after hematopoietic stem cell transplantation. *Psychosomatics* **2005**, *46*, 34–40. [CrossRef]

47. Mo, X.-D.; Jiang, Q.; Xu, L.-P.; Liu, D.-H.; Liu, K.-Y.; Jiang, B.; Jiang, H.; Chen1, H.; Chen, Y.-H.; Zhang, X.-H.; et al. Health-related quality of life of patients with newly diagnosed chronic myeloid leukemia treated with allogeneic hematopoietic SCT versus Imatinib. *Bone Marrow Trans.* **2014**, *49*, 576–580. [CrossRef]
48. Athale, U.; Andolina, J.R.; Redell, M.S.; Hijiya, N.; Patterson, B.C.; Bergsagel, J.; Bittencourt, H.; Schultz, K.R.; Burke, M.J.E.; Kolb, A.; et al. Management of chronic myeloid leukemia in children and adolescents: Recommendations from the Children's Oncology Group CML Working Group. *Pediatr. Blood Cancer* **2019**, *66*, e2782.
49. Bower, H.; Björkholm, M.; Dickman, P.W.; Höglund, M.; Lambert, P.C.; Andersson, T.M.L. Life expectancy of patients with chronic myeloid leukemia approaches the life expectancy of the general population. *J. Clin. Oncol.* **2016**, *34*, 2851–2857. [CrossRef]
50. Ross, D.M.; Branford, S.; Seymour, J.F.; Schwarer, A.P.; Arthur, C.; Yeung, D.T.; Dang, P.; Goyne, J.M.; Slader, C.; Filshie, R.J.; et al. Safety and efficacy of imatinib cessation for CML patients with stable undetectable minimal residual disease: Results from the TWISTER study. *Blood* **2013**, *122*, 515–522. [CrossRef]
51. Mahon, F.X.; Réa, D.; Guilhot, J.; Guilhot, F.; Huguet, F.; Nicolini, F.; Legros, L.; Charbonnier, A.; Guerci, A.; Varet, B.; et al. Intergroupe Français des Leucémies Myéloïdes Chroniques. Discontinuation of imatinib in patients with chronic myeloid leukaemia who have maintained complete molecular remission for at least 2 years: The prospective, multicentre Stop Imatinib (STIM) trial. *Lancet Oncol.* **2010**, *11*, 1029–1035. [CrossRef]
52. National Comprehensive Cancer Network. NCCN Clinical Practice Guidelines in Oncology. Chronic Myelogenous Leukemia, Version 1.2020. Available online: https://www.nccn.org/patients/guidelines/content/PDF/cml-patient.pdf (accessed on 22 June 2020).
53. Saussele, S.; Richter, J.; Guilhot, J.G.; Hjorth-Hansen, H.; Almeida, A.; Janssen, J.J.; Mayer, J.; Koskenvesa, P.; Panayiotidis, P.; Olsson-Strömberg, U.; et al. Duration of deep molecular response has most impact on the success of cessation of tyrosine kinase inhibitor treatment in chronic myeloid leukemia—Results from the EURO-SKI trial. *Blood* **2017**, *130*, 313.
54. Ault, P.; Kantarjian, H.; O'Brien, S.; Faderl, S.; Beran, M.; Rios, M.B.; Koller, C.; Giles, F.; Keating, M.; Talpaz, M.; et al. Pregnancy among patients with chronic myeloid leukemia treated with imatinib. *J. Clin. Oncol.* **2006**, *24*, 1204–1208. [CrossRef]
55. Pye, S.M.; Cortes, J.; Ault, P.; Hatfield, A.; Kantarjian, H.; Pilot, R.; Rosti, G.; Apperley, J.F. The effects of imatinib on pregnancy outcome. *Blood* **2008**, *111*, 5505–5508. [CrossRef]
56. Cortes, J.E.; Abruzzese, E.; Chelysheva, E.; Guha, M.; Wallis, N.; Apperley, J.F. The impact of dasatinib on pregnancy outcomes. *Am. J. Hematol.* **2015**, *90*, 1111–1115. [CrossRef]
57. Hughes, T.P.; Ross, D.M. Moving treatment-free remission into mainstream clinical practice in CML. *Blood* **2016**, *128*, 17–23. [CrossRef] [PubMed]
58. Richter, J.; Söderlund, S.; Lübking, A.; Dreimane, A.; Lotfi, K.; Markevärn, B.; Själander, A.; Saussele, S.; Olsson-Strömberg, U.; Stenke, L. Musculoskeletal pain in patients with chronic myeloid leukemia after discontinuation of imatinib: A tyrosine kinase inhibitor withdrawal syndrome? *J. Clin. Oncol.* **2014**, *32*, 2821–2823. [CrossRef] [PubMed]
59. Rea, D.; Nicolini, F.E.; Tulliez, M.; Guilhot, F.; Guilhot, J.; Guerci-Bresler, A.; Gardembas, M.; Coiteux, V.; Guillerm, G.; Legros, L.; et al. France Intergroupe des Leucémies Myéloïdes Chroniques. Discontinuation of dasatinib or nilotinib in chronic myeloid leukemia: Interim analysis of the STOP 2G-TKI study. *Blood* **2017**, *129*, 846–854. [CrossRef]
60. Hochhaus, A.; Saussele, S.; Rosti, G.; Mahon, F.X.; Janssen, J.J.W.M.; Hjorth-Hansen, H.; Richter, J.; Buske, C.; ESMO Guidelines Committee. Chronic myeloid leukaemia: ESMO Clinical Practice Guidelines for diagnosis, treatment and follow-up. *Ann. Oncol.* **2017**, *28*, iv41–iv51. [CrossRef] [PubMed]
61. Millot, F.; Claviez, A.; Leverger, G.; Corbaciglu, S.; Groll, A.H.; Suttorp, M. Imatinib cessation in children and adolescents with chronic myeloid leukemia in chronic phase. *Pediatr. Blood Cancer* **2014**, *61*, 355–357. [CrossRef]
62. FDA Approves Gleevec for Children with Acute Lymphoblastic Leukemia. Available online: https://www.fda.gov/NewsEvents/Newsroom/PressAnnouncements/ucm336868.html (accessed on 20 January 2020).
63. Suttorp, M.; Claviez, A.; Bader, P.; Peters, C.; Gadner, H.; Ebell, W.; Dilloo, D.; Kremens, B.; Kabisch, H.; Führer, M.; et al. Allogeneic stem cell transplantation for pediatric and adolescent patients with CML: Results from the prospective trial CML-paed I. *Klin. Padiatr.* **2009**, *221*, 351–357. [CrossRef] [PubMed]

64. Suttorp, M.; Millot, F. Treatment of pediatric chronic myeloid leukemia in the year 2010: Use of tyrosine kinase inhibitors and stem-cell transplantation. *Hematol. Am. Soc. Hematol. Educ. Program.* **2010**, *2010*, 368–376. [CrossRef] [PubMed]
65. Champagne, M.A.; Capdeville, R.; Krailo, M.; Qu, W.; Peng, B.; Rosamilia, M.; Therrien, M.; Zoellner, U.; Blaney, S.M.; Bernsteinet, M. Imatinib mesylate (STI571) for treatment of children with Philadelphia chromosome-positive leukemia: Results from a Children's Oncology Group phase 1 study. *Blood* **2004**, *104*, 2655–2660. [CrossRef]
66. Lee, J.W.; Chung, N.G. The treatment of pediatric chronic myelogenous leukemia in the imatinib era. *Korean J. Pediatr.* **2011**, *54*, 111–116. [CrossRef]
67. Muramatsu, H.; Takahashi, Y.; Sakaguchi, H.; Shimada, A.; Nishio, N.; Hama, A.; Doisaki, S.; Yagasaki, H.; Matsumoto, K.; Kato, K.; et al. Excellent outcomes of children with CML treated with imatinib mesylate compared to that in pre-imatinib era. *Int. J. Hematol.* **2011**, *93*, 186–191. [CrossRef]
68. Millot, F.; Guilhot, J.; Nelken, B.; Leblanc, T.; De Bont, E.S.; Be'kassy, A.N.; Gadner, H.; Sufliarska, S.; Stary, J.; Gschaidmeier, H.; et al. Imatinib mesylate is effective in children with chronic myelogenous leukemia in late chronic and advanced phase and in relapse after stem cell transplantation. *Leukemia* **2006**, *20*, 187–192.
69. Miranda, M.B.; Lauseker, M.; Kraus, M.P.; Proetel, U.; Hanfstein, B.; Fabarius, A.; Baerlocher, G.M.; Heim, D.; Hossfeld, D.K.; Kolb, H.J.; et al. Secondary malignancies in chronic myeloid leukemia patients after imatinib-based treatment: Long-term observation in CML Study IV. *Leukemia* **2016**, *30*, 1255–1262. [PubMed]
70. Kalmanti, L.; Saussele, S.; Lauseker, M.; Müller, M.C.; Dietz, C.T.; Heinrich, L.; Hanfstein, B.; Proetel, U.; Fabarius, A.; Krause, S.W.; et al. Safety and efficacy of imatinib in CML over a period of 10 years: Data from the randomized CML-study IV. *Leukemia* **2015**, *29*, 1123–1132. [CrossRef]
71. Drucker, B.J.; Guilhot, F.; O'Brien, S.G.; Gathmann, I.; Kantarjian, H.; Gattermann, N.; Deininger, M.W.; Silver, R.T.; Goldman, J.M.; Stone, R.M.; et al. Five-year follow-up of patients receiving imatinib for chronic myeloid leukaemia. *N. Engl. J. Med.* **2006**, *355*, 2408–2417. [CrossRef] [PubMed]
72. De Lavallade, H.; Apperley, J.F.; Khorashad, J.S.; Milojkovic, D.; Reid, A.G.; Bua, M.; Szydlo, R.; Olavarria, E.; Kaeda, J.; Goldman, J.M.; et al. Imatinib for newly diagnosed patients with chronic myeloid leukaemia: Incidence of sustained responses in an intention-to-treat analysis. *J. Clin. Oncol.* **2008**, *26*, 3358–3363. [CrossRef] [PubMed]
73. Jaeger, B.A.; Tauer, J.T.; Ulmer, A.; Kuhlisch, E.; Roth, H.J.; Suttorp, M. Changes in bone metabolic parameters in children with chronic myeloid leukemia on imatinib treatment. *Med. Sci. Monit.* **2010**, *18*, CR721–CR728. [CrossRef]
74. Giona, F.; Mariani, S.; Gnessi, L.; Moleti, M.L.; Rea, M.; De Vellis, A.; Marzella, D.; Testi, A.M.; Foà, R. Bone metabolism, growth rate and pubertal development in children with chronic myeloid leukemia treated with imatinib during puberty. *Haematologica* **2013**, *98*, e25–e27. [CrossRef] [PubMed]
75. Millot, F.; Guilhot, J.; Baruchel, A.; Petit, A.; Leblanc, T.; Bertrand, Y.; Mazingue, F.; Lutz, P.; Vérité, C.; Berthouh, C.; et al. Growth deceleration in children treated with imatinib for chronic myeloid leukaemia. *Eur. J. Cancer* **2014**, *50*, 3206–3211. [CrossRef]
76. Suttorp, M.; Bornhäuser, M.; Metzler, M.; Millot, F.; Schleyer, E. Pharmacology and pharmacokinetics of imatinib in pediatric patients. *Exp. Rev. Clin. Pharm.* **2018**, *11*, 219–231. [CrossRef]
77. Giona, F.; Putti, M.C.; Micalizzi, C.; Menna, G.; Moleti, M.L.; Santoro, N.; Iaria, G.; Ladogana, S.; Burnelli, R.; Consarino, C.; et al. Long-term results of high-dose imatinib in children and adolescents with chronic myeloid leukaemia in chronic phase: The Italian experience. *Br. J. Haematol.* **2015**, *170*, 398–407. [CrossRef]
78. FDA Approved Dasatinib for Pediatric Patients. Available online: https://www.fda.gov/drugs/resources-information-approved-drugs/fda-approves-dasatinib-pediatric-patients-cml (accessed on 20 January 2020).
79. Gore, L.; Kearns, P.R.; de Martino, M.L.; Lee, C.; De Souza, A.; Bertrand, Y.; Hijiya, N.; Stork, L.C.; Chung, N.-G.; Cardenas Cardos, R.; et al. Dasatinib in Pediatric Patients with Chronic Myeloid Leukemia in Chronic Phase: Results From a Phase II Trial. *J. Clin. Oncol.* **2018**, *36*, 1330–1338. [CrossRef]
80. FDA-approves-nilotinib-pediatric-patients. Available online: https://www.fda.gov/drugs/resources-information-approved-drugs/fda-approves-nilotinib-pediatric-patients-newly-diagnosed-or-resistantintolerant-ph-cml-chronic (accessed on 20 January 2020).
81. Hijiya, N.; Maschan, A.; Rizzari, C.; Shimada, H.; Dufour, C.; Goto, H.; Kang, H.J.; Guinipero, T.; Karakas, Z.; Bautista, F.; et al. Phase 2 study of nilotinib in pediatric patients with philadelphia chromosome-positive chronic myeloid leukemia. *Blood* **2019**, *134*, 2036–2045. [CrossRef] [PubMed]

82. Hochhaus, A.; Saglio, G.; Hughes, T.P.; Larson, R.A.; Kim, D.W.; Issaragrisil, S.; le Coutre, P.D.; Etienne, G.; Dorlhiac-Llacer, P.E.; Clark, R.E.; et al. Long-term benefits and risks of frontline nilotinib vs imatinib for chronic myeloid leukemia in chronic phase: 5-year update of the randomized ENESTnd trial. *Leukemia* **2016**, *30*, 1044–1054. [CrossRef]
83. Giles, F.J.; le Coutre, P.D.; Pinilla-Ibarz, J.; Larson, R.A.; Gattermann, N.; Ottmann, O.G.; Hochhaus, A.; Radich, J.P.; Saglio, G.; Hughes, T.P.; et al. Nilotinib in imatinib-resistant or imatinib-intolerant patients with chronic myeloid leukemia in chronic phase: 48-month follow-up results of a phase II study. *Leukemia* **2013**, *27*, 107–112. [CrossRef]
84. EU Nilotib Approval for Children. Available online: https://www.novartis.com/news/media-releases/novartis-drug-tasignar-nilotinib-secures-eu-approval-first-and-second-line-treatment-ph-cml-cp-children (accessed on 20 January 2020).
85. Nickel, R.S.; Daves, M.; Keller, F. Treatment of an adolescent with chronic myeloid leukemia and the T315I mutation with ponatinib. *Pediatr. Blood Cancer* **2015**, *62*, 2050–2051. [CrossRef]
86. Müller, M.C.; Cervantes, F.; Hjorth-Hansen, H.; Janssen, J.J.W.M.; Milojkovic, D.; Rea, D.; Rosti, G. Ponatinib in chronic myeloid leukemia (CML): Consensus on patient treatment and management from a European expert panel. *Crit. Rev. Oncol. Hematol.* **2017**, *120*, 52–59. [CrossRef] [PubMed]
87. Soverini, S.; Hochhaus, A.; Nicolini, F.E.; Gruber, F.; Lange, T.; Saglio, G.; Pane, F.; Müller, M.C.; Ernst, T.; Rosti, G.; et al. BCR-ABL kinase domain mutation analysis in chronic myeloid leukemia patients treated with tyrosine kinase inhibitors: Recommendations from an expert panel on behalf of European LeukemiaNet. *Blood* **2011**, *118*, 1208–1215. [CrossRef] [PubMed]
88. Millot, F.; Suttorp, M.; de Bont, E.; Kalwak, K.; Nelken, B.; Ducassou, S.; Bertrand, Y.; Baruchel, A. Ponatinib In Childhood Philadelphia Positive Leukemias: The Experience of The International Registry of Childhood Chronic Myleloid Leukemia I-Cml-Ped-Study. *Hemasphere* **2019**, *3*, 161–162. [CrossRef]
89. Lam, M.S. Extemporaneous compounding of oral liquid dosage formulations and alternative drug delivery methods for anticancer drugs. *Pharmacotherapy* **2011**, *31*, 164–192. [CrossRef]
90. Glivec Product Information. Available online: https://www.ema.europa.eu/en/documents/product-information/glivec-epar-product-information_it.pdf (accessed on 18 August 2009).
91. Sprycel Product Information. Available online: https://www.ema.europa.eu/en/documents/product-information/sprycel-epar-product-information_it.pdf (accessed on 18 August 2009).
92. Iclusig Product Information. Available online: https://www.ema.europa.eu/en/documents/referral/iclusig-article-20-procedure-product-information_it.pdf (accessed on 11 July 2009).
93. Tasigna Product Information. Available online: https://www.ema.europa.eu/en/documents/product-information/tasigna-epar-product-information_it.pdf (accessed on 11 July 2009).
94. Andolina, J.R.; Neudorf, S.M.; Corey, S.J. How I treat childhood CML. *Blood* **2012**, *23*, 1821–1830. [CrossRef]
95. De la Fuente, J.; Baruchel, A.; Biondi, A.; de Bont, E.; Dresse, M.F.; Suttorp, M.; Millot, F.; International BFM Group (iBFM) Study Group Chronic Myeloid Leukaemia Committee. Managing children with chronic myeloid leukaemia (CML): Recommendations for the management of CML in children and young people up to the age of 18 years. *Br. J. Haematol.* **2014**, *167*, 33–47. [CrossRef]
96. Hochhaus, A.; Ernst, T.; Eigendorff, E.; La Rosée, P. Causes of resistance and treatment choices of second- and third-line treatment in chronic myelogenous leukemia patients. *Ann. Hematol.* **2015**, *94*, S133–S140. [CrossRef] [PubMed]
97. Cutrignelli, A.; Sanarica, F.; Lopalco, A.; Lopedota, A.; Laquintana, V.; Franco, M.; Boccanegra, B.; Mantuano, P.; De Luca, A.; Denora, N. Dasatinib/HP-β-CD Inclusion Complex Based Aqueous Formulation as a Promising Tool for the Treatment of Paediatric Neuromuscular Disorders. *Int. J. Mol. Sci.* **2019**, *30*, 591. [CrossRef] [PubMed]
98. Melville, N.A. Dasatinib in Children: An Effective Alternative to Imatinib. In Proceedings of the European Hematology Association (EHA) 2017 Congress, Madrid, Spain, 22–25 June 2017.
99. Zwaan, C.M.; Rizzari, C.; Mechinaud, F.; Lancaster, D.I.; Lehrnbecher, T.; Van der Velden, V.H.J.; Beverloo, B.B.; den Boer, M.L.; Pieters, R.; Reinhardt, D.; et al. Dasatinib in Children and Adolescent With Relapsed or Refractory Leukemia: Results of the CA180-018 Phase I Dose-Escalation Study of the innovative Therapies for Children With Cancer Consortium. *J. Clin. Oncol.* **2013**, *31*, 2460–2469. [CrossRef]
100. Allen, L.V. *The Art and Science of Pharmaceutical Compounding*, 2nd ed.; APhA Publications: Washington, DC, USA, 2005; p. 493.

101. Nahata, M.C.; Allen, L.V., Jr. Extemporaneous drug formulations. *Clin. Ther.* **2008**, *30*, 2112–2119. [CrossRef] [PubMed]
102. Allen, L.V. Compounding, stability and beyond-use dates. *Secundum. Artem.* **2003**, *7*. Minneapolis, MN: Paddock Laboratories. Available online: www.paddocklabs.com (accessed on 24 February 2010).
103. Woods, D.J. Extemporaneous formulations: Problems and solutions. *Paediatr. Perinatal. Drug Ther.* **1997**, *1*, 25–29.
104. Batchelor, H.K.; Marriott, J.F. Formulations for children: Problems and solutions. *Br. J. Clin. Pharmacol.* **2015**, *79*, 405–418. [CrossRef]
105. Lopalco, A.; Denora, N.; Laquintana, V.; Cutrignelli, A.; Franco, M.; Robota, M.; Hauschildt, N.; Mondelli, F.; Arduino, I.; Lopedota, A. Taste masking of propranolol hydrochloride by microbeads of EUDRAGIT® E PO obtained with prilling technique for paediatric oral administration. *Int. J. Pharm.* **2020**, *574*, 118922. [CrossRef]
106. Laquintana, V.; Asim, M.H.; Lopedota, A.; Cutrignelli, A.; Lopalco, A.; Franco, M.; Bernkop-Schnürch, A.; Denora, N. Thiolated hydroxypropyl-β-cyclodextrin as mucoadhesive excipient for oral delivery of budesonide in liquid paediatric formulation. *Int. J. Pharm.* **2019**, *572*, 118820. [CrossRef]
107. Lopalco, A.; Curci, A.; Lopedota, A.; Cutrignelli, A.; Laquintana, V.; Franco, M.; Denora, N. Pharmaceutical preformulation studies and paediatric oral formulations of sodium dichloroacetate. *Eur. J. Pharm. Sci.* **2019**, *127*, 339–350. [CrossRef]
108. Hijiya, N.; Suttorp, M. How I treat chronic myeloid leukemia in children and adolescents. *Blood* **2019**, *133*, 2374–2384. [CrossRef]
109. Champagne, M.A.; Fu, C.H.; Chang, M.; Chen, H.; Gerbing, R.B.; Alonzo, T.A.; Cooley, L.D.; Heerema, N.A.; Oehler, V.; Wood, C.; et al. Higher dose imatinib for children with de novo chronic phase chronic myelogenous leukemia: A report from the Children's Oncology Group. *Pediatr. Blood Cancer* **2011**, *57*, 56–62. [CrossRef] [PubMed]
110. Tanizawa, A. Optimal management for pediatric chronic myeloid leukemia. *Ped. Int.* **2016**, *58*, 171–179. [CrossRef] [PubMed]

International Journal of
Molecular Sciences

MDPI

Review

How to Modify Drug Release in Paediatric Dosage Forms? Novel Technologies and Modern Approaches with Regard to Children's Population

Monika Trofimiuk [1,2], Katarzyna Wasilewska [1] and Katarzyna Winnicka [1,*]

1 Department of Pharmaceutical Technology, Medical University of Bialystok, Mickiewicza 2c, 15-222 Bialystok, Poland
2 Department of Clinical Pharmacy, Medical University of Bialystok, Mickiewicza 2a, 15-222 Bialystok, Poland
* Correspondence: kwin@umb.edu.pl

Received: 11 June 2019; Accepted: 28 June 2019; Published: 29 June 2019

Abstract: In the pharmaceutical technology, paediatric population still presents the greatest challenge in terms of developing flexible and appropriate drug dosage forms. As for many medicines, there is a lack of paediatric dosage forms adequate for a child's age; it is a prevailing practice to use off label formulations. Children need balanced and personalized treatment, patient-friendly preparations, as well as therapy that facilitates dosing and thus eliminates frequent drug administration, which can be ensured by modified release (MR) forms. MR formulations are commonly used in adult therapy, while rarely available for children. The aim of this article is to elucidate how to modify drug release in paediatric oral dosage forms, discuss the already accessible technologies and to introduce novel approaches of manufacturing with regard to paediatric population.

Keywords: modified release; drug delivery; paediatric formulation development; paediatric dosage forms

---

1. Introduction

Creating an appropriate dosage form designed exactly for children still appears as an outgoing challenge for pharmaceutical technology and the distinction between adults and children pharmacokinetics should be considered. The statement that children can be treated as "small adults" is obviously incorrect, particularly in determining the therapeutic doses in individual age groups [1–4]. In paediatric pharmacotherapy, many factors regarding a convenient dosage form (e.g., age-suitable formulations in proper strength, off label use, and palatability) have to be included. Creating paediatric dosage forms is associated with many difficulties. Therefore, the main goal raised by regulations of European Medicines Agency (EMA) or paediatric scientific network groups is increasing the safety and efficiency of paediatric therapy by the enhancement the quality of clinical studies for children in various age groups (from birth to 18 years old) with better availability of pharmacokinetic data [5,6]. Scientific and governmental initiatives (The Best Pharmaceuticals for Children—BPCA, Paediatric Investigation Plan—PIP) are focused on the development of paediatric dosage forms adjusted to the child's age and as a consequence of enhancing the efficiency and safety of paediatric therapy [7,8]. The main directives implemented in the appropriate paediatric dosage forms development point basic difficulties connected with paediatric therapy. A special guideline concerning formulation and administration of suitable dosage forms and detailed information on how to use the medicines with regard to children was proposed (State of paediatric medicines in the European Union 10 report) [9]. Due to the special attention focused on the safety of excipients used in the paediatric formulations, the Step and Toxicity of Excipients for Paediatric Patients database has been created (STEP database) [10]. Additionally, EMA gave a support in providing clinical trials with children to promote and harmonize existing paediatric guidelines [7,11–18]. As a result of these acts, the percentage of clinical trials included children (where only one participant was under 18 years at least) has increased from 8.3% in 2007 to 12.4% in 2016. Therefore, it still remains an outgoing problem as there is constantly not enough clinical trials with children compared to the numbers of clinical trials carried out in adults [9]. Challenges associated with conducting clinical trials in children include many childhood diseases, heterogeneity of the population and ethical problems [18]. Main problems related to paediatric therapy are presented in Figure 1.

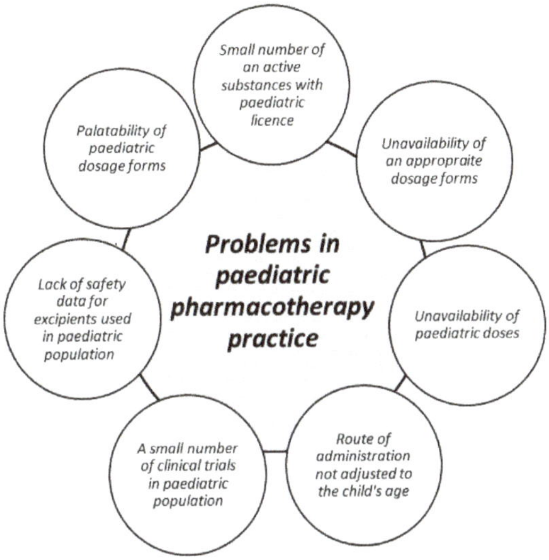

Figure 1. Problems in paediatric pharmacotherapy practice [11,19,20].

Traditional dosage forms (e.g., tablets, capsules, and injections) are often not appropriate for children, hence cutting or crushing tablets, splitting capsule and then mixing with the food (solid or liquid) and dilution of injections are common practice [19–22]. Manipulation with dosage form may cause a risk of damage of formulation structure, hence side effects and changes in pharmacological effect might occur. Despite significant advances in the development of drug dosage forms dedicated especially for children, but unfortunately, unlicensed drugs are still used [1–4].

Oral route of administration is the most natural way of giving medicines for children. Regardless, child age is still the most popular liquid dosage forms. Their main advantages are safety and ease of swallowing, but the most important and most convenient factor is the possibility of drug administration in a wide group of children by volumetrically measuring a dose, precisely adjusted to a child's weight or age. Liquid dosage forms are preferred especially for newborns, infants, and smaller children, avoiding the risk of chocking and enhancing the probability of taking a full dose of the drug. For older children (above 6 years), despite liquid formulations, solid dosage forms (tablets, effervescent formulations, orodispersible tablets, films, pellets, or minitablets) could be safely administered. However, the individual swallowing abilities should be considered when administering oral solid forms for children. Nevertheless, numerous studies proved that pre-school (up to 6 years old) and even infants (6 months old) are able to safely swallow particles smaller than 3 mm (minitablets, pellets) [23–26]. While both liquid and solid forms are available in paediatric therapy, there are not many MR formulations dedicated for children. MR dosage forms ensure drug release in the entire gastrointestinal tract providing constant drug concentration and eliminating the necessity of taking several doses a day, hence improving pharmacotherapy effectiveness [19,27]. Additionally, it is worth emphasizing that utilizing MR technology creates a possibility of increasing drug stability. The most common example is formulating enteric pellets with proton pump inhibitors (omeprazole, pantoprazole), which are unstable in acidic pH. The main reason why MR paediatric forms are lacking includes the clinical aspects like a small range of strength suitable only for particular child's age. Changing clinical parameters in the chronic diseases requires frequent dose changes depending on the response to therapy, disease management, or age-related changes in strength dose, which requires the availability of a wide range of drug doses on the market. From a technological point of view, the production of various doses is simply unprofitable for pharmaceutical companies, considering that some of the MR technologies are rather expensive (Figure 2) [9,20]. The aim of the current article is to overview available paediatric MR oral liquid and solid formulations depending on the dosage forms and utilized technology, as well as to introduce new approaches and possibilities used in the children's pharmacotherapy.

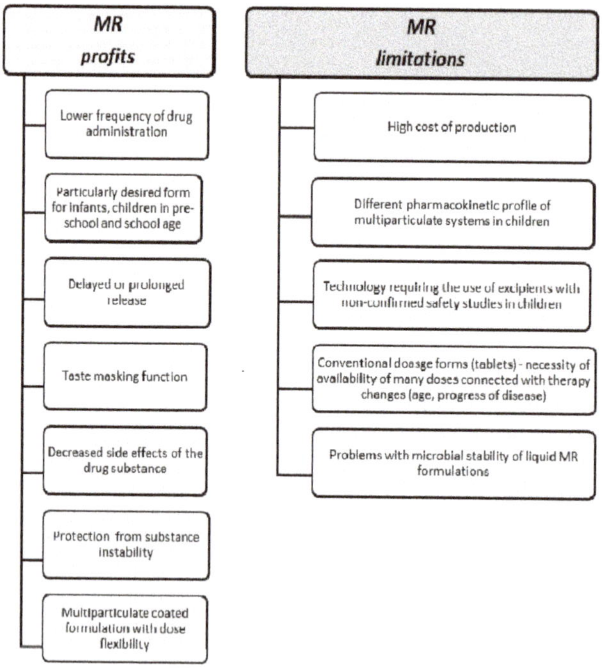

Figure 2. Profits and limitations of use the modified release (MR) formulations in paediatric population.

2. MR Liquid Dosage Forms

Oral liquid formulations (drops, solutions, suspensions, and syrups) are the most popular dosage forms indicated for children of all ages [6]. The main advantage of liquid forms is volumetric dosing, which gives an opportunity to precisely adjust the dose for the specific age groups by measuring an appropriate drug volume. Liquids are easy to swallow and can be administered by child caregivers, without any manipulation before administration, as in the case of tablets (crushing, mixing with food, or fluid). Nevertheless, the limiting factor of using liquids is their physicochemical and microbial instability in an aqueous environment, which entails the need to use preservatives or cosolvents. Additionally, dosing might be a hindering issue in connection with using a not calibrated spoon, oral syringes, or not properly measured drops. Limiting factor of using liquid forms in pediatric preparations is the drug taste as well. Therefore, common practice is to prepare suspensions rather than solutions. For this purpose, excipients such as natural or synthetic flavours or various technologies that minimize the unpleasant feeling of bitterness (e.g., dose sipping technology—straws with coated pellets) can be used [28–31]. Nonetheless, it is evident that conventional liquids are the most frequently chosen formulations, therefore MR oral liquid forms would be preferable for children to eliminate dosing several times per day, especially in chronic diseases. Not only reducing the frequent drug administration is important, but also providing a favorable pharmacokinetic profile of the drug with keeping drug concentration at a constant therapeutic level [32]. For comparison, in case of MR hard capsules, to achieve a prolonged action of the drug, it is possible to open the MR capsule and to administer the content (e.g., powder, pellets, and minitablets) after mixing with the fluid or food. However, this practice can result in dose errors or MR disturbance if the medication will be chewed not swallowed [19,33]. The following modifications are used for formulating prolonged drug release in liquid dosage forms: Drug/resin complexes, in situ gel formation, microencapsulation, and MR microparticles [34,35]. Despite liquid MR formulations seem to be suitable for applying to children, currently the number of MR liquid forms for children available in the market is limited. Table 1 presents an overview of MR oral liquid formulations with paediatric license and products with finished clinical trials being under the approval registration process by the Food and Drug Administration (FDA).

Table 1. Modified release oral liquid dosage forms—an overview of available formulations with paediatric license and drugs with finished clinical trials being under approval registration process by the Food and Drug Administration (FDA).

Product (Manufacturer)	Drug	Dosage Form	Polymer (MR Technique)	Paediatric Licence	Indication	References
Delsym® (Reckitt Benckiser LLC.)	dextromethorphan	extended-release suspension	ion exchange resin *drug/polymer complexation*	≥4 year	Cough	[36]
Dyanavel XR (TrisPharma)	amphetamine	extended-release suspension	sodium polystyrene sulfonate *drug/polymer complexation*	≥6 years	attention-deficit hyperactivity disorder (ADHD)	[37]
MST® Continus® (Napp Pharmaceuticals Limited)	morphine	prolonged-release suspension	cationinc exchange resin *drug/polymer complexation*	≥1 year	Pain	[38]
Quillivant XR (Pfizer)	methylphenidate	extended-release suspension	sodium polystyrene sulfonate *drug/polymer complexation*	≥6 years	ADHD	[39]
Tussionex®(UCB)	Hydrocodone chlorpheniramine	extended-release suspension	ion exchange resin *drug/polymer complexation microparticles*	≥6 years	common cold, flu	[40]
Zmax® (Pfizer)	azithromycin	extended-release suspension	glyceryl behenate and poloxamer 407 *microspheres*	≥6 month	bacterial infections	[41]
under approval registration process by FDA	bupivacaine	NA *	SABER™ Delivery System in situ *gel formation*	NA *	local anesthetic	[42–44]
under approval registration process by FDA	methylphenidate	NA *	ORADUR™ in situ *gel formation*	NA *	pain	[45,46]
under approval registration process by FDA	mirabegron	sustained-release suspensions	microspheres with lauryl sulfate salt/complex	NA *	urinary incontinence	[47,48]
under approval registration process by FDA (Remoxy®)	oxycodone	NA *	ORADUR™ in situ *gel formation*	NA *	Pain	[49–51]

* NA: no data available.

2.1. Drug-Resin Complexes

Utilizing ion-exchange resins for getting a complex with drug is one of the techniques enabling modified drug release in oral liquids (Figure 3) [52]. Nowadays, the most often used resins are cationic exchange-resin with a free sulphuric acid group on the crosslinked polystyrene matrix (extended release utilized in liquids with, e.g., chlorpheniramine and dextromethorphan) and the anionic exchange-resin with amino groups, which are available in the market with paediatric licence (Table 1). Developed complexes are often incorporated into microcapsules (inside the particle or in the coat), lipospheres (lipid microspheres, size 0.01–100 μm), or directly suspended in suspending vehicles. Obtained suspensions are usually administered once (Dyanavel XR®, Quilivant XR®) or twice daily (Delsym®, MST® Continous®) [36,37,39]. Dyanavel XR® utilizes an ion exchange resin, where the amphetamine is bound to the sodium polystyrene sulfate resin through an ionic binding reaction. Dyanavel XR® contains immediate release and extended release components as complexes coated with pH independent polymers: Povidone and polyvinyl acetate [37]. Quillivant XR® is a powder forming an extended-release oral suspension after reconstitution with water. It contains approximately 20% immediate-release and 80% extended-release methylphenidate in drug-polistyrex complex form [39]. MST® Continous® is the example of syrup containing MR granules with morphine sulphate complexed with Dowex 50WX8 cationic exchange resin and suspended in a sugar free medium. It is recommended to take the suspension every 12 h [38].

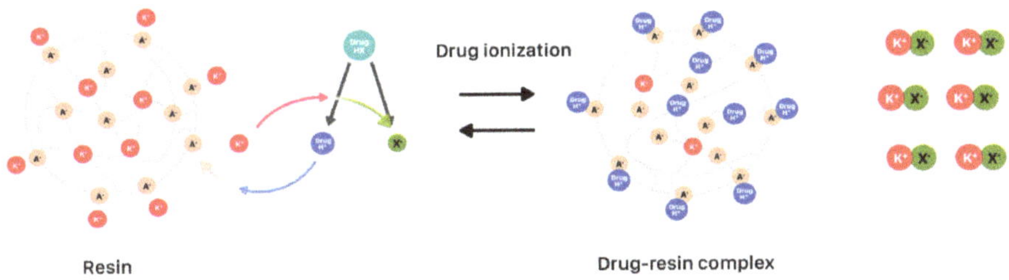

Figure 3. Drug release modification utilizing drug-resin complexes.

2.2. Microparticles—Spray Drying Technique

The other method of obtaining MR in oral liquids is the spray-drying technique. It creates the possibility of designing microparticles (microspheres, microcapsules) so that drug is incorporated or enclosed in a polymeric shell and then suspended in the liquid (Figure 4) [53,54]. This technique brings the great area on the achieving desired dissolution profile, improved drug stability, and also provides taste masking effect. Depending on the polymers used in the process, various sizes and properties of microparticles might be created. Spray drying is the process by which a dry powder product is formed from the starting solution or suspension [55–58]. Zmax® is an example of a single-dose, prolonged-release formulation of microspheres for oral suspension containing azithromycin (Table 1). Zmax® was approved as a one-dose-only treatment for mild-to-moderate acute bacterial sinusitis and community-acquired pneumonia. Azithromycin microspheres (50–300 μm) are produced with glyceryl behenate and poloxamer 407 utilization by a combining hot melt extrusion with spray congealing technology. Microspheres were formed by suspending azithromycin in a molten carrier matrix and spraying by a spinning-disk atomizer to form droplets, which congeal into solid microspheres upon cooling [41].

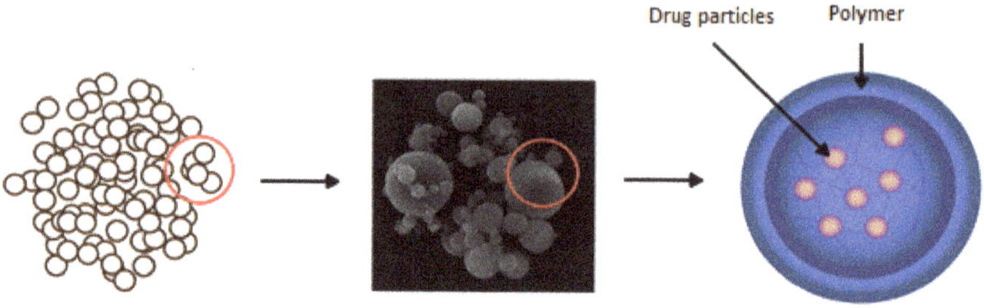

Figure 4. Drug release modification using spray-dried microparticles.

2.3. *In situ Gel Formation*

Modified drug release in liquid formulations could be obtained by the in situ gel formation, which depends on temperature, pH, ions, or UV irradiation. Gel formation from liquid allows achieving a sustained or controlled release profile (Figure 5) [59–61]. Physicochemical characteristics like solubility or gelling properties of polymers are crucial for obtaining desired MR. Polymers used for in situ gel formation include gellan gum, xyloglucan, pluronics, tetronics, alginic acid, carbomer, hypromellose, pectins, chitosan, polycaprolactone or poly(DL-lactic acid), or poly(DL-lactide-co-glycolide) [62–64]. The in situ forming gel technique is used in SABER™ technology (sucrose acetate isobutyrate extended release, SABER™ Delivery System), where the biodegradable gel initially appears in low-viscosity fluid form. SABER™ systems are dedicated as carriers for drugs in dosage forms administered orally (the clinical trials with SABER™ delivery system for bupivacaine). After application, its viscosity increases, making an adhesive gel and MR of drug could be obtained. Orally administered drug diffuses rapidly, leaving in situ drug depot [42,44]. Another system utilizing gel forming matrices is ORADUR™. Gel matrices are able to accumulate high concentrations of drug, with the goal of once and twice-daily dosing. These are also designed to provide controlled long-term treatment preventing the abuse, e.g., the preparations with oxycodone (Remoxy®) and methylphenidate [49–51]. Unfortunately, formulations based on ORADUR™ and SABER™ technologies despite completed clinical trials (phase 3) have still not been approved for marketing (Table 1).

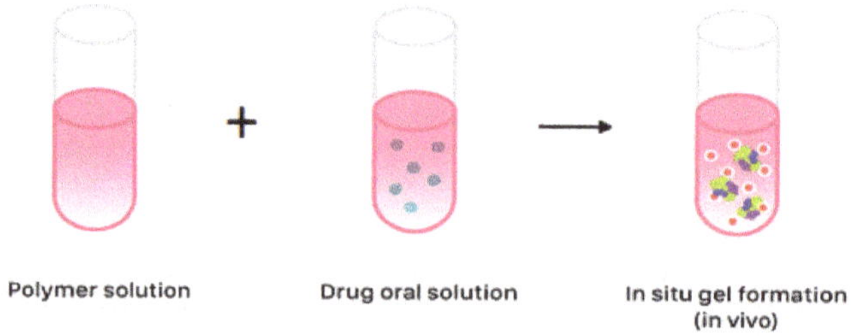

Figure 5. In situ gel formation from liquid for obtaining sustained or controlled release profile of drug in liquid dosage form.

3. MR Solid Dosage Forms

3.1. *Matrix and Coated Tablets*

Traditional MR tablets are formulated by drug embedding in a hydrophilic (swelling), lipid, or insoluble matrix. A determinant factor for maintaining MR from all tablets types is obtained by incorporation to ensuring a sufficiently long way of drug diffusion. As hydrophilic carriers, water-soluble polymers are utilized: Cellulose derivatives (methylcellulose, hypromellose, and hydroxyprophylmethylcellulose) or sodium alginate. A common feature of these polymers is the formation of highly viscous gel in an aqueous environment hindering the drug diffusion. By suspending drug in the gastrointestinal insoluble carrier, shell tablets are obtained. Among excipients forming insoluble shell inorganic compounds (calcium sulphate, and di- and triphosphate) or organic (ethylcellulose, cellulose acetate) can be distinguished [65–68]. An interesting example of matrix tablet (intended for children ≥12 years) is Lamictal® XR (Table 2). The tablets are coated with an enteric layer ensuring MR. Simultaneously, there are apertures drilled from the core to the outer layer on both faces of the tablet's structure (DiffCORE™) to provide a controlled release of drug in the acidic environment of the stomach (Figure 6). Such a combination is designed to control the dissolution rate of lamotrigine over a period of approximately 12 to 15 h, leading to a gradual increase of lamotrigine level in serum [69].

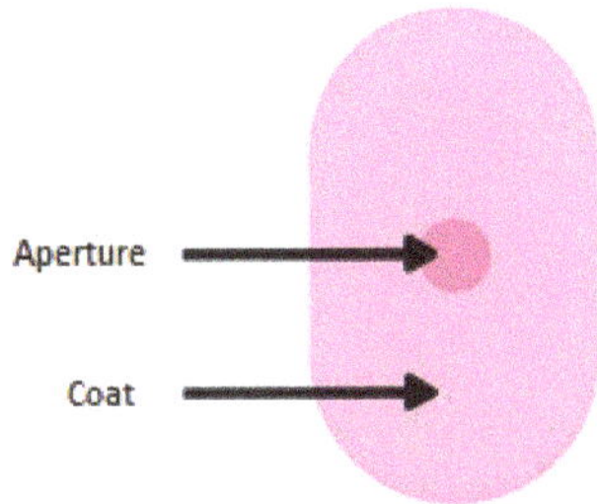

Figure 6. Scheme illustration of DiffCORE™ system.

Table 2. Modified release solid dosage forms—an overview of available formulations with paediatric licence.

Product (Manufacturer)	Drug	Dosage Form	Polymer/MR Technique	Paediatric Licence	Indication	References
Aciphex® Sprinkle™ (Eisai Management Co., Ltd.)	rabeprazole	capsules with granules	ethylcellulose *(delayed release)*	≥1 year	reflux	[70]
Adderall XR® (Shire US Inc)	mixed salts of single-entity amphetamine product	capsules with granules	hypromellose, methacrylic acid copolymer *(immediate and delayed release)*	≥6 years	attention deficit hyperactivity disorder (ADHD)	[71]
Adzenys XR-ODT™ (Neos Therapeutics)	amphetamine	extended release orally disintegrating tablets (ODT)	methacrylic acid, ethylcellulose *(ion resin technology)*	≥6 years	ADHD	[72,73]
Azulfidine® (Pfizer)	sulfasalazine	tablets	cellulose acetate phthalate *(delayed release)*	≥6 years	mild to moderate ulcerative colitis Crohn disease – off labell use	[74]
Concerta® (Janssen-Cilag)	methylphenidate	tablets	cellulose acetate *(matrix, extended release)*	≥6 years	ADHD	[75]
Coreg CR® (GSK)	carvedilol	capsules	methacrylic acid copolymers *(controled release)*	≥6 years	hypertension	[76]
Cotempla XR-ODT® (Neos Therapeutics)	methylphenidate	orodispersible tablets	methacrylic acid, ethylcellulose, polystyrene sulfate *(immediate and extended release)*	≥6 years	ADHD	[77,78]
Creon® (Solvay Pharmaceuticals)	pancreatic enzymes	Minitablets (in capsules)	dibuthyl phthalate, hypromellose phthalate *(delayed-release)*	≥6 years	chronic pancreatitis, cystic fibrosis	[79]
Finlepsin® (Teva Pharmaceuticals)	carbamazepine	tablets	Eudragit RS 30D, Eudragit L 30 D *(extended release)*	≥6 years	epilepsy	[80]
Focalin™ XR (Elan Holdings Inc.)	dexmethylphenidate	capsules	ammonio methacrylate copolymer *(extended release)*	≥6 years	ADHD	[81]
GranuPAS® (Lucane Pharma)	para-aminosalicylic acid	gastro-resistant granules in sachet	methacrylic acid – ethyl acrylate copolymer (1:1) *(extended release)*	≥1 year	tuberculosis	[82]
Kapvay ® (Concordia Pharmaceuticals Inc.)	clonidine	tablets	hypromellose (diffusion from gel matrix structure *-extended release*)	≥6 years	ADHD	[83]
Keppra XR® (UCB)	levetiracetam	tablets	hypromellose, polyvinyl alcohol- (diffusion from gel matrix structure *-extended release*)	≥12 years	epilepsy	[84]
Lamictal® XR (GSK)	lamotrigine	tablets	methacrylic acid copolymers *(extended release)*	≥12 years	epilepsy	[69]
Losec MUPS (AstraZeneca AB)	omeprazole	gastro-resistant tablets with coated pellets	methacrylic acid – ethyl acrylate copolymer (1:1) dispersion *(delayed-release)*	≥1 year	reflux	[85]
Metadate CD® (UCB Manufacturing, Inc.)	methylphenidate	capsules with granules	hypromellose, polyethylene glycol, ethylcellulose *(immediate and extended release)*	≥6 years	ADHD	[86]
Moxatag™ (MiddleBrook Pharmaceuticals, Inc.)	amoxicilin	tablets	prolonged-release pulsatile delivery technology MUPS	≥12 years	tonsillitis, pharyngitis	[87,88]

Table 2. *Cont.*

Product (Manufacturer)	Drug	Dosage Form	Polymer/MR Technique	Paediatric Licence	Indication	References
Orfiril Long (Desitin Arzneimittel GmbH)	natrii valproas	Minitablets (*in sachet or capsule*)	ethylcellulose, ammonium methacrylate copolymer (*extended release*)	≥6 years	epilepsy	[89]
Pancrease MT® (McNeil)	pancreatic enzymes	enteric-coated minitablets in capsule	methacrylic acid ethyl acrylate copolymers (*delayed release*)	from birth	chronic pancreatitis, cystic fibrosis	[90]
Pentasa® (Ferring GmbH)	mesalazine	granules	Ethylcellulose (*prolonged release*)	≥6 years	Crohn's disease	[91]
Prevacid® SoluTab™ (Takeda)	lansoprazole	orodispersible tablets	methacrylic acid copolymer (*delayed relese*)	≥1 year	reflux	[92,93]
Ritalin® LA (Novartis)	methylphenidate	capsules	ammonio methacrylate copolymer, gelatin, methacrylic acid copolymer (*delayed relese*)	≤ 6 years	ADHD	[94]
Tegretol® XL (Novartis)	carbamazepine	tablets	ethylcellulose dispersion (*matrix, extended release*)	≤ 6 years	epilepsy	[95]
TOPROL-XL® (Aralez Pharmaceuticals)	metoprolol	tablets	cellulose compounds (*extended release*)	≥6 years	hypertension	[96]
Viramune® (Boehringer Ingelheim International GmbH)	nevirapine	tablets	hypromellose (*prolonged release*)	≥3 years	human immunodeficieny virus (HIV) infection	[97]

3.2. Multiparticulate MR Solid Dosage Forms (MultiP)

Single-unit formulations contain drug in a single tablet or capsule form, whereas MultiP dosage forms comprise of quantity of particles combined into one dosage unit. They may exist as pellets, granules, minitablets, microparticles (microspheres, microcapsules), or nanoparticles with drugs being entrapped in or compacted in the matrix, as well as layered around cores and placed per se in sachets or capsules. MultiP provide many advantages over single-unit systems because of their small size and large surface, which allows leaving the stomach within a short period of time, which results in better distribution and bioavailability improvement. Another advantage of MultiP is the decreased risk of dose dumping due to damaged/broken coating, as well as reduced local irritation as MultiP are more uniformly dispersed in the gastrointestinal tract. Pellets reduce retention in a throat compared to the capsules or powders and improve physicochemical stability. MultiP dosage forms may offer a flexible dosing that allows covering a broad range of doses for different age groups [98,99].

An example of such a system is Multi-Unit Pellet System (MUPS)—utilizing coated pellets for controlled release, usually filled into a capsule or compressed to a tablet form. MUPS technology has been adopted by the pharmaceutical industry as an alternative to conventional immediate or MR tablets. MUPS consisting of pellets ensures divisible dosage form without imparing the drug release characteristic of the individual units. In addition, compared to other carrier systems, MUPS preparations entail a lower risk of irritation and toxicity, dose stability, minimal fluctuations of drug concentration in plasma, and the ability to administer drugs with a narrow therapeutic index. Another advantage is the possibility of using different taste masking techniques. The leading preparation being manufactured utilizing this technology is Losec MUPS with omeprazole for children (approved over 1 year of age and $\geq$ 10 kg) [85,100]. Formerly, attempts were made to obtain omeprazole as liquid form, however, due to its instability (rapid decomposition at wide range of pH), such a formulation could not be manufactured and launched in the market [101,102].

Another example of MultiP with paediatric licence is Moxatag™—prolonged release pulsatile delivery technology (PULSYS™) with multiple pellets inside. The typical PULSYS™ drug delivery format is a tablet containing pellets with different release profiles. The pellets with amoxicillin are formulated in a proportion that delivers optimal antibiotic levels. Manufacturers assume that this product can be presented in the future as sprinkle granules for the youngest patients [87,88]. Prevacid® is an instance of delayed release capsules with pellets (also available as delayed release orally disintegrating tablet—Prevacid® SoluTab™) containing lansoprazole (Table 2). The pharmacokinetics of lansoprazole was studied in paediatric patients with gastroesophageal reflux disease (GERD) in two separate clinical studies (one children group aged from 1 to 11 years and the second from 12 to 17 years). Each capsule contains enteric-coated pellets consisting of methacrylic acid copolymer providing delayed release of lansoprazole [92,93].

The MR MultiP can also be used in the treatment of hypertension in children above 6 years. The combined (immediate and control) release profile (Coreg CR®) or extended release (Toprol-XL®) were designed for carvedilol and metoprolol, respectively (Table 2). Coreg CR® is in hard gelatin capsules form filled with immediate or controlled release microparticles. Controlled release of carvedilol is ensured by coating multiparticles using methacrylic acid copolymer [76]. TOPROL-XL® is available as extended release tablets and has been formulated to provide a controlled release of metoprolol for once-daily administration. The tablets comprise a multiple unit system containing metoprolol succinate in a multitude of controlled release pellets. Each pellet acts as a separate drug delivery unit and is designed to deliver metoprolol continuously over the dosage interval [96,103].

The group of patients that requires special attention related to effective pharmacotherapy and has a particular need for taking MR formulations are children suffering from ADHD. MR drug dosage forms allow for extended delivery with less variability throughout the day, improved tolerability and less frequent administration ensuring convenience and adherence, which is important in this patients' group therapy. Therefore, some pharmaceutical preparations utilizing MR have been introduced for treatment of children with ADHD (Table 2). An interesting example of multiparticulate drug delivery system is SODAS® (Spheroidal Oral Drug Absorption System). Based on the production of controlled release beads, it is possible to provide a number of release profiles, including immediate and sustained release, giving rise to a fast onset of action, which is maintained for 24 h (Figure 7) [104]. Ritalin LA® (methylphenidate hydrochloride) is an extended-release capsule with a bi-modal release profile based on SODAS® system with paediatric license (Table 2). The preparation contains 50% of immediate release beads and 50% extended release beads covered by the polymer overcoat. The first peak in its bimodal profile occurs after 1 to 3 h and the second peak is approximately 6 h post dosage, therefore it is designed to be effective throughout the school day [94]. The second example bases on SODAS® technology is Focalin® XR (Table 2). Similar to the above description, each capsule contains 50% immediate release beads of methylphenidate and 50% extended release beads covered by a polymer, with the difference that the beads can be sprinkled in food [81].

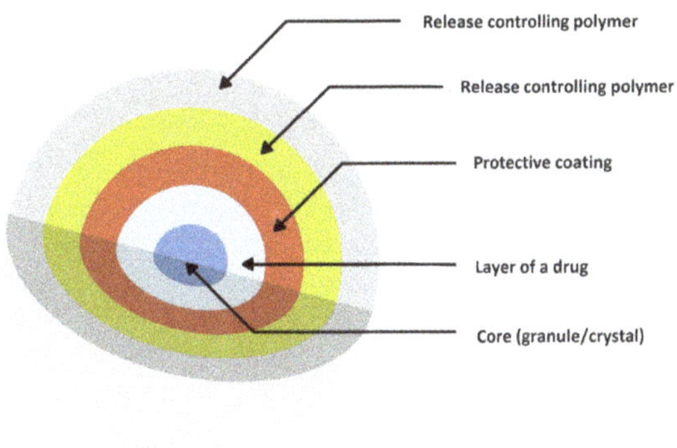

Figure 7. Scheme illustration of SODAS® delivery system modified from Elan drug technologies [105].

The other example of MR formulation designed for children with ADHD is Adderall® XR administered once daily (Table 2). Each capsule contains a 50:50 ratio of immediate and delayed-release beads. Diffucaps system (utilized in Metadate CD®) comprises both immediate release (30%) and extended release (70%) beads. Diffucaps is multiparticulate system, where drug profiles are created by layering a drug onto a neutral core (e.g., sugar spheres, crystals, or granules) followed by the application of a rate-controlling, functional membrane (Figure 8). The physicochemical characteristics of coating materials (being water soluble/insoluble, pH dependent/independent) are conditional on individual drug features. Obtained beads are small in size, approximately 1 mm or less. By incorporating beads with differing drug release properties, combined release profiles can be achieved. Metadate CD® designed to provide efficacy throughout the school day is available in six capsule strengths (with possibility of sprinkle over food) [86].

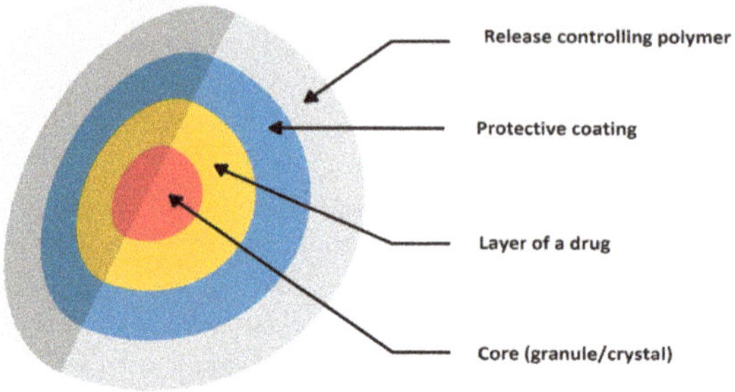

Figure 8. Scheme illustration of Diffucaps® bead technology modified from Weil [106].

3.3. Minitablets

Minitablets constitute a dosage form that ensures more dose flexibility and ease of drug administration in various children age groups than conventional tablets or capsules. Their main advantage is its small size ranging from 1 to 3 mm, with an average mass from 5 to 25 mg and possibility of adjusting single dose by counting the proper amount of minitablets. Minitablets are produced in the same way as conventional ones, by compression using tableting technology with single or multi-punch. They may appear as individual dosage form or could be delivered in capsules or sachets (Figure 9). The clinical researches have demonstrated that 2 mm tablets can be easily used in six-month old infants and 4 mm in children above one year of age, while orodispersible 2 mm tablets can be administered already for preterm neonates. Moreover, Klingmann et al. proved that children above six-months show greater acceptance of minitablets than syrups. Furthermore, they might provide combined release patterns [25,107–109]. Examples of commercially available MR minitablets are: Orfiril Long® and Pancrease MT® (Table 2). Orfiril Long® is provided in hard capsule or single sachet [89]. Pancrease MT® is also provided in minitablets form enclosed inside the capsule [90].

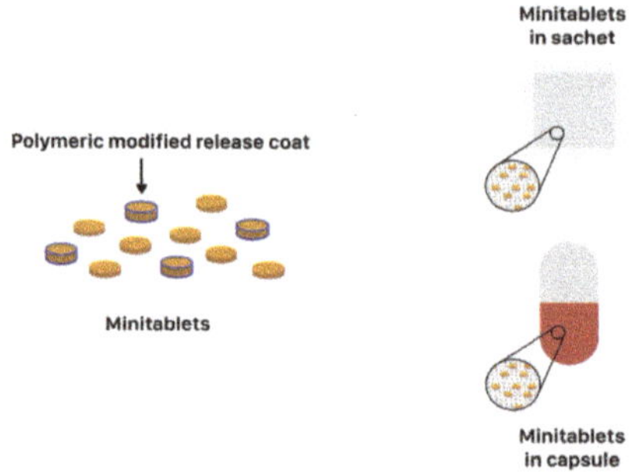

Figure 9. Scheme illustration of minitablets placed in capsule and sachet.

3.4. MR Orodispersible Formulations

MR may be also ensured by rapidly disintegrating forms like orodispersible tablets and films. They constitute a relatively new and dynamically developing group of MR formulations. Preparing tablets with such a short disintegration time is a particularly beneficial feature, especially for young patients [110–112]. Prevacid® SoluTab™ is an instance of delayed release orally disintegrating tablet with compressed MR pellets containing lansoprazole (Table 2). An interesting example of immediate and extended release profiles combined in single dosage form are two commercially available medications namely Adzenys XR®-ODT with amphetamine (FDA approval in January 2016) and Cotempla® XR-ODT with methylphenidate group (approval in June 2017). They are indicated for ADHD treatment in children from 6 to 17 years of age. These formulations are available in wide range of doses:

Adzenys®—3.1 mg, 6.3 mg, 9.4 mg, 12.5 mg, 15.7 mg, 18.8 mg, and Cotempla®—8.6 mg, 17.3 mg, and 25.9 mg allowing dosing in a wide age group. These products are obtained by XR-ODT (extended-release orally disintegrating tablet) technology and they dissolve quickly in the mouth (according to the FDA guidelines—up to 30 s or less) so that it can be easily swallowed. The technology utilized in the tablet uses two different types of microparticles: immediately released in 25% (Cotempla®) or 50% (Adzenys®) and slowly released with the other 75% and 50%, respectively, throughout the day. Two different polymers' coatings are applied to the XR microparticles: Interior polymer coating as diffusion barrier (ethylcellulose) and exterior polymer coating being pH dependent (methacrylic acid). The technology allows for a drug to be incorporated into orodispersible dosage form using ion resin technology [72,73,77,78,113].

Orodispersible films are defined as thin polymeric films supposed to disintegrate in the oral cavity within seconds (there is no detailed monography in any Pharmacopoeia; FDA indicated 30 s or less as disintegration time). Their size and shape resemble postage stamp, with a thickness ranges from 12 to 100 μm and a surface from 2 cm^2 to 8 cm^2 (in the literature, the most frequently encountered dimension is 3×2 cm^2, 2×2 cm^2) [113–115]. They are usually manufactured by solvent casting, hot melt extrusion, semisolid casting method, rolling method or electrospinning. There are a number of preparations dedicated specifically for children in this form, but with immediate release, e.g., Pedia-Lax® Quick Dissolve Strip, Orajel™ Kids Sore Throat Relief Strips, IvyFilm Kiddies® [116]. In contrast to fast dissolving films, MR release might be obtained by the mucoadhesive effect, which underlies buccal films preparation allowing for prolonged release at the application place. Buccal films are particularly addressed for pre-school and school children since they are thin, adaptable to the mucosal surface and able to offer an exact and flexible dose. Abruzzo et al. have designed buccal films for propranolol hydrochloride (β-blocker used in paediatric patients primarily for the treatment or prevention of cardiac arrhythmias and hypertension) administration. Polymeric layer was prepared by casting and drying of film-forming polymers' solutions (polyvinylpyrrolidone or polyvinylalcohol with addition of gelatin or chitosan). As a second layer applied onto the primary one in order to obtain prolonged drug delivery and mask its bitter taste, ethylcellulose was utilized. The formulation is intended for children ≥ 2 years of age and body weight around 12 kg [117].

4. Excipients Utilized in MR—Safety of Use in Children

The safety of children's pharmacotherapy depends not only on the drug substance itself, but also on ingredients forming medicines (excipients). The choice of excipients is a crucial factor in the development of medicinal products for paediatric use. Excipient safely and commonly used in adults' therapy could be harmful for children, e.g., ethanol or propylene glycol cause neurotoxicity; some preservatives like benzyl alcohol, sodium benzoate may lead to allergic reactions. Additionally, a questionable issue is the utilization of parabens in the paediatric population. The most common polymers used for obtaining MR formulations are cellulose derivatives (especially hypromellose and ethylcellulose) utilized as coating polymers, taste masking agents, or e.g. a microparticles matrix (Table 3) [10,68,118–122]. Ethylcellulose as a biocompatible and gastro resistant polymer is used for preparaing sustained release syrup with hydrocodone and chlorpheniramine indicated for children above six-years (Tussionex®, Table 1) [40]. Ethylcellulose (in aqueous suspension form) was also used for formulating MR microspheres with mirabegron by the spray drying technique, so oral sustained-release suspensions were obtained [51,52]. Emami et al. used hypromellose for formulating MR suspensions with theophylline [123]. Most cellulose derivatives are generally recognized as safe (GRAS) to use in children pharmacotherapy (Table 3). However, for most excipients used in pediatric formulations (e.g., ethylcellulose acetate, methacrylic acid copolymers, and lauryl sulfate), safety data are still limited. The EMA guidelines serve as a database for assessing the safety profile of excipients, so the presented data must be actualized, related to the age group, and relevant to the maximum daily exposure uptake [10,68,118].

Table 3. Safety data of excipients used in paediatric modified release preparation [10,68].

Excipient		Paediatric Safety Data Use	Main Function in Formulation
Cellulos derivatives	cellulose acetate	NA *	MR
	cellulose acetate phthalate	NA *	MR
	carmellose sodium	yes	suspending agent
	ethylcellulose	yes	MR taste masking
	hypromellose	yes	suspending agent MR taste masking
	methylcellulose	yes	suspending agent
ion exchange resin		NA *	drug/polymer complexation
methacrylic acid copolymers		NA *	MR
sodium polystyrene sulfate		NA *	drug/polymer complexation MR
sodium alginate		NA *	MR
calcium sulfate		NA *	MR
lauryl sulfate		NA *	MR
polyvinyl alcohol		NA *	MR

* NA: no data available.

5. Novel Technologies 3D Printing for MR Formulations

Pharmaceutical applications of 3D printing have increased over the past years. Printing technologies are cutting edge methods in tablets and films manufacturing. Inject printing is experimentally used for drug printing on different matrices, flexographic printing is employed to coat the drug loaded substrate with a polymeric film. An increasing number of researchers are employing 3D printing technologies to develop oral dosage forms with MR. There is a new approach using a non-contact printing system that incorporates both piezo-electric and solenoid valve-based inkjet printing technologies to deliver both drug and excipients onto the matrix. The main ideas of using this type of technology revolve around the fact that printing technologies would allow to develop pharmaceuticals in a tailored manner to meet some of the envisaged personalization needs of patients for potential use in the paediatric population [124,125]. Recently, 3D printing was utilized to create a multi-active solid dosage form, containing five different drugs within the same capsule, which were autonomously controlled with two separate release profiles, called Polypill® [126]. It would be especially useful for all patients who are taking medicines many times a day. The feasibility of 3D printing coupled with hot melt extrusion to prepare paediatric medicines that can be consumed easily by children from 2–11 years old was introduced. The medicines were designed in such a way to imitate 'candy-like' chewable tablets. For the purposes of the study, Starmix® (HARIBO PLC, UK) formulations were printed using indometacine as model drug and hypromellose acetate succinate as the polymeric excipient [127]. The culmination of 3D printing applications in oral dosage forms is the FDA approval (in August 2015) of 3D printed drug product called Spritam® (levetiracetam)—tablets for oral suspension manufactured by using the ZipDose® technology based on a powder bed (liquid 3D printing patented technology). Spritam® became the first 3D-printed drug approved by FDA as a prescription adjunctive therapy for treating patients with epilepsy for children from the age of four-years old (or 20 kg up). The technology enables immediate disintegration of the drug with a sip of water, making it easy for the patients to administer the drug, even in high doses. ZipDose® technology creates a porous formulation using 3D process that binds powders without compression. The method enables delivery of high drug doses of up to 1 g. Drugs formulated using ZipDose® technology are specially designed for people with swallowing difficulties (drug dosage form disintegrate within approximately 11 s according to manufacturer) and those who skip regular drug doses, resulting in ineffective treatment outcomes—the children population perfectly fit in [128,129].

6. Conclusions

Pharmacotherapy of children's population is an important matter in a field of modern pharmaceutical technology. The main problems regarding children treatment result from the diversity of the paediatric population, as well as a relatively small number of appropriate dosage forms, including modern ones. Creating children-made formulation is challenging but an essential task in formulating an appropriate dosage form, which should be adjustable for a wide range of ages, palatable, easy to administer, but first of all safe and effective. Key aspects in modern formulations involve development of novel MR formulations (minitablets, pellets, MR oral liquid formulations) when considering chronic diseases that affects children and minimizing the dose frequency. Simultaneously, safety of excipients and child's acceptability should be kept in mind. The limited number of MR formulation in the market (especially for children under six-years) arise from the high cost of technologies and lack of relevant clinical trials in the paediatric population. Therefore, new regulations and additional funding opportunities, as well as innovative collaborative research initiatives should be constantly developed.

Author Contributions: Conceptualization, M.T., K.W. (Katarzyna Wasilewska) and K.W. (Katarzyna Winnicka); Writing-Review & Editing, M.T., K.W. (Katarzyna Wasilewska) and K.W. (Katarzyna Winnicka); Visualization, M.T., K.W. (Katarzyna Wasilewska); Supervision, K.W. (Katarzyna Winnicka); Project Administration, M.T.; Funding Acquisition, K.W. (Katarzyna Wasilewska) and K.W. (Katarzyna Winnicka).

Funding: This research was funded by Medical University of Bialystok grant N/ST/MN/16/002/2215, N/ST/MN/17/001/2215 and N/ST/MN/18/002/2215.

Conflicts of Interest: The authors declare no conflict of interest.

Abbreviations

MR	Modified Release
EMA	European Medicines Agency
BPCA	The Best Pharmaceuticals for Children
PIP	Paediatric Investigation Plan
STEP	Step&Toxicity of Excipients for Paediatric Patients Databaseb
FDA	Food and Drug Administration
ADHD	Attention-Deficit Hyperactivity Disorder
NA	No Data Available
HIV	Human Immunodeficiency Virus
MultiP	Multiparticulates
MUPS	Multi-Unit Pellet System
GERD	Gastro-Esophageal Reflux Disease
GRAS	Generally Recognized as Safe

References

1. Gore, R.; Chugh, P.K.; Tripathi, C.D.; Lhamo, Y.; Gautam, S. Paediatric off-label and unlicensed drug use and its implications. *Curr. Clin. Pharmacol.* **2017**, *12*, 18–25. [CrossRef] [PubMed]
2. McIntyre, J.; Conroy, S.; Avery, A.; Corns, H.; Choonara, I. Unlicensed and of label prescribing of drugs in general practice. *Arch. Dis. Child.* **2000**, *83*, 498–501. [CrossRef] [PubMed]
3. Frattarelli, D.A.; Galinkin, J.L.; Green, T.P.; Johnson, T.D.; Neville, K.A.; Paul, I.M.; Van Den Anker, J.N. American Academy of Paediatrics Committee on Drugs.: Off-label use of drugs in children. *Paediatrics* **2014**, *133*, 563–567.
4. Salunke, S.; Tuleu, C. Formulating better medicines for children—Setting the pace for the future. *Int. J. Pharm.* **2013**, *457*, 308–309. [CrossRef] [PubMed]

5. List of drugs for which paediatric studies are needed. Federal Register Notices; 2003; p. 68. Available online: https://bpca.nichd.nih.gov/prioritization/status/Documents/Federal_Register_01-21-2003.pdf (accessed on 5 April 2019).
6. Reflection paper: Formulations of choice for the paediatric population, EMEA/CHMP/PEG/194810/2005. Available online: http://www.ema.europa.eu/docs/en_GB/document_library/Scientific_guideline/2009/09/WC500003782.pdf (accessed on 5 April 2019).
7. Regulation (EC) No 1901/2006. Available online: https://ec.europa.eu/health//sites/health/files/files/eudralex/vol-1/reg_2006_1901/reg_2006_1901_en.pdf (accessed on 5 April 2019).
8. Salunke, S.; Liu, F.; Batchelor, H.; Walsh, J.; Turner, R.; Ju, T.R.; Tuleu, C. European Paediatric Formulation Initiative (EuPFI)-formulating ideas for better medicines for children. *AAPS PharmSciTech.* **2017**, *18*, 257–262. [CrossRef] [PubMed]
9. State of paediatric medicines in the EU 10 years of the EU paediatric regulation, report from the Commission to the European Parliament and the Council, COM (2017) 626. Available online: https://ec.europa.eu/health/sites/health/files/files/paediatrics/docs/2017_childrensmedicines_report_en.pdf (accessed on 5 April 2019).
10. Safety & Toxicity of Excipients for Paediatrics, STEP database. Available online: http://www.eupfi.org/step-database-info/ (accessed on 5 April 2019).
11. Breslow, L.H. The best pharmaceuticals for children act of 2002: The rise of the voluntary incentive structure and congressional refusal to require paediatric testing. *Harvard J. Legis.* **2003**, *40*, 133–193. [PubMed]
12. The BPCA priority list of needs in paediatric therapeutics. Federal Register Notices; 2014; 79. Available online: http://www.gpo.gov/fdsys/pkg/FR-2014-08-25/html/2014-20156.htm (accessed on 5 April 2019).
13. Best Pharmaceuticals for Children Act (BPCA) Paediatric Formulation Initiative (PFI) Working Meeting. 6–7 December 2005. Available online: https://bpca.nichd.nih.gov/collaborativeefforts/documents/pfi_meeting_12-06-2005.pdf (accessed on 5 April 2019).
14. Best Pharmaceuticals for Children Act Paediatric Formulations Initiative Workshop. 1–2 November 2011. Available online: https://bpca.nichd.nih.gov/collaborativeefforts/Documents/pfi_workshop_11-1-2011.pdf (accessed on 5 April 2019).
15. Guideline on Pharmaceutical Development of Medicines for Paediatric Use, EMA/CHMP/QWP/805880/2012 Rev. 2. 2013. Available online: http://www.ema.europa.eu/docs/en_GB/document_library/Scientific_guideline/2013/07/WC500147002.pdf (accessed on 5 April 2019).
16. Better Medicines for Children from Concept to Reality PROGRESS REPORT ON THE PAEDIATRIC REGULATION (EC) N°1901/2006. Available online: https://ec.europa.eu/health/sites/health/files/files/paediatrics/2013_com443/paediatric_report-com%282013%29443_en.pdf (accessed on 5 April 2019).
17. Joseph, P.D.; Craig, J.C.; Caldwell, P.H.Y. Clinical trials in children. *Br. J. Clin. Pharmacol.* **2015**, *79*, 357–369. [CrossRef] [PubMed]
18. Swain, T.R. Clinical trials for children: Some concerns. *Indian J. Pharmacol.* **2014**, *46*, 145–146. [CrossRef]
19. Batchelor, H.K.; Marriott, J.F. Formulations for children: Problems and solutions. *Br. J. Clin. Pharmacol.* **2015**, *79*, 405–418. [CrossRef] [PubMed]
20. Lopez, F.L.; Ernest, T.B.; Tuleu, C.; Gul, M.O. Formulation approaches to paediatric oral drug delivery: Benefits and limitations of current platforms. *Expert Opin. Drug Deliv.* **2015**, *12*, 1727–1740. [CrossRef]
21. Richey, R.H.; Shah, U.U.; Peak, M.; Craig, J.V.; Ford, J.L.; Barker, C.E.; Nunn, A.J.; Turner, M.A. Manipulation of drugs to achieve the required dose is intrinsic to paediatric practice but is not supported by guidelines or evidence. *BMC Pediatr.* **2013**, *13*, 81. [CrossRef] [PubMed]
22. Richey, R.H. The Manipulation of Dosage Forms of Medications, With the Aim of Achieving the Required Dose, for Administration to Children. Ph.D. Thesis, University of Liverpool, Liverpool, UK, 2013. Available online: https://livrepository.liverpool.ac.uk/15475/4/RicheyRob_May2013_15475.pdf (accessed on 5 April 2019).
23. Spomer, N.; Klingmann, V.; Stoltenberg, I.; Lerch, C.; Meissner, T.; Breitkreutz, J. Acceptance of uncoated mini-tablets in young children: Results from a prospective exploratory cross-over study. *Arch. Dis. Child.* **2012**, *97*, 283–286. [CrossRef] [PubMed]
24. Klingmann, V.; Spomer, N.; Lerch, C.; Stoltenberg, I.; Frömke, C.; Bosse, H.M.; Breitkreutz, J.; Meissner, T. Favorable acceptance of mini-tablets compared with syrup: A randomized controlled trial in infants and preschool children. *J. Pediatr.* **2013**, *163*, 1728–1732. [CrossRef] [PubMed]
25. Klingmann, V.; Seitz, A.; Meissner, T.; Breitkreutz, J.; Moeltner, A.; Bosse, H.M. Acceptability of uncoated mini-tablets in neonates-a randomized controlled trial. *J. Pediatr.* **2015**, *167*, 893–896. [CrossRef] [PubMed]

26. Kluk, A.; Sznitowska, M.; Brandt, A.; Sznurkowska, K.; Plata-Nazar, K.; Mysliwiec, M.; Kaminska, B.; Kotłowska, H. Can preschool-aged children swallow several minitablets at a time? Results from a clinical pilot study. *Int. J. Pharm.* **2015**, *485*, 1–6. [CrossRef]
27. Feldman, M.; Bélanger, S. Extended-release medications for children and adolescents with attention-deficit hyperactivity disorder. *Paediatr. Child Health.* **2009**, *14*, 593–597. [CrossRef] [PubMed]
28. Van Riet-Nales, D.A.; Schobben, A.F.A.M.; Vromans, H.; Egberts, T.C.G.; Rademaker, C.M.A. Safe and effective pharmacotherapy in infants and preschool children: Importance of formulation aspects. *Arch. Dis. Child.* **2016**, *101*, 662–669. [CrossRef]
29. Milap, C.N.; Loyd, V.A. Extemporaneous drug formulations. *Clin. Ther.* **2008**, *30*, 2112–2119.
30. Riet-Nalesa, D.A.; Ferreira, J.A.; Schobben, A.F.A.M.; Neef, B.J.; Egberts, T.C.G.; Rademaker, C.D.A. Methods of administering oral formulations and child acceptability. *Int. J. Pharm.* **2015**, *491*, 261–267. [CrossRef]
31. Loyd, V.A. Dosage form design and development. *Clin. Ther.* **2008**, *30*, 2102–2111.
32. Flament, M.P. Extended release dosage: Recent advances and potential in paediatric medicine. *Ther. Deliver.* **2016**, *7*, 197–199. [CrossRef] [PubMed]
33. Ivanovska, V.; Rademaker, C.M.A.; Dijk, L.; Mantel-Teeuwisse, A.K. Paediatric drug formulations: A review of challenges and progress. *Paediatrics* **2014**, *134*, 361–372. [CrossRef] [PubMed]
34. Khan, A.Y.; Talegaonkar, S.; Iqbal, Z.; Ahmed, F.J.; Khar, R.K. Multiple emulsions: An overview. *Curr. Drug Deliv.* **2006**, *3*, 429–443. [CrossRef] [PubMed]
35. Kathpalia, H.; Phadke, C. Novel oral suspensions: A review. *Curr. Drug Deliv.* **2014**, *11*, 338–358. [CrossRef] [PubMed]
36. Full Prescribing Information Delsym®. Available online: https://www.fda.gov/ucm/groups/fdagov-public/@fdagov-drugs-gen/documents/document/ucm085592.pdf (accessed on 5 April 2019).
37. Full Prescribing Information Dyanavel XR®. Available online: http://dyanavelxr.com/pdfs/pi.pdf (accessed on 5 April 2019).
38. Full Prescribing Information MST® Continous®. Available online: https://www.medicines.org.uk/emc/product/1015/smpc#PHARMACOKINETIC_PROPS (accessed on 5 April 2019).
39. Full Prescribing Information Quillivant XR®. Available online: https://www.fda.gov/downloads/Drugs/DrugSafety/DrugShortages/UCM602794.pdf (accessed on 5 April 2019).
40. Tussionex® Drug Information: Description, User Reviews, Drug Side Effects, Interactions—Prescribing Information. Available online: https://www.rxlist.com/tussionex-drug.htm (accessed on 5 April 2019).
41. Full Prescribing Information Zmax®. Available online: https://www.accessdata.fda.gov/drugsatfda_docs/label/2012/050797s016lbl.pdf (accessed on 5 April 2019).
42. Saber/Oradur Technology on Durect Platform Information. Available online: http://www.durect.com/science-technologies/long-acting-injectables/saber-long-acting-injectables/ (accessed on 5 April 2019).
43. Clinical trials of bupivacaine. Available online: https://clinicaltrials.gov/ct2/show/NCT01052012 (accessed on 5 April 2019).
44. Ekelund, A.; Peredistijs, A.; Grohs, J.; Ellis, D.; Verity, N.; Rasmussen, S. SABER-Bupivacaine Reduces Postoperative Pain and Opioid Consumption Following Arthroscopic Subacromial Decompression. Personal Communication. 2016. Available online: http://www.durect.com/files/1914/6376/6227/EFORT2016_PostopPainReduction_Shoulder.pdf (accessed on 5 April 2019).
45. ORADUR®-Methylphenidate ER Information. Available online: http://www.durect.com/pipeline/development/oradur-methylphenidate-er/ (accessed on 5 April 2019).
46. Clinical Trials of Methylphenidate. Available online: https://clinicaltrials.gov/ct2/show/NCT02450890 (accessed on 5 April 2019).
47. Kasashima, Y.; Uchida, S.; Yoshihara, K.; Yasuji, T.; Namiki, N. Oral sustained-release suspension based on a lauryl sulfate salt/complex. *Int. J. Pharm.* **2016**, *515*, 677–683. [CrossRef] [PubMed]
48. Available online: https://www.ema.europa.eu/en/documents/variation-report/betmiga-h-c-2388-p46-0008-epar-assessment-report_en.pdf (accessed on 5 April 2019).
49. REMOXY® (ORADUR®-Oxycodone) ER Capsules Information. Available online: http://www.durect.com/pipeline/development/remoxy/ (accessed on 5 April 2019).
50. Clinical Trials of Oxycodone. Available online: https://clinicaltrials.gov/ct2/show/NCT01559701 (accessed on 5 April 2019).

51. Pain Therapeutics Announces Feedback From Recent Meeting with FDA on Remoxy. Available online: https://www.drugs.com/nda/remoxy_er_190205.html (accessed on 5 April 2019).
52. Singh, I.; Rehni, A.K.; Kalra, R.; Joshi, G.; Kumar, M.; Aboul-Enein, H.Y. Ion exchange resins: Drug delivery and therapeutic applications. *FABAD J. Pharm. Sci.* **2007**, *32*, 91–100.
53. Singh, M.N.; Hemant, K.S.Y.; Ram, M.; Shivakumar, H.G. Microencapsulation: A promising technique for controlled drug delivery. *Res. Pharm. Sci.* **2010**, *5*, 65–77.
54. Kojima, Y.; Ohta, T.; Shiraki, K.; Takano, R.; Maeda, H.; Ogawa, Y. Effects of spray drying process parameters on the solubility behavior and physical stability of solid dispersions prepared using a laboratory-scale spray dryer. *Drug Dev. Ind. Pharm.* **2013**, *39*, 1484–1493. [CrossRef]
55. Liu, W.; Chen, X.D.; Selomuyla, C. On the spray drying of uniform functional microparticles. *Particuology* **2015**, *22*, 1–12. [CrossRef]
56. Vehring, R. Pharmaceutical particle engineering via spray drying. *Pharm. Res.* **2008**, *25*, 999–1022. [CrossRef]
57. Cal, K.; Sollohub, K. Spray Drying Technique. I: Hardware and Process Parameters. *J. Pharm. Sci.* **2009**, *99*, 575–586. [CrossRef] [PubMed]
58. Madan, M.; Bajaj, A.; Lewis, S.; Udupa, N.; Baig, J.A. In situ forming polymeric drug delivery systems. *Indian J. Pharm. Sci.* **2009**, *71*, 242–251. [CrossRef] [PubMed]
59. Binu, C.; Surajpal, V. Preparation and evaluation of novel in situ gels containing acyclovir for the treatment of oral herpes simplex virus infections. *ScientificWorldJournal* **2014**, 1–7.
60. Makwana, S.B.; Patel, V.A.; Parmar, S.J. Development and characterization of in-situ gel for ophthalmic formulation containing ciprofloxacin hydrochloride. *Results Pharma. Sci.* **2016**, *6*, 1–6. [CrossRef] [PubMed]
61. Alami-Milani, M.; Zakeri-Milani, P.; Valizadeh, H.; Salehi, R.; Salatin, S.; Naderinia, A.; Jalvehgari, M. Novel pentablock copolymers as thermosensitive self-assembling micelles for ocular drug delivery. *Adv. Pharm. Bull.* **2017**, *7*, 11–20. [CrossRef] [PubMed]
62. Sarada, K.; Firoz, S.; Padmini, K. In-situ gelling system: A review. *Int. J. Curr. Pharm. Res.* **2015**, *5*, 76–90.
63. Jaya, R.K.K.; Selvadurai, M.; Sokkalingam, A.D. A review: Polymeric in-situ gel system. *Res. Rev. J. Pharm. Pharm. Sci.* **2013**, *2*, 1–7.
64. Van Tomme, S.R.; Storm, G.; Hennink, W.E. In situ gelling hydrogels for pharmaceutical and biomedical applications. *Int. J. Pharm.* **2008**, *355*, 1–18. [CrossRef]
65. Orubu, E.S.F.; Tuleua, C. Medicines for children: Flexible solid oral formulations. *Bull. World Health Organ.* **2017**, *95*, 238–240. [CrossRef]
66. Nokhodchi, A.; Raja, S.; Patel, P.; Asare-Addo, K. The role of oral controlled release matrix tablets in drug delivery systems. *Bioimpacts.* **2012**, *2*, 175–187.
67. Mondal, N. The role of matrix tablet in drug delivery system. *Int. J. App. Pharm.* **2017**, *10*, 1–6. [CrossRef]
68. Development of Paediatric Medicines: Points to Consider in Formulation. Available online: https://www.who.int/childmedicines/partners/SabineKopp_Partners.pdf (accessed on 5 April 2019).
69. Full Prescribing Information Lamictal® XR. Available online: https://www.fda.gov/downloads/drugs/developmentapprovalprocess/developmentresources/ucm215664.pdf (accessed on 5 April 2019).
70. Full Prescribing Information Aciphex® Sprinkle™. Available online: https://www.accessdata.fda.gov/drugsatfda_docs/label/2014/020973s035204736s005lbl.pdf (accessed on 5 April 2019).
71. Full Prescribing Information Adderall XR®. Available online: https://www.accessdata.fda.gov/drugsatfda_docs/label/2013/021303s026lbl.pdf (accessed on 5 April 2019).
72. Adzenys XR-ODT® Tablets Information. Available online: https://www.adzenysxrodt.com/ (accessed on 5 April 2019).
73. Full Prescribing Information Adzenys XR-ODT®. Available online: https://www.accessdata.fda.gov/drugsatfda_docs/label/2017/204326s002lbl.pdf (accessed on 5 April 2019).
74. Full Prescribing Information Azulfidine®. Available online: https://www.accessdata.fda.gov/drugsatfda_docs/label/2009/007073s124lbl.pdf (accessed on 5 April 2019).
75. Full Prescribing Information Concerta®. Available online: https://www.accessdata.fda.gov/drugsatfda_docs/label/2017/021121s038lbl.pdf (accessed on 5 April 2019).
76. Full Prescribing Information Coreg CR®. Available online: https://www.accessdata.fda.gov/drugsatfda_docs/label/2009/022012s010s013lbl.pdf (accessed on 5 April 2019).
77. Cotempla XR-ODT® tablets information. Available online: https://www.cotemplaxrodt.com/ (accessed on 5 April 2019).

78. Full Prescribing Information Cotempla XR-ODT®. Available online: https://www.accessdata.fda.gov/drugsatfda_docs/label/2017/205489s000lbl.pdf (accessed on 5 April 2019).
79. Full Prescribing Information Creon. Available online: https://www.accessdata.fda.gov/drugsatfda_docs/label/2009/020725s000lbl.pdf (accessed on 5 April 2019).
80. Full Prescribing Information Finlepsin. Available online: http://www.ecopharm.bg/images/product/product_13_44_file2.pdf (accessed on 5 April 2019).
81. Full Prescribing Information Focalin™ XR. Available online: https://www.accessdata.fda.gov/drugsatfda_docs/label/2005/021802lbl.pdf (accessed on 5 April 2019).
82. GranuPAS® Granules Information. Available online: https://www.ema.europa.eu/en/medicines/human/EPAR/granupas-previously-para-aminosalicylic-acid-lucane (accessed on 5 April 2019).
83. Full Prescribing Information Kapvay®. Available online: https://www.accessdata.fda.gov/drugsatfda_docs/label/2010/022331s001s002lbl.pdf (accessed on 5 April 2019).
84. Full Prescribing Information Keppra®. Available online: https://www.accessdata.fda.gov/drugsatfda_docs/label/2009/021035s078s080,021505s021s024lbl.pdf (accessed on 5 April 2019).
85. Losec MUPS Capsules Information. Available online: https://www.medicines.org.uk/emc/files/pil.1493.pdf (accessed on 5 April 2019).
86. Full Prescribing Information Metadate CD®. Available online: https://www.accessdata.fda.gov/drugsatfda_docs/label/2009/021259s021lbl.pdf (accessed on 5 April 2019).
87. Full Prescribing Information Moxatag™. Available online: https://www.accessdata.fda.gov/drugsatfda_docs/label/2008/050813lbl.pdf (accessed on 5 April 2019).
88. Moxatag™ Tablets Information. Available online: http://www.moxatag.com/hcp_delivery.htm (accessed on 5 April 2019).
89. Orfiril Long Minitablets Information. Available online: https://ec.europa.eu/health/documents/community-register/2018/20180531140837/anx_140837_en.pdf (accessed on 5 April 2019).
90. Full Prescribing Information Pancrease MT®. Available online: https://www.accessdata.fda.gov/drugsatfda_docs/nda/2010/022523Orig1s000ClinPharmR.pdf (accessed on 5 April 2019).
91. Full Prescribing Information Pentasa. Available online: https://www.accessdata.fda.gov/drugsatfda_docs/label/2018/020049s031lbl.pdf (accessed on 5 April 2019).
92. Full Prescribing Information Prevacid®, Prevacid® Solutab™. Available online: https://www.accessdata.fda.gov/drugsatfda_docs/label/2012/020406s078-021428s025lbl.pdf (accessed on 5 April 2019).
93. Full Prescribing Information Prevacid®, Prevacid® Solutab™. Available online: https://www.accessdata.fda.gov/drugsatfda_docs/label/2006/020406s058,021281s017,021428s006lbl.pdf (accessed on 5 April 2019).
94. Full Prescribing Information Ritalin® LA. Available online: https://www.accessdata.fda.gov/drugsatfda_docs/label/2010/021284s010lbl.pdf (accessed on 5 April 2019).
95. Full Prescribing Information Tegretol® -XR. Available online: https://www.accessdata.fda.gov/drugsatfda_docs/label/2009/016608s101,018281s048lbl.pdf (accessed on 5 April 2019).
96. Full Prescribing Information Toprol-XL®. Available online: https://www.accessdata.fda.gov/drugsatfda_docs/label/2009/019962s038lbl.pdf (accessed on 5 April 2019).
97. Full Prescribing Information Viramune®. Available online: https://www.accessdata.fda.gov/drugsatfda_docs/label/2011/020636s039_020933s030lbl.pdf (accessed on 5 April 2019).
98. Martínez-Terán, M.E.; Hoang-Thi, T.H.; Flament, M.P. Multi-particulate dosage forms for paediatric use. *Pediatr. Ther.* **2017**, *7*, 314. [CrossRef]
99. Shajahan, A.A.; Chandewar, V.; Jaiswal, S.B. A flexible technology for modified-release drugs: Multiple-unit pellet system (MUPS). *J. Control Release* **2010**, *147*, 2–16.
100. Full Prescribing Information Prilosec (Omeprazole Magnesium) Delayed-Release Oral Suspension. Available online: https://www.fda.gov/downloads/AdvisoryCommittees/CommitteesMeetingMaterials/PaediatricAdvisoryCommittee/UCM610733.pdf (accessed on 5 April 2019).
101. Milić, J.; Radojković, B.; Jančić-Stojanović, B.; Drašković, J.; Miraševič, S.; Čalija, B. Investigation of omeprazole stability in oral suspensions for paediatric use prepared extemporaneously from omeprazole capsules. *Arh. Farm.* **2017**, *67*, 14–25.
102. Garg, S.; Svirskis, D.; Al-Kabban, M.; Farhan, S.; Komeshi, M.; Lee, J.; Liu, Q.; Naidoo, S.; Kairuz, T. Chemical stability of extemporaneously compounded omeprazole formulations: A comparison of two methods of compounding. *Int. J. Pharm. Compd.* **2009**, *13*, 250–253. [PubMed]

103. Moodley, K.; Pillay, V.; Choonara, Y.E.; du Toit, L.C.; Ndesendo, V.M.K.; Kumar, P.; Cooppan, S.; Bawa, P. Oral drug delivery systems comprising altered geometric configurations for controlled drug delivery. *Int. J. Mol. Sci.* **2012**, *13*, 18–43. [CrossRef] [PubMed]
104. Venkata, P.D.B. Spheroidal oral drug absorption system (SODAS). *J. Glob. Pharma Technol.* **2011**, *3*, 1–5.
105. Elan Drug Technologies. Spheroidal Drug Absorption System (SODAS®). Available online: http://www.elandrugtechnologies.com/oral_controlled_release/sodas (accessed on 5 April 2019).
106. Weil, A.J. Cyclobenzaprine extended-release: The difference is in the formulation. *Pharm. Times*. 2009. Available online: https://www.pharmacytimes.com/p2p/cyclobenzaprine-extended-release (accessed on 5 April 2019).
107. Kluk, A.; Sznitowska, M. Application properties of oral gels as media for administration of minitablets and pellets to paediatric patients. *Int. J. Pharm.* **2014**, *460*, 228–233. [CrossRef] [PubMed]
108. Preis, M. Orally Disintegrating films and mini-tablets—innovative dosage forms of choice for paediatric use. *APS PharmSciTech.* **2015**, *16*, 234–241. [CrossRef] [PubMed]
109. Klingmann, V. Acceptability of mini-tablets in young children: Results from three prospective cross-over studies. *AAPS PharmSciTech.* **2017**, *18*, 263–266. [CrossRef] [PubMed]
110. Zajicek, A.; Fossler, M.J.; Barrett, J.S.; Worthington, J.H.; Ternik, R.; Charkoftaki, G.; Lum, S.; Breitkreutz, J.; Baltezor, M.; Macheras, P.; et al. Report from the paediatric formulations task force: Perspectives on the state of child-friendly oral dosage form. *AAPS J.* **2013**, *15*, 1072–1081. [CrossRef] [PubMed]
111. Van Riet-Nales, D.A.; Kozarewicz, P.; Aylward, B.; de Vries, R.; Egberts, T.C.; Rademaker, C.M.; Schobben, A.F. Paediatric drug development and formulation design—An European perspective. *AAPS PharmSciTech.* **2017**, *18*, 241–249. [CrossRef] [PubMed]
112. FDA, CDER, Guidance for industry—orally disintegrating tablets. 2008. Available online: https://www.fda.gov/downloads/Drugs/.../Guidances/ucm070578.pdf (accessed on 5 April 2019).
113. *The European Pharmacopoeia*, 9th ed.; Council of Europe: Strasburg, France, 2016.
114. Visser, J.C.; Woerdenbag, H.J.; Hanff, L.M.; Frijlink, H.W. Personalized medicine in paediatrics: The clinical potential of orodispersible films. *AAPS PharmSciTech.* **2017**, *18*, 267–272. [CrossRef] [PubMed]
115. Wasilewska, K.; Winnicka, K. How to assess orodispersible film quality? A review of applied methods and their modifications. *Acta Pharm.* **2019**, *69*, 155–176. [CrossRef]
116. Hoffmann, E.M.; Breitenbach, A.; Breitkreutz, J. Advances in orodispersible films for drug delivery. *Expert Opin. Drug Deliv.* **2011**, *8*, 299–316. [CrossRef] [PubMed]
117. Abruzzo, A.; Nicoletta, F.P.; Dalena, F.; Cerchiara, T.; Luppi, B.; Bigucci, F. Bilayered buccal films as child-appropriate dosage form for systemic administration of propranolol. Formulation and characterization of fast dissolving buccal films: A review. *Int. J. Pharm.* **2017**, *5*, 257–265. [CrossRef] [PubMed]
118. Buckley, L.A.; Salunke, S.; Thompson, K.; Baer, G.; Fegley, D.; Turner, M.A. Challenges and strategies to facilitate formulation development of paediatric drug products: Safety qualification of excipients. *Int. J. Pharm.* **2018**, *536*, 563–569. [CrossRef] [PubMed]
119. Walsh, J. Excipients for the formulation of medicines for children. *Eur. Ind. Pharm.* **2012**, *13*, 14–16.
120. Nahata, M.C. Safety of "inert" additives or excipients in paediatric medicines. *Arch. Dis. Child. Fetal. Neonatal. Ed.* **2009**, *94*, 392–393. [CrossRef] [PubMed]
121. Christiansen, N. Ethanol exposure through medicines commonly used in paediatrics. *Arch. Dis. Child. Educ. Pract. Ed.* **2015**, *100*, 101–104. [CrossRef] [PubMed]
122. Trissel, L.A. *Stability of Compounded Formulations*, 5th ed.; American Pharmacists Association: Washington, DC, USA, 2012.
123. Emami, J.; Varshozas, J.; Ahmadi, F. Preparation and evaluation of a liquid sustained-release drug delivery system for theophylline using spray-drying technique. *Res. Pharm. Sci.* **2007**, *2*, 1–11.
124. Preis, M.; Breitkreutz, J.; Sandler, N. Perspective: Concepts of printing technologies for oral film formulations. Printing technologies in fabrication of drug delivery systems. *Int. J. Pharm.* **2015**, *305*, 78–584.
125. Daly, R.; Harrington, T.S.; Martin, G.D.; Hutchings, I.M. Inkjet printing for pharmaceutics—A review of research and manufacturing. *Int. J. Pharm.* **2015**, *30*, 554–567. [CrossRef] [PubMed]
126. Khaled, S.A.; Burley, J.C.; Alexander, M.R.; Yang, J.; Roberts, C.J. 3D printing of five-in-one dose combination polypill with defined immediate and sustained release profiles. *J. Control Release* **2015**, *10*, 308–314. [CrossRef] [PubMed]

127. Scoutaris, N.; Ross, S.A.; Douroumis, D. 3D Printed "Starmix" drug loaded dosage forms for paediatric applications. *Pharm Res.* **2018**, *16*, 34. [CrossRef] [PubMed]
128. Spritam® Drug Information. Available online: https://www.spritam.com/#/patient/about-spritam/what-is-spritam (accessed on 5 April 2019).
129. Full Prescribing Information Spritam®. Available online: https://www.accessdata.fda.gov/drugsatfda_docs/appletter/2016/207958Orig1s002ltr.pdf (accessed on 5 April 2019).

International Journal of
Molecular Sciences

MDPI

Review

Making Medicines Baby Size: The Challenges in Bridging the Formulation Gap in Neonatal Medicine

Fiona O'Brien [1], David Clapham [2], Kamelia Krysiak [1], Hannah Batchelor [3], Peter Field [4], Grazia Caivano [5], Marisa Pertile [5], Anthony Nunn [6] and Catherine Tuleu [4,*]

1. School of Pharmacy, Royal College of Surgeons in Ireland, 111 St Stephens Green Dublin 2, Ireland; fionaobrien@rcsi.ie (F.O.); kameliakrysiak@rcsi.ie (K.K.)
2. 14 Tailors, Bishops Stortford CM23 4FQ, UK; david.clapham@ntlworld.com
3. College of Medical and Dental Sciences, Institute of Clinical Sciences, University of Birmingham, Birmingham B15 2TT, UK; H.K.Batchelor@bham.ac.uk
4. University College London School of Pharmacy, 29-39 Brunswick Square, London WC1N 1AX, UK; peter.field@ucl.ac.uk
5. Chiesi Farmaceutici S.p.A. Largo Francesco Belloli 11/A—43122 Parma, Italy; g.caivano@chiesi.com (G.C.); m.pertile@chiesi.com (M.P.)
6. Department of Women's and Children's Health, University of Liverpool, Liverpool Women's Hospital, Liverpool L8 7SS, UK; a.j.nunn@liverpool.ac.uk

* Correspondence: c.tuleu@ucl.ac.uk; Tel.: +44-207-7535857

Received: 5 April 2019; Accepted: 24 May 2019; Published: 31 May 2019

Abstract: The development of age-appropriate formulations should focus on dosage forms that can deliver variable yet accurate doses that are safe and acceptable to the child, are matched to his/her development and ability, and avoid medication errors. However, in the past decade, the medication needs of neonates have largely been neglected. The aim of this review is to expand on what differentiates the needs of preterm and term neonates from those of the older paediatric subsets, in terms of environment of care, ability to measure and administer the dose (from the perspective of the patient and carer, the routes of administration, the device and the product), neonatal biopharmaceutics and regulatory challenges. This review offers insight into those challenges posed by the formulation of medicinal products for neonatal patients in order to support the development of clinically relevant products.

Keywords: neonates; formulation; product development; formulation development; oral; parenteral; topical; inhaled; intra nasal; biopharmaceutics; administration; excipient; NICU; device; medication error; dosage form

---

1. Introduction

Neonates are not small adults, and neither can they be classified as small children when it comes to medicinal products and their formulation development.

Neonates include term, post-term and preterm babies. The neonatal period for term and post-term newborn infants is defined as the day of birth plus 27 days. The neonatal period for preterm newborn infants is defined as the day of birth through to the expected date of delivery plus 27 days [1].

Each year, some 15 million babies arrive in the world, more than one in 10 babies are born prematurely, according to the report 'Born Too Soon: The Global Action Report on Preterm Birth' (2012) [2]. Even if born at term and ready to grow outside of the mother's womb, most organs and their functions are still immature. This immaturity of organ and function is more profound and impactful in preterm infants. For example, neonates have reduced gastric emptying, intestinal transit time and surface area, and transporter immaturity, which have relevance for oral drug delivery. Additionally, the skin barrier may not be fully formed, and respiratory function may be immature. A host of other physiological factors such as gastro intestinal (GI) pH, body surface to volume ratio, body fat to lean tissue ratio are also different and are known to change rapidly with time [3].

The underlying complexity of pharmacology and biopharmaceutics in neonates, especially in preterm babies, can lead to altered and variable pharmacokinetics and pharmacodynamics even compared to that in young babies [4–6]. In turn, this can lead to potential lack of efficacy or reduced safety of a medicinal product leading to additional special requirements for development of products for these age groups. This includes (amongst others) formulation aspects, dosage form choice, drug administration considerations and storage and handling advice.

The recent European Medicines Agency (EMA) report to the European Commission on the experience acquired as a result of the first 10 years application of the Paediatric Regulation, acknowledged that neonates still represent a particularly neglected paediatric subpopulation in the development of medicines despite the regulatory push [7].

In paediatric patients, there is still significant off-label use of medicines due to a lack of medicines developed and authorised for the specific needs of the very young [8,9]. This remains an even more significant problem in the neonatal population, due to the difficulty in conducting the necessary clinical trials in vulnerable subsets with lower patient numbers. Considering the extra challenges they present, pharmaceutical companies are not incentivised to formulate medicines for the neonatal population. In fact, trials open for recruiting neonates were included in only a quarter of all agreed paediatric investigation plans (PIPs), often at the request of the EMA Paediatric Committee (PDCO), due to an array of reasons: Lack of neonate specific indications, recruitment/enrollment challenges, lack of incentives [10]. There is a 'Catch 22' situation because to protect neonates, trials have been deferred so that safety and efficacy data are obtained in older age groups meanwhile the unmet needs gap widens, necessitating off-label use [11]. An international consortium of experts has produced a white paper to facilitate successful neonatal clinical trials of medicines and includes useful information on neonatal dosage forms and formulations [12].

The situation is not helped by a relative lack of relevant guidelines on the development of medicines for neonates (let alone those born prematurely). In terms of international guidance, it is only in the latest revision of the ICH E11 Guideline on Clinical Investigation of Medicinal Products in the Paediatric Population (2018) [1] that neonates are specifically mentioned as an age classification and paediatric subgroup. In these guidelines, a rather general and brief set of considerations are made in terms of formulation and around polypharmacy via parenteral routes of administration in the hospital setting. The capability to administer small volumes in relation to dosing error is also mentioned. This is insufficient considering the much wider and complex needs of neonates [13] and lack of clarity about how this limited guidance is to be translated into patient-centric product development, which meets with regulatory approval.

The common ground in current guidelines is that there are pointers to the lack of safety data on excipients and that available data generally does not apply to neonates requiring further justification including provision of non-clinical safety data [14,15].

The aim of this review is, therefore, to provide insights and factors to consider in order to assist those developing products for neonates, but with little or no neonatal medicine knowledge or paediatric formulation development background, to overcome the range of challenges posed by this patient group and so enhance the provision of clinically relevant products. This includes factors that differentiate the needs of preterm and term neonates from those of the older paediatric subsets, in terms of environment of care, ability to measure and administer the dose (from the patient, the routes of administration, the device and the product perspectives), formulation development, neonatal biopharmaceutics and regulatory challenges.

2. Formulation Considerations

2.1. Environment of Care

Neonates who receive treatment in hospital, will most often be located in the Neonatal Intensive Care Unit (NICU). The NICU admits high-risk premature and full-term neonates with serious medical or surgical conditions.

Factors which may necessitate neonatal admission include (modified from [16])

- Premature birth <37 weeks gestation
- Delayed birth >42 weeks gestation
- Birth weight <2500 g
- Concern about size, e.g., intrauterine growth restriction (IUGR)
- Medication or resuscitation required in the delivery room
- Birth defects, e.g., congenital heart defects, intraventricular haemorrhage, macrosomia, retinopathy of prematurity (ROP)
- Respiratory problems including RDS (respiratory distress syndrome) and BDP (bronchopulmonary dysplasia)
- Infection (including neonatal sepsis)
- Seizures
- Hypoglycemia
- Requiring additional support (extra oxygen or monitoring, body temperature control support, intravenous (IV) therapy, or medications) or specialized treatments (blood transfusion)
- Feeding issues
- Jaundice

Often multiple conditions exist concomitantly leading to a need for polypharmacy.

Neonates admitted to the NICU often require periods in specialised incubators to maintain optimum environmental conditions and may also be attached to electrocardiogram (ECG), oxygen saturation and blood pressure monitors (Figure 1). In addition, they may require respiratory support through oxygen supplementation or require phototherapy for jaundice [17]. The majority of pharmaceutical medicinal products administered to neonates are liquid oral or parenteral formulations [18]. In this population, the environment of care is critical with the vast majority of IV infusions delivered in an intensive care setting where the additional environmental requirements of light, temperature and oxygen may impact the physicochemical stability of medication being delivered [19]. The effect of the environment of care should be a critical consideration in the formulation design of a parenteral product intended for this patient population and in use stability studies will be discussed in more detail below.

Neonates in incubators also provide some challenges for the provision of suitable nutrition and hydration requiring enteral or parenteral feeding. Given the limited number of access points and the requirement for polypharmacy that is common in this patient group, there are more opportunities for stability challenges, interactions and incompatibilities.

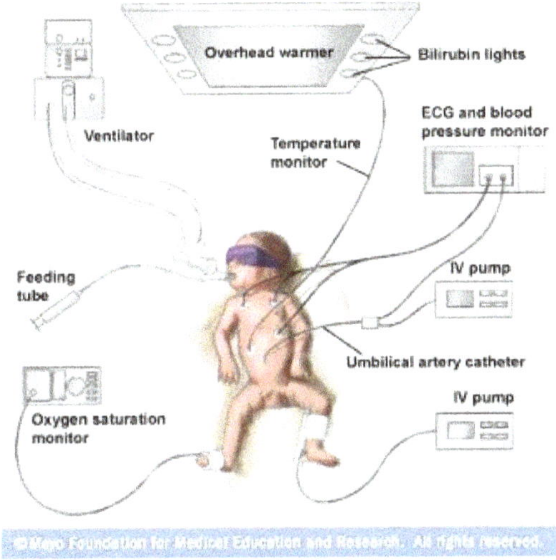

Figure 1. Schematic of NICU support for neonate (used with permission of Mayo Foundation for Medical Education and Research, all rights reserved).

2.2. *Ability to Dose: Patient (Developmental Age)/Physiological/Administration Routes Factors to Consider*

Some drugs can be administered to these patient groups orally, topically, via inhalation or indeed by any of the usual administration routes and the specific issues associated with the ability to dose via these routes is discussed under each of these sections below. However, the main route of administration in the neonatal patient is parenteral and in particular the IV route for seriously ill preterm and term neonates [20].

2.2.1. Parenteral Delivery

The neonatal IV infusion is a demanding process, involving vulnerable patients and complex IV administration apparatus. Venous access in neonates can be via a catheter with tip placed in the vena cava, a central venous catheter (CVC) (threaded through from a peripheral vein and known as a 'PICC' or peripherally-inserted central catheter, inserted through the umbilical vein (UVC) or placed surgically) or via a peripheral cannula or catheter (accessing smaller veins in the hands, arms and feet). A UVC remains in situ for up to two weeks after birth and PICC may last for several weeks whilst cannulas and catheters accessing the small peripheral veins may only be patent for hours or days [21].

All cannulas and catheters require scrupulous care to avoid blockage and infection. Since blood flow in central versus peripheral veins is greater, an administered drug is rapidly diluted [22]. Some irritant (chemotherapy, amiodarone, vasopressors [22] and hyperosmolar (glucose >12.5%, total parenteral nutrition (TPN)) [23] medications, may be indicated for central administration only. Many medications are compatible with a peripheral route of administration. However, some may cause phlebitis (e.g., dopamine, dobutamine, sodium bicarbonate, calcium gluconate) [24].

The choice of available catheters is summarised in Table 1 alongside characteristics and issues associated with each.

Tubing of small internal diameter is often used to reduce the dead space in apparatus for IV drug administration. Even so, drugs may be exposed to adverse temperatures and light from the point of administration until they reach the vascular system. This should be taken into account when assessing in-use stability as well as pharmacokinetics and clinical outcomes.

Table 1. Characteristic of available types of the vascular catheters (adapted from [21,22,24–26].

Type of Catheter	*Characteristics*	*Issues*
	Peripheral Venous Catheters	
Peripheral venous catheter	Application: Most IV drugs, isotonic IV fluids, blood transfusions Low flow rates Physicochemical irritation with some drugs results in phlebitis Dwell time: Most need to be removed within three days due to complications	Difficult to insert in the neonates due to the small and hard to visualize vessels
	Central Venous Catheters (CVC)	
Umbilical venous catheter (UVC)	Application: For diagnostic and therapeutic purposes—infusion of medication, TPN, hypertonic IV fluids, central venous pressure and venous blood gas monitoring, blood transfusions Dwell time: Up to 14 days	Suitable for neonates only as the umbilical vein remains for up to two weeks after birth UVC usually inserted within 12 hours of birth if indicated, for parenteral nutrition and/or inotropic support.
Peripherally inserted central catheter (PICC)	Application: Medication and IV fluid administration, TPN, blood sampling Suitable for irritant drugs Not suitable for large volume administration in emergency situations (for 28G 20 cm long catheter the max flow is 1 mL/min) Available multi-lumen catheters Made of polyurethane or silicone	Links the benefits of peripheral and central catheter PICC inserted at any time and used for all drugs (in conjunction with UVC helps reduce risk of drug incompatibilities).

An important formulation consideration associated with IV administration in preterm and term neonates is the volume of fluid that can be tolerated. Neonates, especially those delivered preterm, will have critical fluid and electrolyte requirements. At full term, a neonate has a fluid allowance of 100–140 mL/kg/day resulting in administration of 10–20 mL/hr, which includes all routes of administration [27]. In the case of severely fluid restricted neonates, IV fluids and drugs must be concentrated and so may need to be infused at flow rates as low as 0.02 mL/hr. Formulations for IV drugs and infusions should be designed with such requirements in mind and should take account of the sickest babies requiring multiple therapies, each with their own method of administration to be considered.

Generally, drugs administered intravenously should not be required to be administered in a fixed volume. It is preferable to investigate and report the minimum and maximum concentration at which the drug is sufficiently stable and to note any restrictions that this may pose for vascular access. For example, the need to use the central venous routes and slow rates of administration if higher concentrations might irritate or damage veins. The consequences of inadvertent extravascular administration should also be considered.

When drugs require dilution for administration the carrier fluid studied should include dextrose (5% and 10% *w/v*) as well as sodium chloride 0.9% w/v to maintain isotonicity. If dextrose solutions are suitable, this will help with sodium restriction and provide additional energy. Water for Injection should not usually be considered a suitable infusion fluid because of the potential risk of infusing hypotonic solutions, but information on stability and osmolarity may be useful if dextrose is inappropriate and sodium balance a problem.

Many neonatal patients in a critical care setting receive between 15 and 20 IV medications daily, the majority of these are unlicensed or used off label [4]. Lack of knowledge around the physicochemical incompatibilities of IV drugs in NICU and PICU settings often necessitates the use of a dedicated IV catheter in neonates and infants who have limited IV access [28,29]. Drug incompatibilities are often an underestimated aspect of clinical practice and are concerning in the neonatal population where a lower capability to compensate for adverse drug reactions may lead to higher morbidity and death [30,31]. This concern is exacerbated in neonates by the frequent requirements for polypharmacy, multiple infusions delivered through a single catheter due to limited vascular access, low infusion rates exposing drugs to longer interaction and the possibility of incomplete dissolution or precipitation of drug due to low volumes of drug solutions [32]. Realistically, limited venous access can result in little choice but to co-administer drugs.

A recent report highlighted the extent of the problem and reported that among medicines tested as a co-infusion of two drugs listed as a common NICU medication, only 4% of the combinations were fully compatible [28]. The compatibility of IV injections with other commonly administered drugs and infusion fluids should be studied when mixing at a 'Y'-site when the drug is likely to be used in the intensive care situation.

2.2.2. Oral Delivery

Medicine administration via the oral route to children less than two years of age can be difficult for both parents and children [13,33]. The main issue in administering oral medicines to neonates lies in their ability to effectively swallow the medicine. Typically, most oral processes are present from birth (rooting, lip, lateral tongue, mouth opening, biting, and emerging chewing behaviours). However, even oral syrups are not always fully swallowed when administered to neonates and infants [34]. This is further exacerbated in preterm neonates, where issues around swallowing and ADME (absorption, distribution, metabolism, and excretion) factors may be less well understood. Clinical evidence suggests, that the oral absorption process of a drug undergoes substantial changes after birth [35,36]. The impact of these physiological changes on oral drug therapy is currently unclear.

Historically oral liquid formulations have been used in neonates. Examples of commonly used oral medicines for neonates include vitamin D drops, analgesic suspensions (ibuprofen, paracetamol), antibiotics, glucose gel for treatment of hypoglycaemia and anti-reflux medicines. The first approved medicine to treat neonatal diabetes (Amglidia®) is a suspension formulation of glibenclamide [37].

EMA guidance [20] suggests, that the following oral formulation options are suitable from birth: Powders and granule (administered as a liquid) and oral liquid preparations (solution, suspension and oral drops).

It is important that appropriate strength medications are available for neonates to ensure that appropriate doses/volumes can be accurately measured and administered. The maximum recommended single dosing volume is 5 mL for children aged below five years and 10 mL for children aged below 10 years [38]. For neonates, dose volumes as low as 0.1 mL may be required.

The importance of accurate dosing to the youngest children has previously been highlighted [39]. Dosing devices for this age group mainly involves the use of an oral syringe or dropper though both have significant issues in terms of accuracy [40,41]. Therapeutic nipple shield [42,43] and medicine dispensing pacifiers [44,45] have been suggested for home use, but these are as yet experimental. In hospitalised patients, enteral tubes can be used for the administration of oral medicines where liquids (ideally solutions rather than suspensions) are the preferred dosage forms. Emulsion formulations can also theoretically be delivered via this route. One example is enteral nutrition or milk. There is a risk that the PK of a highly lipophilic drug, delivered via this route, may be altered if co-administered with an emulsion formulation or milk. Although during development compatibility of drug-enteral feed will be evaluated for new medicines, it is impossible to consider every feed available to review the likelihood of an interaction. Thus, there are circumstances where a drug–feed interaction may lead to PK variability as a result of a novel feed. Many medicines currently used in neonates are unlicensed for use in this population, and there is little or no information on their compatibility with milk or enteral feeds which may lead to a change in the anticipated PK. This is addressed in more detail in section (biopharmaceutical consideration).

Palatability considerations for neonates are more commonly associated with volume and texture of medicines in addition to taste [46,47]. However, some taste preferences seem to be innate (e.g., sweetness), and in extreme cases, bitter taste can lead to vomiting, so taste cannot be ignored even in neonates. New products that are designed for children are likely to be those that offer flexible dosing to target not only neonates but also older infants and children. Therefore, a product that meets the needs of all age groups will be developed which will address issues of palatability for infants and children.

2.2.3. Rectal Delivery

Rectal administration is more commonly used from infancy onwards [48]. This route suffers in neonates from unpredictable absorption which seems to be the limiting factor [36]. This is highly correlated with faecal incontinence/retention of the drug dose, which is inversely related to age/maturation. Solid dosage forms are usually better retained in the rectum than liquids. However, flexible and accurate dosing, which is an important product feature, cannot be delivered when suppositories are split at the point of administration [49]. This does not completely prevent the use of the rectal route to reduce the need for parenteral administration, or overcome oral administration restrictions (e.g., physiologically reduced gastrointestinal absorption, nausea, vomiting, seizures, nil by mouth), but safety and efficacy with appropriate bioavailability studies and patient size-appropriate product are paramount. In fact, recent studies (pain, patent ductus arteriosus closure) [50–52] demonstrated similar effectiveness for rectal delivery (mainly small enemas) as compared with IV or oral delivery even in very low birth weight preterm infants as well as being cheap and safe. Therefore, in resource-limited settings, rectally formulated drugs for pre-referral use could have great potential, e.g., neonatal septicaemia, pneumonia or malaria [53]. Recently, rectal antibiotic (Ceftriaxone) to reduce treatment delays in neonatal sepsis presented as rectodispersible capsules have been proposed [54].

2.2.4. Pulmonary Delivery

The respiratory route is not much used therapeutically in preterm and term neonates with one major exception.

Respiratory distress syndrome (RDS) is a life-threatening condition, which occurs almost exclusively in preterm neonates with a deficiency [55,56], dysfunction or inactivation of pulmonary surfactant. The physiological role of surfactant is to allow the lungs to expand and avoid collapse (atelectasis) during the expiratory phases. Lack of surfactant results in difficulty in breathing, with low oxygenation, increased breathing effort and the need for respiratory support.

The administration of exogenous surfactant can alleviate the symptoms of RDS by supplementing the endogenous pool of surfactant, thereby enabling the biofilm to be replenished, dramatically reducing mortality and morbidity. This is often administered via invasive methods, such as endotracheal intubation and mechanical ventilation (MV). The gold standard version of this approach [57,58] is called INSURE (intubation, surfactant administration, extubation). Less invasive approaches, using thinner catheters (such as LISA technique—less invasive surfactant administration [59,60]), have been designed to supply exogenous surfactant to spontaneously breathing neonates, but these are still partially invasive. Therefore, there is a great interest by neonatologists in the development of a truly non-invasive procedure for surfactant administration, such as nebulisation. Some interesting pilot studies have been published on this issue [61–63].

These various intubation techniques also allow administration of other therapeutic interventions to treat a wide range of breathing difficulties should these be required. It is important that any formulations administered in this way, are of the correct concentration to be able to deliver the required dose in a small volume, and that the rheological profile is adequate to permit flow through the narrow tubes involved.

For neonates who can breathe spontaneously the major route of drug delivery is via nebulisation. A wide range of breathing difficulties can be treated, including respiratory syntical virus (RSV) bronchiolitis or asthma-like reactive airways disease. Typical drugs administered include beta agonists, steroids, ribavirin and sometimes adrenalin.

An advantage of using the pulmonary route is that efficacy can be achieved with reduced systemic drug levels and hence, decreased side effects [64]. For example, a recent review [63] has shown that pulmonary administration of corticosteroids can effectively prevent bronchopulmonary dysplasia in preterm infants with RSD, without the adverse effects on growth and neurodevelopmental outcome associated with systemic delivery. An emerging practice consists of adding budenoside to surfactant [65].

The formulation for nebulisation is usually a sterile aqueous solution or suspension, mainly based on the solubility and stability proprieties of the active drug substance. Many times, in order to achieve the suitable physico-chemical characteristics (i.e., osmolality, viscosity, pH etc.) and stability, it may be necessary to include excipients in the formulation. However, only a few excipients are approved for the inhalation route, and usually, there is little if any specific safety data available for neonates or the wider paediatric population [15], let alone after pulmonary exposure. Nebuliser solutions should be presented in an age-related unit dose of appropriate concentration if possible. Where this is not possible, the device used to measure the dose must not be a syringe designed for injection, thus reducing the risk of the nebuliser solution being inadvertently injected. The most appropriate nebuliser system to achieve good inhalation performance in terms of output rate, dose delivery to the lung, aerodynamic particle/droplet size distribution (i.e., fine particle or respirable fraction below 5μm or 3.3μm) and respirable drug delivery rate must be chosen. It is worth highlighting that the lung deposition of a nebulised product can be influenced by the breathing pattern (i.e., tidal volume, breath frequency and inhalation/exhalation ratio). The breathing pattern depends on the patient's age, and so, neonates (preterm and term) and children have significantly different and variable breathing profiles, making reliable dosing difficult when the efficiency of drug delivery systems are breathing pattern dependant [66,67]. This is further complicated if the infant is crying as this also changes breathing patterns.

It is possible to model the influence of various anatomical, physical, and physiological factors on aerosol delivery in preterm neonates on the efficiency of the delivery of an aerosolized drug to the bronchial tree using 3D models such as the (PrINT model) [68].

Nebulised delivery may also be important in older paediatric patients during times when the child's inspiratory force is low such as during an acute exacerbation of their condition. However, for both babies and older paediatric patients, some studies have shown that drug delivery from a Pressurised Metered Dose Inhaler (pMDI) and spacer system with suitably sized facemask can be at least as effective as nebulisation [69,70]. Delivery via pMDI and a spacer system is recommended in NICE guidance [71] for routine treatment of older paediatric patients.

2.2.5. Nasal Delivery

Although the nasal aperture is small in diameter in neonates and the nasal mucosa is often delicate and coated with mucus, it nevertheless offers a potential route for employing local drug administration to effect systemic drug delivery, thus reducing the need for injectable administration. In emergency care, this could also decrease the need for additional painful procedures such as insertion of IV cannulas for medication administration. For example, intranasal fentanyl has been used in the palliative care of term neonates and infants [72]. As such, it has attracted significant research interest in the last few years for use in this age group. Even if intubation is required, nasal administration could provide a less impactful means of aiding that procedure. Nasal midazolam was used and was more efficient than nasal ketamine, to adequately sedate neonates requiring intubation in the delivery room [73]. A typical dose volume was 0.1 mL/kg in each nostril.

Potential advantages of this route of administration, are that it is less invasive than IV whilst maintaining rapid onset since medications are directly absorbed (parenterally) through the nasal mucosa into systemic circulation, it also results in higher bioavailability compared to oral medications, as first-pass hepatic metabolism is bypassed. There is also some evidence that brain penetration can be enhanced as a consequence of nasal delivery [74–76].

Compared to buccal administration, (another method of avoiding first-pass metabolism), where salivation and swallowing issues after administration in neonates could have a substantial impact on dosage accuracy, more consistent absorption and bioavailability have been obtained following intranasal administration [77].

Bypassing the gastrointestinal tract with nasal drug delivery allows for the delivery of some macromolecules with low permeability across the gastrointestinal tract or susceptibility to chemical/proteolytic degradation. Recent studies in neonatal mice highlighted the potential of intranasal immunization of neonates with live vaccines [78]. As vaccination at birth would provide early protection for neonates and infants, expanding and improving the available means of neonatal vaccination is a global health priority.

There is a well-established link between nasal dosing and enhanced blood–brain barrier (BBB) permeability. This may or may not be of therapeutic benefit. If central activity is not desired, then nasal dosing may lead to enhanced toxicity. The BBB is a dynamic physiological barrier which regulates the passage of hydrophilic molecules into the central nervous system (CNS) via a physical barrier formed by tight junctions (TJ) between the endothelial cells and also a system of influx and efflux transporters and enzymes. Recent reviews and papers on the development of the BBB in preterm and term neonates suggest that although the physical barrier is formed very early, many other factors that affect brain penetration favour increased drug levels in the brain and cerebrospinal fluid (CSF) and that these factors change rapidly with maturation [79–81]. As yet, there is little or no information on the effect of actives and excipients on the development of the BBB in these age groups. It is known, however, that some penetration enhancers used in intranasal formulations for adults work via breaking tight junctions which is clearly not safe in neonates [82]. Conversely, cyclodextrins (e.g., hydroxypropyl-β-cyclodextrin), which are increasingly being used as vehicles to transport lipophilic drug through BBB [83], may have a neuroprotective effect and are used therapeutically in the treatment of neonatal hypoxia-ischemia [84].

Alterations in BBB development and in TJ expression could lead to anomalies later in life as well as to increased predisposition to some diseases. Neonatal hypoxia-ischemia (HI) causes severe brain damage and remains a major cause of neonatal morbidity and mortality [85]. At present, treatment options for neonatal HI brain damage are very limited and have only modest efficacy [86]. A study conducted in several rodent models of ischemic brain injury demonstrated the therapeutic potential of mesenchymal stem cells (MSCs) transplantation that improves functional outcome and also restores brain structure [87]. These findings in rodents indicate that the nasal route was an efficient route for stem cell transplantation after brain injury in the neonates [88].

During the development of a new formulation for nasal administration to the neonate, as well as all the usual factors (e.g., pH, osmolality, chemical irritation and overall acceptability) the formulator needs to keep in mind:

- The ratio between the ideal volume per nostril and the concentration of solution/ suspension to be administered. In practice, the maximum volume for single administration into one nostril is 0.1 mL in neonates and 0.5 mL in older children [89]. There is no agreement about the volume that can be given to preterm neonates. Larger doses can be given by using these dose volumes (or half the total volume provided this does not exceed the safe total volume) in both nostrils.
- The need for a 'baby size' device able to dose accurately very low volumes of liquids without causing physical damage to the nasal mucosa.
- The potential irritancy of highly concentrated solutions, especially if these are hypertonic.
- The choice of excipients. For example, a penetration enhancer may be required to aid the absorption of polar drugs. Many of these could cause irritation of nasal epithelium of neonates, and for most common penetration enhancers, no safety data are available in the neonatal population.

2.2.6. Dermal and Transdermal Delivery

In neonatal practice, many of the topical treatment decisions are made by specialist nurses. It is therefore very important for formulators to provide a high level of support to enable safe and effective use of topical formulations.

Formulators should be aware that the skin barrier may not be fully formed in preterm infants both at birth and for up to four weeks afterwards [90]. There are also changes in the type and proportion of lipids present and the development of ancillary skin structures such as sweat ducts and hair follicles [91]. This has implications for potential toxicity of active pharmaceutical ingredients

(API's) and formulations usually delivered for a localised effect via the topical route. In the absence of a fully competent skin barrier, higher systemic exposure of both API and common topical formulation excipients could be expected. The skin barrier may also be breached in term neonates and infants due to conditions such as cradle cap, nappy rash and eczema/dermatitis. Even if the skin is fully formed and functional, the lipid composition changes rapidly and the stratum corneum tends to be thinner in young babies and more hydrated. Additionally, occlusive dressings including nappies/diapers with impermeable plastic coverings can increase absorption. Besides permeability, the simple BSA/kg ratio puts neonates at a higher risk of increased absorption. Therefore, any adverse effects tend to be exacerbated by the relatively higher surface to volume ratio in children leading to a risk of unwanted high systemic exposure if large areas of skin are treated.

Formulators, therefore, need to ensure that APIs delivered topically in neonates have excellent systemic safety (limitation to the skin alone cannot be assumed). It is insufficient to only undertake topical safety studies as systemic absorption should be assumed and therefore systemic safety studies are required. Toxicity of ingredients in subpopulations such as neonates requires a careful risk assessment to avoid the use of excipients with unclear safety in target population. It is also necessary to select formulation excipients that are intrinsically safe and have a low potential to cause irritation or other skin reactions, e.g., sodium lauryl sulfate [92].

Finally, it is important to ensure that the pH and tonicity of the formulation is well matched to the specific requirements of the skin at the various stages of development and an understanding that this can change rapidly [93,94].

Coupled with the state of development of the barrier function of the skin, it is important to remember that skin in the neonate is often fragile and that it can be damaged by mechanical abrasion. Thus, the rheological profile and other cosmetic attributes of the formulation need to be taken into account. It may well be desirable to use a specific neonatal formulation that is more fluid/easier to spread than is common for creams and ointments used in older children. Similar considerations apply to the potential skin damage that can be caused by adhesives used in transdermal patches.

In some cases, the topical route is deliberately used to achieve systemic delivery [91]. In this case, it is still important to remember that the barrier properties of the skin may be rapidly changing during early development as this might affect the systemic blood levels achieved from a particular dose with implications for the topical dose that needs to be given.

Indeed, dose flexibility is a challenge in the neonate due to the above-mentioned developmental factors specific to the skin and due to the rapidly changing weight of the neonate. For standard semi-solid formulations, this can be accommodated to some extent by the area that is treated, but a wide range of strengths is still likely to be required. This is often achieved by diluting a formulation extemporaneously with a base. If this is done the stability implications (chemical, physical and microbiological) need to be considered along with the issues of ensuring homogeneity of the diluted formulation. For 'unit dose' topical formulations such as patches, dose control is achieved either by cutting patches or masking them (usually off-label) to reduce the contact area. Both practices are subject to significant risks of poor dose control unless the developer fully validates this practice.

Once a fully competent skin barrier is formed, it is possible to use the full range of topical and transdermal delivery [91] including creams, ointments, sprays, lotions, baths and, less frequently, transdermal patches with suitable adjustment to the dose. Some patches have paediatric labelling supported by clinical trials, whereas others are used off-label. There is little literature on the utility of innovative delivery methods such as needleless injectors, iontophoresis, sonophoresis and microneedle patches in the neonate. However, there is evidence of them being tested for use in older paediatric patients in order to attain more reliable drug levels, either locally in the skin or systemically, that are bioavailable from creams and ointments, especially where dermal permeation of the API is low or slow.

2.3. *Ability to Administer: Product Factors to Consider*

The lack of appropriate dosage forms frequently results in off-label and/or unlicensed use of modified adult formulations for administration in neonate and paediatric patients [95]. It has been reported that between 71% and 100% of patients in the NICU receive at least one off-label or unlicensed medicine [96].

The dose of a medicine can vary 100-fold between that for a preterm baby and an adult. Whilst a liquid medicine may be acceptable for all ages, using just one concentration for all ages would mean that an appropriate dose volume for an adult (say 10 mL) would be impossible to measure accurately for a preterm neonate (say 0.1 mL). Thus, more than one concentration would be required in this example.

Fortunately, many medicines can be dosed according to weight bands (e.g., 5–10 kg, 11–20 kg) or by age bands making it easier to produce an acceptable standard concentration of a drug. Formulations available from manufacturers should follow this rule, making it more feasible to manipulate with drug doses for neonatal application. However, doses of potent medicines for neonates necessitates individualised dose calculations based on body weight (e.g., mg/kg) or body surface area (e.g., mg/m^2). This could lead to the preparation of very low drug concentrations, several manipulations in preparation of the end formulation and the management of multi-infusion IV administration [97]. This has been reported to lead to a significant risk of medication errors and adverse drug events, especially in the NICU [98].

To overcome these issues associated with individualised concentration calculations based on low body weight, formulations of standard concentrations [97,99], specifically of high alert medication, have been implemented into NICU settings [100–102]. These can then be used with intelligent infusion pumps programmed with the standard concentrations allowing automated calculation of the individual flow rate based on body weight. Standardised concentrations also make it easier for pharmacies to prepare or source drugs which are 'ready to infuse' rather than them being prepared at the bedside with the attendant risks. They also reduce the burden of complex patient-individualised calculations on the prescriber and nurse administering the medication [97,99,103] and standard concentrations of different strengths allow for flexibility to compensate for a patient's differing fluid requirements, though the existence of more than one concentration introduces the potential for product selection error.

As the flow rates of standard concentrations are linked to patient weight, the flow rate decreases with decreasing patient's weight. This can result in very low flow rates in extremely low birth weight neonates where rates may be as low as 0.02 mL/h. Low flow rates have been associated with long delays in drug delivery times and hence the onset of effect, delays in transitioning to new infusion rates and delays in reducing effects when the infusion is stopped [27]. Such unpredictable delivery is unacceptable in the administration of life-sustaining medicines to critically ill patients [104–106]. Lack of awareness of these issues among clinical staff may lead to inappropriate clinical decision-making [27,104]. Observed lack of response to treatment due to delivery delay may cause precocious dose increase leading to overdosing and toxicity. When examined together with low flow rates, several other factors may contribute to prolonged and unpredictable delivery such as, the dead space of the administration system [27,104], the adsorption of the drug to tubing, backflow of the infusion [105], change of the flow rate of other infusions or a carrier fluid [27], the type of pump used, the placement of the infusion tubing and the line architecture [27].

When designing new formulations or reformulations of existing drugs, clinicians and drug formulators should work together to maximise an optimal drug delivery under special neonatal drug administration conditions (Table 2). Information and advice provided to the end user should facilitate a practical dosing regimen taking into account the potency of the drug and potential adverse effects and the setting in which the drug is to be measured and administered (NICU, hospital ward, home).

Using standard concentrations of medicines can help avoid medication errors when patients move between care settings [107]. Due to the implementation of standard concentrations to the neonatal care, the number of individually prepared drug doses reduced dramatically. Formulation of medication strengths that could represent standard concentration used in the NICU would eliminate the step of drug manipulation to prepare standard concentrations.

Table 2. Factors to consider in neonatal parenteral drug formulation and administration.

Chemical and physical compatibility of drug formulation used in multi-drug administration [28] including generic brands
Chemical and physical compatibility of drug formulation used in combination with neonatal TPN [108,109]
Compatibility of drug with diluents typically used in the NICU and stability after dilution
Compatibility of drug formulation while mixing at Y-site junction at different mixing ratios [108,109]
Stability of drug formulation over extended period of time (e.g., over 24 h infusion)
Stability of drug formulation exposed to different environmental conditions (high temperature, strong light, high oxygen levels) [110]
Stability and compatibility of excipients used in drug formulation
Stability and compatibility of excipients used in drug formulation with IV administration set and container
Compatibility of drug formulation with IV administration set and container
Strength(s)/concentration of drug that can cover neonatal weight- or age-bands as well as fluid restricted patients
Performance of medical equipment delivering drug—volumetric and smart pumps, syringe drivers
Design of IV administration set minimising drug delivery delays

2.3.1. In use Stability Issues

As well as all the administration factors discussed above, it is clear that formulations need to be stable in order to deliver the expected dose to the patient.

Although all the usual ICH stability requirements will apply, the therapeutic environment will often present some additional 'in use' stability challenges that should be considered by the formulator when developing products for this age group. This is particularly true for preterm infants and those being treated in high dependency incubators or intensive care environments and when those infants are treated almost exclusively via the parenteral route. They are complex to anticipate and therefore to reproduce in vitro despite regulatory expectations. Yet, conditions in NICU may be more controlled than in other environments. Clinical colleagues can provide information on likely treatment scenarios that will allow the most common potential interactions to be mimicked. Drug concentrations, diluents, formulations, mixing ratios and environmental conditions (light and temperature) are some factors pertinent to the design of compatibility studies. Some of the major issues are true pharmacy issues that need to be considered within the clinical context:

a) Photostability—useful reviews have been published [111,112]. Light intensity is often higher in the neonatal environment. Phototherapy is often used to treat postnatal jaundice either with lamps having the required spectral power distribution to break down the bilirubin in the skin or even exposure to direct sunlight. The implications of this treatment to the stability of any drug that partitions to the skin or eye need to be considered along with the implications for photosafety particularly if the metabolic immaturity of the neonate might be expected to affect clearance of any photoproducts that are generated. When bilirubin is present in the skin, protein binding will also be high. This can either exacerbate or mitigate photostability issues, depending on the mechanism of photodegradation. It is extremely difficult to predict what effect this protein binding might have *a priori*.

If the infant is being treated parenterally, it should be remembered that solution concentration tends to be low (though they can also be high if the neonate is fluid restricted) and infusion rates can also be low leading to long residence times in the administration lines. There is often a transit time of several hours between the infusion pump and the patient [27,106]. Added to this is the potentially very large surface to volume ratio due to the solution flowing through a long narrow bore tube. All of these factors can lead to significant levels of photodegradation that may not be fully predicted by some ICH Q1B photostability tests. It is therefore important to undertake a well-designed in use photostability test to allow for suitable advice to be provided to the healthcare practitioner.

Another factor that needs to be considered here is that the material of construction of any transfer line may be quite variable. Thus, some plastics will contain quite high levels of UV stabilisers whilst others may have much lower levels or even none at all. Clearly, this can lead to different levels of photoprotection for the content of the line.

If the product is known to be photolabile, then it may be wise to advise that any holding vessel (such as a syringe or infusion bag) and particularly any transfer lines are shielded from light exposure either with a suitable coloured cover or with a light impermeable barrier such as aluminium foil. One downside of doing this is that it will render visual inspection more difficult.

b) Other environmental factors—due to potential difficulties in breathing and less developed thermoregulation preterm infants may be in an oxygen-rich environment that is maintained at a higher temperature than would usually be the case in the hospital environment. This has obvious potential implications for formulations that are subject to oxidation or thermally driven hydrolysis if they are stored in such an environment for extended periods. Again, this may be further exacerbated by low concentration. The formulator will need to consider whether or not these conditions might have an effect on their product and provide suitable handling advice accordingly.

c) Potential Interactions—due to the limited number of access points for parenteral delivery to the neonate, it is often necessary to provide several drugs via the same entry point. In addition, neonates may be provided with nutrition via the same route. Thus, it is important to consider potential incompatibilities (both chemical and physical) between the various medications and the potential for API to adsorb to components of the parenteral or enteral feeding formulation used. In some cases, high concentrations of dextrose may be administered in this age group, both of which could have effects on product stability. On top of that, there is a risk that the API may interact with the material of the transfer tube leading to adsorption of the API. If the concentration is low, there may be a significant loss of therapeutic activity.

d) It may be a challenge to provide a sufficient range of formulation strengths to cover all needs of the paediatric population. Dilution may be required to allow accurate measurement using routinely available syringes. Large dilution factors may be required if only a single product concentration (often designed for adults or older children) is available. If dilution is to be avoided it may be necessary to develop and provide an appropriate dose measurement device and/or advice on providing an accurate dosage. Another method for controlling dose (volume) delivery is to change the infusion rate as discussed in the parenteral section.

If the formulation contains a stabiliser of some sort (e.g., antioxidant, photostabiliser) then dilution may render that stabiliser ineffective with obvious potential consequences for the formulation stability. The extent of the impact will be determined by a knowledge of the stability of the unprotected API and the level of stabiliser that remains. If the parenteral product is an emulsion or suspension, then dilution may lead to physical instability potentially leading to blockage of cannulas or even phlebitis. On the positive side, dilution may also reduce intake of an excipient that might be harmful to the neonate, e.g., sodium metabisulfite in parenteral inotropes.

The diluent used is also of importance. In some instances, photodegradation of the diluent can lead to degradation of an otherwise photo stable API (e.g., [113]). Formulators should be aware of this possibility when considering which diluent to recommend.

Photostability (and possibly oxidative stability) is also of concern for formulations that are applied topically in such a high-intensity light environment. Clearly, there is a risk that stability could be poor, leading to suboptimal dosing and/or the generation of relatively high levels of photo products in situ. The fact that the API concentration could be low should be considered when assessing the likely impact.

2.3.2. Excipients

Some of the challenges in the development of formulations for neonates have recently been reviewed by Kogermann et al. (2017) [114]. Not surprisingly, this review focuses largely on pharmaceutical excipients since mainly liquid dosage forms are used in these vulnerable patients and these may require preservatives to limit microbial growth and/or an array of excipients to achieve a solution of the active or to formulate a palatable liquid. Given that most actives are not specifically developed for neonatal conditions, dosage forms may contain excipients deemed safe in adults or older children that have not been studied for neonates [14,15].

The immaturity of organs and physiological systems of the neonate is important for the 'handling' of excipients as well as active drug substances. Little is known or published about the acute or chronic effects of excipients in the very young and that which has been published indicates their vulnerability to toxic effects considered safe for older children and adults. The exposure limits for many excipients have been published, but often apply to adults and should not be applied to neonates unless specifically indicated. Non-clinical work in appropriate juvenile animal models may be required for excipients used for the first time in the very young, and there should be appropriate justification for the use of any excipient supported by studies of high quality.

Epidemiological studies have demonstrated the exposure of neonates to excipients and the variation between products marketed in different countries [115]. Studies have also shown that excipient exposure can exceed current safety limits and that exposure may be from several medicines [116]. Measuring blood levels and the kinetics of excipients is possible in the very young and can be contemplated as part of clinical studies if necessary [117].

The formulator should first consider whether any excipient is required when developing a formulation for neonates. For example, it should not be necessary to include an antimicrobial preservative in a liquid product if designed for single use. Even an oral liquid medicine might be presented as a sterilised unit dose preparation without preservative excipients. Colours are generally not needed for this population and if intended for administration via enteral tubes, sweetening and taste masking may not be required. A risk-based approach [118] taking into account PK, ADME and safety data relevant to neonates may be taken in discussion of potential formulations with clinical and toxicology colleagues. For example, an excipient toxicity profile acceptable for single dose administration for a life-threatening illness may be quite different from that for long-term administration for a condition for which effective remedies are already available. Of course, the overall cost of formulation development for neonates and the viability of the commercial product will also require consideration during this pre-formulation discussion.

The European Paediatric Formulation Initiative (EuPFI) STEP database [119] provides comprehensive information on excipients for children and is freely available. The European Commission, through the EMA, is revising the labelling requirements for excipients in authorised products. Supporting the thresholds for labelling requirements are extensive reviews of excipient properties, uses and toxicology [120]. Whilst not intended to provide direct guidance to the formulator, that for propylene glycol [121] serves as a good example of the way in which toxicity can be assessed and safe exposure limits calculated.

As with any product development, compatibility studies of any proposed excipients with the API are an important part of early product development.

2.3.3. A Shift towards Solid Dosage Forms?

The mainstay of oral product administration for neonates are liquid dosage forms. Despite being more common for the paediatric market, they tend to contain excipients with elevated toxicological risks in neonates compared to adults (e.g., ethanol, propylene glycol, benzyl alcohol, polysorbate, parabens, etc.) due to the ongoing organ development and incomplete maturation [122]. REF oral solid dosage form is by far dominating the adult market [123]. Although they may sometimes contain excipients with some toxicological concerns of their own (for example, some may be allergens or irritants), in general, excipients used in solid dosage forms have a more benign safety profile than some used for liquid medicines [124]. As discussed above (Section 2.2) neonates are not always able to swallow liquids let alone solid dosage forms even though there are many inappropriate products on the market such as capsules and tablets licensed from birth that will have to be manipulated for administration [125]. It is highly unlikely that this is based on clinical data and hence, would no longer be acceptable following the implementation of new regulations worldwide.

The attraction of solid dosage forms has led to the emergence since 2008 of flexible solid oral dosage form (FSOD) which has been actively promoted by the World Health Organization (WHO) to overcome challenges of ensuring access to suitable medicines for children [126].

FSOD are solid forms that do not have to be swallowed whole, such as dispersible tablets, effervescent tablets, orodispersible tablets, and sprinkle capsules. They are flexible in administration, though not necessarily in dose. They may provide a convenient dose unit that can be manipulated to provide a suitable form for delivery to a neonate.

For example, some tablets for older children can be cut accurately into two, four or even eight approximately equal pieces. Such tablets were developed for fixed-dose combination of zidovudine and lamivudine in fast-disintegrating subunits (per 5 kg of BW) [127]. These can be easily administered after dispersion in a liquid or milk.

Dispersible tablets offer more dose flexibility. However, the minimum adequate volume of dispersion linked to the reproducibly of accurate dose withdrawal needs to be considered concomitantly [128]. In theory, such a dispersion could be dosed via an enteral tube. The suspension should also of sufficiently low viscosity and be fine enough to pass through the enteral tubes if this is foreseen.

As effervescent tablets have high-sodium content, their use should be carefully considered for neonates. They are also clearly contraindicated in cases where the patient in hypernatraemic, who need to be sodium restricted. Dispersions probably need to be degassed prior administration.

Some tablets have been developed to be administered in a Therapeutic Nipple Shield to safely deliverer medications and nutrients to breastfeeding infants [129]. In one proof of concept study, formula or freeze-dried milk/drug tablets were formulated to demonstrate both reliable drug delivery to babies and the ability of milk to dissolve poorly water-soluble drugs [130,131]. Both are interesting concepts considering babies' exclusive milk-based diet. There is also some evidence that milk can mask the taste of some medicines. However, the risk of interfering with feeding by rendering the taste of the milk foul needs to be mitigated with decent palatability of the medicine itself.

Until recently, few studies have been undertaken on the acceptability of mini tablets for patients under six years of age. Recently, however, a study in 6 to 12 month-old children that also included some neonates (n = 151) demonstrated complete swallowing of one mini-tablet (82.2%) compared to 72.2% for a liquid syrup [132] (Figure 2). However, it is to be noted that this mini tablet was uncoated so would have dispersed if it were inhaled inadvertently rather than swallowed. This was for safety reasons but may have somewhat confounded the study. On the positive side, there was no evidence of choking. As this has been a concern for some time, it would appear that the acceptance of this dosage form has come a long way. Data on acceptability and swallowability of several hundred mini tablets in slightly older babies, infants and children (six months to six years) are awaited. In parallel orodispersible minitablets that can be dispersed in the mouth or in baby-friendly beverages to particles that are easy to swallow are proposed to reduce the risk of choking while allowing dosing via nasogastric tubes (NGTs) [133].

Other orodispersible dosage forms, such as thin polymeric film, have been studied in older age groups (0.5–6 years old) [134]. They are attractive as they overcome the need for swallowing a solid entity. However, there is no precedent of use in neonates. This is also true of micropellet formulations such as sprinkles. It is hoped that work with older babies will provide usability and safety evidence and translate in product development in the near future that could benefit neonates too.

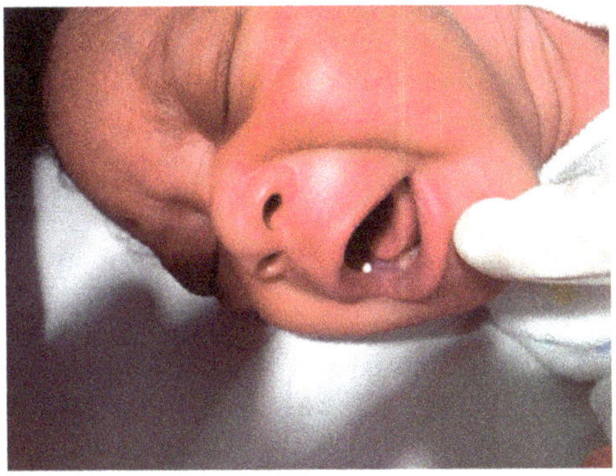

Figure 2. Newborn child with uncoated mini-tablet in the cheek pouch before swallowing (with permission to use from Thabet et al. 2018 [132]).

2.4. Ability to Administer: Device Factors to Consider

2.4.1. Accuracy of Small Volumes

Measuring small volumes (e.g., bolus IV injections, oral liquids, low rate IV infusions) with sufficient accuracy can be problematic if routinely available administration devices such as oral/enteral syringes and injection syringes are to be used [40]. Using a liquid product with a concentration designed to administer standard volumes to adults or older children may mean that volumes of 0.1 mL or less might be required for neonates [39,40,135–139]. This issue can be even more acute if clinical practice standardises on a single oral/enteral syringe design rather than using any specific device that may be provided by the manufacturer. To mitigate this issue, formulators should ensure that they develop the right product strength(s) so that there is no need to measure small volumes, especially that no volume is less than 0.1 mL and at the same time, the strength allows to cover body weights from 0.5 kg to 5 kg (a 10-fold range).

Calculating the required dose volume may be difficult, and dilution steps may add to calculation errors and be undertaken in an unsuitable environment. Decimal fractions involving hundredths of a mL can be confusing.

For injections the volume presented in a container should not be greater than ten times the dose for the smallest child and for drugs given by other routes of administration the risks of miscalculation or inaccurate measurement should be risk assessed and steps taken to reduce the risk [20].

It would be helpful if the device industry could develop administration devices that can accurately measure and reproducibly deliver very small volumes.

2.4.2. Enteral Tubes Administration

Neonates may require enteral feeding tubes to allow safe administration of enteral feeds, fluids and medicines. Older children with swallowing difficulties (sometimes related to administration of medicines only [140]) may have enteral feeding tubes inserted at different sites in the upper gastrointestinal tract for the long-term administration of food, fluids and medicines. The likelihood of administration via an enteral tube is one of the factors that will determine the investigations required to demonstrate that the drug preparation can be delivered effectively via the tube and without adverse effects. The flush volume required to ensure the whole dose is delivered must be carefully considered in relation to fluid requirements. If the product is likely to be administered via an enteral (e.g., nasogastric, nasojejunal) tube, issues such as viscosity of formulation (to permit flow of the product through neonatal tubes [e.g., 6FR/8FR] and avoid blockage), size of particles, adsorption to commonly used enteral tubes and interaction with common formula/breast milk should be investigated [20]. A recent Q&A has been produced by the Quality Working Party of EMA and sets out the potential problems of administration of medicines through an enteral tube and the steps to be taken to investigate and minimise them [141].

2.4.3. Parenteral Catheters and Administration Sets

An important formulation development consideration is to assess the risk of interactions between certain drugs and the materials of the devices used to administer them. Neonatal catheters and cannulas can be made of a variety of different materials including silicone, polyurethane, polyvinyl chloride and polyethylene [142]. Other components of the administration set may well be constructed from a different polymer to that of the catheter being used. Thus, the nature of the incompatibility may be complex. For this reason, the formulator should be aware of the composition of medical devices and potential incompatibilities.

The most commonly observed interactions are chemisorption, mechanical trapping of drug molecules onto or within the device and leaching. The occurrence of an interaction depends on the chemical nature of a drug and of the device material [143], the number of binding sites on the material's surface (surface area, device dimensions) [143], the drug concentration [144], and the time of contact between the drug and the material's surface which is influenced by the flow rate and the length and diameter of any tubing. The pH of the formulation may influence binding by changing the ionisation state of the API and any binding sites on the material of construction of the administration device. It is also important to consider any physical entrapment due to sharp bends in the tubing or via in-line filters. In the latter case, the pore size and material of construction could be important. Where medications are used at low neonatal concentrations, the loss due to interactions may be relatively significant [104]. A well-studied example is insulin [143,145–147]. Insulin adsorbs to glass and plastic polymers, binding more strongly at low flow rates to polyvinyl chloride (PVC) than to polyethylene (PE) tubing [143]. The opioid analgesic fentanyl and some benzodiazepines (diazepam, clonazepam) used in neonatal population for sedation and seizure control have been shown to interact with PVC IV lines and bags whilst there was no interaction with PE [144,147].

Another issue to consider here is that of the potential risk posed by extractables and leachables from any polymeric components. This issue is not restricted to parenteral delivery but may be expected to have a significant impact in this patient group. Leachables may be an issue not only for administration sets but also for primary, secondary and tertiary packaging. Table 3 below shows the likelihood of packaging component interactions.

Table 3. Assessment of drug product leachables associated with pharmaceutical packaging/delivery systems (modified from "FDA/CDER/CBER risk-based approach to consideration of leachables" (USP—General chapter <1664>) [148]).

Degree of Concern Associated with the Route of Administration	Likelihood of Packaging Component–Dosage Form Interaction		
	High	Medium	Low
Highest	Inhalation aerosol and spray	Injections and injectable suspensions, inhalation solution	Sterile powders and powders for injection, inhalation powders
High	Transdermal ointment and patches	Ophthalmic solutions and suspension, nasal aerosol and spray	
Low	Topic solutions and suspensions, topical and lingual aerosol, oral solutions and suspensions		Oral tablets and oral (hard and soft gelatin) capsules, topical powders, oral powders

Term (and to an even greater extent preterm) neonates are at higher risk of increased sensitivity to toxicants due to their organ immaturity and altered metabolic function [149]. This should be taken into account when establishing the safety threshold of any extractables and leachables. Official guidance on how to do this is available [150]. This includes the calculation of 'safety factors' for neonates and children but does not cover preterm infants.

With the advent of ever more sensitive analytical techniques, the number of compounds that may need to be evaluated, albeit at trace levels, makes this aspect of product development very challenging. It can, however, have significant effects in clinical practice.

As an example, some drugs, such as amiodarone and etoposide and some formulation excipients, such as those used in lipid emulsions, are known as leaching promoters [104,151]. Diethylhexyl phthalate (DEHP) is a plasticiser that is commonly used in PVC polymers that may be used in administration sets [152]. Animal studies have shown that high levels of phthalates may cause a range of issues in animals, including cancers, endocrine disruption and kidney injury [153]. The US Department of Health and Human Services issued a report in 2006 in which the risks of DEHP to human reproductive and development was evaluated [154]. In the report, neonatal patients and in particular pre-term neonates were highlighted by the FDA and the European Commission as being particularly at risk of exposure to levels of DEHP where there may be toxicological concerns [155]. However, DEHP free administration sets may not always be readily available. Lala et al. have described a case where an urgent need for an IV amiodarone infusion (which was infrequently used by their unit) was delayed several hours whilst attempts were made to source DEHP free components to adapt administration apparatus to avoid leaching of DEHP [104].

2.4.4. IV Polypharmacy

Patients cared for in intensive care units frequently require multiple IV infusions and present with limited vascular access. To avoid mixing different drug solutions in one IV container and to reduce the contact time between these drugs Y-site, T- or multi-lumen connectors are used to separate drug administration [31,156,157]. Some examples are shown in Figure 3 below.

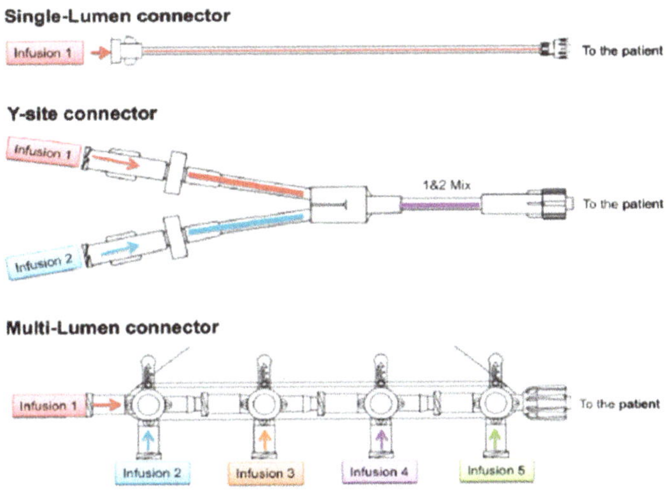

Figure 3. Different types of connectors (adapted from IV Sets and Access Devices Product Catalog—B. Braun Medical Inc., effective August 2017).

In practice, multi-lumen and y-site connectors are often used to simultaneously deliver multiple compatible drugs through one port of entry. The use of these devices can allow for multiple drug infusions to converge and be administered through a single venous port, if compatibility studies are available. This is illustrated in Figure 4. However, the optimum geometry, effect of mixing, flow resistance and risk of inadvertent boluses due to drug flow through these parts have not yet been sufficiently studied and compared.

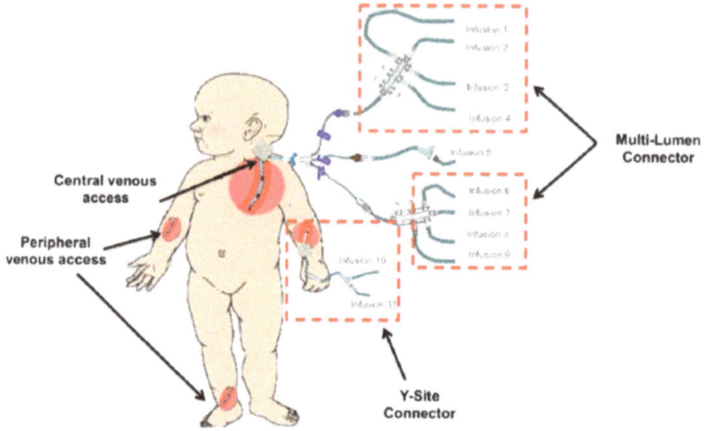

Figure 4. Example of venous access and multiple drug administration devices.

2.4.5. Inhalation Devices

The device used is an integral part of dose delivery of respiratory medicines, and hence, devices have been discussed to a significant extent under the respiratory product development section. Metered dose inhalers (MDI) need to be assessed for the safety of any extractables and leachables, e.g., from polymeric components of valves used in MDI. Aqueous formulations present less of a risk. Nevertheless, there have been some notable innovations in the use of devices to aid drug delivery in unusual settings such as the delivery of aerosolised therapy to a sleeping infant (Figure 5) and around device development [158,159].

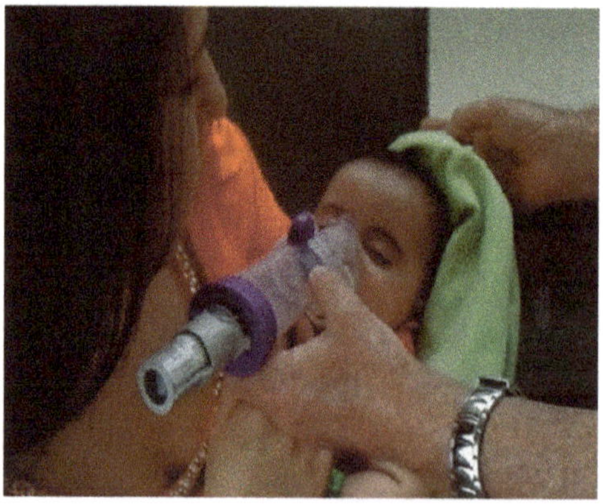

Figure 5. Photograph illustrating the method of aerosol administration to a sleeping infant showing the Respimat inhaler, InspiraChamber and SootherMask. (reproduced from [160] with permission from BMJ Publishing Group Ltd.).

2.5. Biopharmaceutical Considerations

The rate and extent of absorption, distribution, metabolism and excretion of drugs in children are different from that encountered in adults with the greatest differences from the adult being observed in neonates [161]. The impact of a number of biopharmaceutics factors on product development has been discussed inter alia during the discussion about the development of products for the various routes of administration.

Rectal absorption in full-term neonates has been reported as excellent. Intrarectal pH was shown to be significantly lower in neonates (6.5) compared with infants (6.9), and illness had no effect [162]. Neonates' erratic absorption seems to be the limiting factor for wider use [36].

When medicines are administered via inhalation, breathing patterns can have an effect on the bioavailability and crying (which is more prevalent in neonates compared to adults) can reduce the amount of dose absorbed.

The skin barrier may not be fully formed in preterm infants, there are changes in the type and proportion of lipids present, and the development of ancillary skin structures such as sweat ducts and hair follicles differs to that of the skin of a term neonate. Skin may also be thin, fragile, hydrated and prone to occlusion. This has implications for potential toxicity of APIs and formulations usually delivered for a localised effect via the topical route.

Altered or reduced distribution is likely via most parenteral routes.

A recent review describes these differences in detail [13], the most relevant information has been extracted and is presented in Table 4.

Table 4. Summary of differences between neonatal and adult physiology that affect absorption/distribution of drugs (extracted from [13]).

Route of Administration	Impact on Absorption/Distribution	Reasons
Oral	Altered absorption	Neonatal pH is elevated in the stomach (increased for basic drugs and reduced for acidic drugs) Immature ontogeny of transporter expression
	Reduced absorption	Slower gastric emptying Reduced relative surface area in the intestine
	Increased absorption	Slower intestinal transit Reduced intestinal P-glycoprotein expression
Rectal	Decreased surface area	Reduced relative surface area of rectum
Respiratory	Decreased absorption	Immature lung branching and development Reduced lung capacity and inspiratory flow
Nasal	No data shown	Potential for irritation in the nasal mucosa in neonates
Dermal and transdermal	Increased absorption	Higher BSA/kg ratio Thinner stratum cornea layer More hydrated stratum corneum Higher relative surface area to bodyweight
IV	Reduced distribution	Reduced blood volume
Intramuscular	Reduced distribution	Reduced muscle mass
	Altered distribution	Variable muscle blood flow
Subcutaneous	Reduced distribution	Reduced subcutaneous fat

The gastrointestinal environment is very different in neonates compared to adults due to the ontogeny of transporters, pH and permeability of the intestinal wall [6,13,163–165]. Furthermore, the impact of feeding on drug absorption needs to be considered due to the very different feeding pattern in neonates as well as the co-administration of medicines with foods to improve acceptance and palatability [166,167]. In neonates and infants, attention has been paid to the mixing of medicines with liquid feeds due to the impact on the osmolality of the resulting fluid which may affect GI transit [168,169] although the actual physiological impact is still unclear [170].

Many medicines are mixed with food to improve palatability, yet the impact of this mixing on the absorption of the drug is unknown [171]. The dissolution of drug products in breast and formula milk has been evaluated in vitro to identify whether any effects can be predicted [172]. Other paediatric-relevant dissolution testing systems have also been introduced [173] as well as media that represents the gastric and intestinal fluid from children [163].

Doses used in paediatric populations are often derived from weight-based and surface-area-based dosing equations that are often based on adult data and then scaled based on size and age as an approximation for drug activity in children. However, paediatric growth and development is not a linear process, and there are risks associated with simple scaling to determine doses to use in neonates.

Comprehensive physiologically based pharmacokinetic modelling (PBPK) systems have been introduced that replicate the known parameters of the paediatric GI tract to better predict oral dosing [174–182]. However, the major limitation in both in vitro and in silico models is the lack of accurate knowledge about the neonatal and infant intestinal parameters [178,183,184]. These models have focused on oral absorption to date, and there is a need to develop appropriate tools for other routes of drug administration in neonates.

Clinical examples of differences in absorption in neonates compared to adults are scarce. Recent work suggested that the processes underlying changes in oral drug absorption rate typically reach adult levels within one week of birth [35].

2.6. Regulatory Challenges

Useful background information on the European Paediatric Regulation, together with progress reports, is published by the European Commission [7,185]. The US FDA has a similar regulatory framework and many other countries model their approach to the assessment of products on these two sets of regulations [186]. The clarity that these regulations and associated guidelines have provided has advanced the number of new medicines that are available for children. However, the 2017 review of 10 years of the implementation of the European regulations shows there is still a long way to go [185].

Many of the medicines used for neonates have only been developed and authorised for adults and older children. That is, they are used off-label. They may sometimes have been studied extensively perhaps 'in-house', but information has often been gathered and published by academia and never submitted by the pharmaceutical industry to regulatory scrutiny [187,188]. Frequently the dosage form that is authorised is not acceptable for neonates or younger children and must be modified for administration via an extemporaneous preparation by the pharmacist or manipulation at the point of administration such as splitting or crushing a solid oral dosage form [189]. Modification of the dosage form may generate an 'unlicensed' medicine with attendant uncertainty and risks. Although there is a strong drive to develop licensed medicines for all paediatric subpopulations, as noted throughout this paper this can be extremely challenging for neonates and in some cases industry verified manipulations can provide some increased level of assurance of suitability for use in the interim before a suitable neonate formulation can be licensed [188].

Whilst regulation demands a paediatric investigation plan (PIP) for all new medicines, this does not apply to long-established molecules. There are incentives to study and seek authorisation for age-appropriate dosage forms of off-patent medicines used by children (the Paediatric-Use Marketing Authorisation (PUMA) [190]), but it has not been considered successful in providing the medicines required. The incentives do not appear to be enough to balance the cost of development studies, and despite the acknowledged benefits of using a licensed product, the inevitable cost of such medicines can limit market entry. Enforcement of the use of licensed medicines appears not to be systematic and unlicensed medicines continue to be used.

Since it appears that extemporaneous modification is likely to persist the European Directorate for the Quality of Medicines (EDQM) is establishing a formulary of paediatric medicines prepared extemporaneously. As of January 2019, two monographs are available for public consultation. This seeks to improve access to medicines of demonstrable quality when suitable marketed products are not available [191].

3. Burden of Proof

Given all the product, environmental, biopharmaceutics, dose delivery and device factors discussed above, along with the multiplicity of different practices in the clinical setting based upon the age of the neonate, their stage of development and their clinical need, the potential number of pharmaceutical development studies and in particular the compatibility and stability studies are huge. There is a complex matrix of confounding factors such as polypharmacy/coadministrations, low or high concentrations, long residence time and high risk of adsorption to the medical devices used for the administration, e.g., tubing, extractables and leachables assessment, interaction with packaging and unique environmental conditions many of which either alone or in combination can lead to unfavourable clinical outcomes.

The principal aim of a formulator is to design a formulation that provides the intended efficacy for the target patients, ensuring a safe and robust quality profile for the final drug product. However, it would perhaps be unrealistic to test every possible combination. Although this is part of current practices of drug product development, it is important in the context of neonates to reinstate that product developers provide evidence and advice that will cover the majority of anticipated use cases covering both what is acceptable and what is contraindicated from a product quality, safety and efficacy point of view. A lot of this advice can be extrapolated from studies performed for other paediatric subpopulations or adults. However, specific pharmaceutical development studies taking into account the specific requirements of neonates may well be required.

If clinical practice demands that manipulation of formulations or co-dosing etc. that is outside the range of studies performed by the product developer or reported and justified in the literature, then it is the responsibility of the team treating that patient to consider the wisdom of the proposed action. Clearly the pharmacist supporting that team can be of significant importance in evaluating the product quality aspects. The level of practical work supporting those decisions will depend on the circumstances. Thus a 'one off' intervention may perhaps be supported on a risk/benefit judgement based on general medical and pharmaceutical principles. If the intervention is used more frequently, then the pharmacy should undertake and publish appropriate compatibility and stability studies. If the intervention is to become commonplace, then formal studies perhaps in association with the original product developer and/or academia will be warranted.

It is essential to assign to appropriate shelf life and/or in-use period for the final product under specific storage conditions (i.e., temperature, humidity and light). Conditions in the NICU may, in fact, represent the best case-controlled environment for the delivery of medications to neonate and drug formulators should investigate other environments in which uncontrolled environmental conditions may present challenges not seen in the NICU. It is vital to demonstrate the in-use usability of the entire product including storage requirements in the likely use environment, access to the packaging, reconstitution procedure, dose measurement, suitability and ease of use of any device required, clarity of instructions. All these factors can/should be assessed in a simulated use evaluation supervised by clinical and/or pharmacy staff using a representative sample of professional staff (i.e., nurse, physician, pharmacy technicians) and where relevant non-professional carers such as parents and guardians. With the increased involvement of staff and carers in the treatment of neonates, evaluation of the role of human factors could be particularly relevant during development.

Studies should aim to mirror clinical practice where possible bearing in mind that the needs for each neonate could vary significantly and might require customized treatment regimens, especially as new treatments become available. Usability study should also include the stability of the formulation over the extended "worst case" time periods being cognisant of regulatory policies on microbiological stability. Limited venous access in neonates can in clinical practice result in little choice but to co-administer drugs. Formulators should be aware of this potential and should extend stability studies to examine Y-site stability of co-administered drugs and also TPN [109,157].

4. Conclusions

Development of medicines is challenging per se, and this is even more relevant in the development of medicines for children, let alone preterm babies and neonates. When those children have such rapidly changing physiology and biopharmaceutical characteristics accompanied by critical clinical conditions and requirements, such as is the case for neonates (and in particular preterm neonates), then product development is very challenging. Neonates are still 'therapeutic orphans' in terms of access to appropriate drugs and formulations that have been studied and approved by regulators. To design an appropriate formulation for neonates, it is important to understand their physiological status, development and care environment as well as methods of drug administration and the limitations these factors place on formulation development. Given that neonates will usually form a very small fraction of the population that might benefit from a drug, there may be other constraints that limit the ability to provide unique neonatal formulations. A good understanding of the various constraints will allow the formulator to provide for neonates whilst having due regard for the needs of the older population. If the neonate is considered early in the formulation design process, some delays in clinical trials in this population may be avoided.

In recent years, great strides have been made in understanding physiological development in paediatrics in general and neonates in particular. Nonetheless, it is clear that as yet some fundamental information is not available to inform the pharmaceutical development process. For example, further research is needed on the safety of excipients in this population, the development of the CNS and in particular the effect of API and excipients on the maturation of the BBB, the robustness of the skin barrier. The maturation of drug elimination processes and the effect of a rapidly changing fat/lean ratio on drug distribution. Guidance is also needed on compatibility studies that are required and/or desirable in the light of the polypharmacy that is common in the treatment of neonates. Finally, the place of oral therapy with minitablets needs clarification. Further research in these areas will help inform even better product development for this patient group.

Nevertheless, this review has provided information, advice and guidance on factors for the product developer and, in particular, the formulator when seeking to meet the requirements of this highly vulnerable patient group. In general, guidance is given for what is required for treatment in ICU settings but where possible, we have set that in the context of paediatric product development as a whole.

Funding: This research received no external funding.

Conflicts of Interest: The authors declare no conflict of interest.

References

1. European Medicines Agency. Ich e11(r1) Guideline on Clinical Investigation of Medicinal Products in the Pediatric Population. Available online: https://www.ema.europa.eu/documents/scientific-guideline/ich-e11r1-guideline-clinical-investigation-medicinal-products-pediatric-population-revision-1_en.pdf (accessed on 19 March 2019).
2. March of Dimes; PMNCH; Save the Children; WHO. Born too Soon: The Global Action Report on Preterm Birth. Available online: https://www.who.int/pmnch/media/news/2012/201204_borntoosoon-report.pdf (accessed on 19 March 2019).
3. Allegaert, K.; van de Velde, M.; van den Anker, J. Neonatal clinical pharmacology. *Paediatr. Anaesth.* **2014**, *24*, 30–38. [CrossRef] [PubMed]
4. Allegaert, K.; Samardzic, J.; Bajcetic, M.; van den Anker, J.N. Developmental pharmacology and therapeutics in neonatal medicine. In *Neonatology: A Practical Approach to Neonatal Diseases*, 2nd ed.; Buonocore, G., Bracci, R., Weindling, M., Eds.; Springer: Cham, Switzerland, 2018; pp. 693–707.
5. Allegaert, K.; Simons, S.; van den Anker, J. Research on medication use in the neonatal intensive care unit. *Expert Rev. Clin. Pharmacol.* **2019**, *12*, 343–353. [CrossRef] [PubMed]
6. Van den Anker, J.; Reed, M.D.; Allegaert, K.; Kearns, G.L. Developmental changes in pharmacokinetics and pharmacodynamics. *J. Clin. Pharmacol.* **2018**, *58*, S10–S25. [CrossRef] [PubMed]
7. European Commission. State of Paediatric Medicines in the eu. 10 Years of the eu paediatric Regulation. Report from the Commission to the European Pparliament and the Council. Available online: https://ec.europa.eu/health/sites/health/files/files/paediatrics/docs/2017_childrensmedicines_report_en.pdf (accessed on 19 March 2019).
8. Cuzzolin, L.; Atzei, A.; Fanos, V. Off-label and unlicensed prescribing for newborns and children in different settings: A review of the literature and a consideration about drug safety. *Expert Opin. Drug Saf.* **2006**, *5*, 703–718. [CrossRef] [PubMed]

9. Gore, R.; Chugh, P.K.; Tripathi, C.D.; Lhamo, Y.; Gautam, S. Pediatric off-label and unlicensed drug use and its implications. *Curr. Clin. Pharmacol.* **2017**, *12*, 18–25. [CrossRef] [PubMed]
10. Tomasi, P.A.; Egger, G.F.; Pallidis, C.; Saint-Raymond, A. Enabling development of paediatric medicines in europe: 10 years of the eu paediatric regulation. *Paediatr. Drugs* **2017**, *19*, 505–513. [CrossRef]
11. Yen, E.; Davis, J.M.; Milne, C.-P. Impact of regulatory incentive programs on the future of pediatric drug development. *Ther. Innov. Regul. Sci.* **2019**. Published online 14 April 2019. [CrossRef] [PubMed]
12. Ward, R.M.; Benjamin, D.; Barrett, J.S.; Allegaert, K.; Portman, R.; Davis, J.M.; Turner, M.A. Safety, dosing, and pharmaceutical quality for studies that evaluate medicinal products (including biological products) in neonates. *Pediatr. Res.* **2017**, *81*, 692–711. [CrossRef] [PubMed]
13. Linakis, M.W.; Roberts, J.K.; Lala, A.C.; Spigarelli, M.G.; Medlicott, N.J.; Reith, D.M.; Ward, R.M.; Sherwin, C.M.T. Challenges associated with route of administration in neonatal drug delivery. *Clin. Pharmacokinet.* **2016**, *55*, 185–196. [CrossRef] [PubMed]
14. Allegaert, K.; Cosaert, K.; van den Anker, J.N. Neonatal formulations: The need for a tailored, knowledge driven approach. *Curr. Pharm. Des.* **2015**, *21*, 5674–5679. [CrossRef] [PubMed]
15. Valeur, K.S.; Holst, H.; Allegaert, K. Excipients in neonatal medicinal products: Never prescribed, commonly administered. *Pharmaceut. Med.* **2018**, *32*, 251–258. [CrossRef] [PubMed]
16. Stanford Children's Health. The Neonatal Intensive Care Unit (nicu). Available online: https://www.stanfordchildrens.org/en/topic/default?id=the-neonatal-intensive-care-unit-nicu-90-P02389 (accessed on 21 March 2019).
17. Rennie, J.M.; Roberton, N.R.C. *Textbook of Neonatology*, 3rd ed.; Churchill Livingstone: Edinburgh, UK, 1999.
18. Zhou, D. Understanding physicochemical properties for pharmaceutical product development and manufacturing ii: Physical and chemical stability and excipient compatibility. *J. Validation Tech.* **2009**, *15*, 36–47.
19. Allegaert, K.; van den Anker, J. Neonatal drug therapy: The first frontier of therapeutics for children. *Clin. Pharmacol. Ther.* **2015**, *98*, 288–297. [CrossRef] [PubMed]
20. European Medicines Agency. Guideline on Pharmaceutical Development of Medicines for Paediatric Use. Available online: https://www.ema.europa.eu/documents/scientific-guideline/guideline-pharmaceutical-development-medicines-paediatric-use_en.pdf (accessed on 19 March 2019).
21. Concepcion, N.D.P.; Laya, B.F.; Lee, E.Y. Current updates in catheters, tubes and drains in the pediatric chest: A practical evaluation approach. *Eur. J. Radiol.* **2017**, *95*, 409–417. [CrossRef] [PubMed]
22. Farrelly, J.S.; Stitelman, D.H. Complications in pediatric enteral and vascular access. *Semin. Pediatr. Surg.* **2016**, *25*, 371–379. [CrossRef] [PubMed]
23. Koletzko, B.; Goulet, O.; Hunt, J.; Krohn, K.; Shamir, R.; for the Parenteral Nutrition Guidelines Working Group. 1. Guidelines on paediatric parenteral nutrition of the european society of paediatric gastroenterology, hepatology and nutrition (espghan) and the european society for clinical nutrition and metabolism (espen), supported by the european society of paediatric research (espr). *J. Pediatr. Gastroenterol. Nutr.* **2005**, *41*, S1–S4. [PubMed]
24. Dongara, A.R.; Patel, D.V.; Nimbalkar, S.M.; Potana, N.; Nimbalkar, A.S. Umbilical venous catheter versus peripherally inserted central catheter in neonates: A randomized controlled trial. *J. Trop. Pediatr.* **2017**, *63*, 374–379. [CrossRef] [PubMed]
25. Detaille, T.; Pirotte, T.; Veyckemans, F. Vascular access in the neonate. *Best Pract. Res. Clin. Anaesthesiol.* **2010**, *24*, 403–418. [CrossRef] [PubMed]
26. Pedreira, M.L. [obstruction of peripherally inserted central catheters in newborns: Prevention is the best intervention]. *Revista paulista de pediatria: orgão oficial da Sociedade de Pediatria de São Paulo* **2015**, *33*, 255–257. [CrossRef]
27. Sherwin, C.M.; Medlicott, N.J.; Reith, D.M.; Broadbent, R.S. Intravenous drug delivery in neonates: Lessons learnt. *Arch. Dis. Child.* **2014**, *99*, 590–594. [CrossRef] [PubMed]
28. Kalikstad, B.; Skjerdal, A.; Hansen, T.W. Compatibility of drug infusions in the nicu. *Arch. Dis. Child.* **2010**, *95*, 745–748. [CrossRef] [PubMed]
29. Gaetani, M.; Frndova, H.; Seto, W.; Parshuram, C. Concurrent intravenous drug administration to critically ill children: Evaluation of frequency and compatibility. *J. Crit. Care* **2017**, *41*, 198–203. [CrossRef]
30. Bradley, J.S.; Wassel, R.T.; Lee, L.; Nambiar, S. Intravenous ceftriaxone and calcium in the neonate: Assessing the risk for cardiopulmonary adverse events. *Pediatrics* **2009**, *123*, e609–e613. [CrossRef]

31. Manrique-Rodriguez, S.; Sanchez-Galindo, A.; Mora-Garcia, T.; Fernandez-Llamazares, C.M.; Echarri-Martinez, L.; Lopez-Herce, J.; Rodriguez-Gomez, M.; Bellon-Cano, J.M.; Sanjuro-Saez, M. Development of a compatibility chart for intravenous y-site drug administration in a pediatric intensive care unit. *J. Infus. Nurs.* **2012**, *35*, 109–114. [CrossRef] [PubMed]
32. Perez, M.; Décaudin, B.; Maiguy-Foinard, A.; Barthélémy, C.; Lebuffe, G.; Storme, L.; Odou, P. Dynamic image analysis to evaluate subvisible particles during continuous drug infusion in a neonatal intensive care unit. *Sci. Rep.* **2017**, *7*, 9404. [CrossRef] [PubMed]
33. Rouse, C.; Mistry, P.; Rayner, O.; Nickless, J.; Wan, M.; Southern, K.W.; Batchelor, H.K. A mixed methods study of the administration of flucloxacillin oral liquid; identifying strategies to overcome administration issues of medicines with poor palatability. *Int. J. Pharm. Pract.* **2017**, *25*, 326–334. [CrossRef] [PubMed]
34. Sheppard, J.J.; Mysak, E.D. Ontogeny of infantile oral reflexes and emerging chewing. *Child Dev.* **1984**, *55*, 831–843. [CrossRef] [PubMed]
35. Somani, A.A.; Thelen, K.; Zheng, S.; Trame, M.N.; Coboeken, K.; Meyer, M.; Schnizler, K.; Ince, I.; Willmann, S.; Schmidt, S. Evaluation of changes in oral drug absorption in preterm and term neonates for biopharmaceutics classification system (bcs) class i and ii compounds. *Br. J. Clin. Pharmacol.* **2016**, *81*, 137–147. [CrossRef] [PubMed]
36. Anderson, B.J.; van Lingen, R.A.; Hansen, T.G.; Lin, Y.C.; Holford, N.H.G. Acetaminophen developmental pharmacokinetics in premature neonates and infants—A pooled population analysis. *Anesthesiology* **2002**, *96*, 1336–1345. [CrossRef]
37. European Medicines Agency. First Medicine to Treat Neonatal Diabetes. Press Release 23/02/2018. Available online: https://www.ema.europa.eu/en/news/first-medicine-treat-neonatal-diabetes (accessed on 19 March 2019).
38. European Medicines Agency. Reflection Paper: Formulations of Choice for the Paediatric Population. Available online: https://www.ema.europa.eu/documents/scientific-guideline/reflection-paper-formulations-choice-paediatric-population_en.pdf (accessed on 19 March 2019).
39. Ainscough, L.P.; Ford, J.L.; Morecroft, C.W.; Peak, M.; Turner, M.A.; Nunn, A.J.; Roberts, M. Accuracy of intravenous and enteral preparations involving small volumes for paediatric use: A review. *Eur. J. Hosp. Pharm.* **2018**, *25*, 66–71. [CrossRef]
40. Gurung, K.; Arenas-Lopez, S.; Wei, L.; Tuleu, C. Accuracy of enteral syringes for liquid medicines prescribed in children. *Arch. Dis. Child.* **2014**, *99*, e3. [CrossRef]
41. Brown, D.; Ford, J.L.; Nunn, A.J.; Rowe, P.H. An assessment of dose-uniformity of samples delivered from paediatric oral droppers. *J. Clin. Pharm. Ther.* **2004**, *29*, 521–529. [CrossRef] [PubMed]
42. Maier, T.; Scheuerle, R.L.; Markl, D.; Bruggraber, S.; Zeitler, A.; Fruk, L.; Slater, N.K.H. Zinc delivery from non-woven fibres within a therapeutic nipple shield. *Int. J. Pharm.* **2018**, *537*, 290–299. [CrossRef] [PubMed]
43. Hart, C.W.; Israel-Ballard, K.A.; Joanis, C.L.; Baniecki, M.L.; Thungu, F.; Gerrard, S.E.; Kneen, E.; Sokal, D.C. Acceptability of a nipple shield delivery system administering antiviral agents to prevent mother-to-child transmission of hiv through breastfeeding. *J. Hum. Lact.* **2015**, *31*, 68–75. [CrossRef] [PubMed]
44. Ducki, S.; Martin, C.; Bohn, L. Oral administration of medicines to infants: The dummy which relieves [2]. *Arch. Pediatr.* **2002**, *9*, 1298–1299.
45. Hansen, K.; Yee, L.; Lee, J.; Horeczko, T.; Saidinejad, M.; Padlipsky, P.S.; Gausche-Hill, M.; Tanen, D.A. Parent and nurse satisfaction using pacidose®oral medication delivery device in the pediatric emergency department: A pilot study. *J. Pediatr. Nurs.* **2018**, *42*, 100–103. [CrossRef] [PubMed]
46. WHO. Development of Paediatric Medicines: Points to Consider in Formulation. Who Technical Report Series, no. 970, 2012, annex 5. Available online: http://apps.who.int/medicinedocs/en/m/abstract/Js19833en/ (accessed on 19 March 2019).
47. Paroche, M.M.; Caton, S.J.; Vereijken, C.; Weenen, H.; Houston-Price, C. How infants and young children learn about food: A systematic review. *Front. Psychol.* **2017**, *8*, 33. [CrossRef]
48. Jannin, V.; Lemagnen, G.; Gueroult, P.; Larrouture, D.; Tuleu, C. Rectal route in the 21st century to treat children. *Adv. Drug Deliv. Rev.* **2014**, *73*, 34–49. [CrossRef]
49. Kim, T.W.; Rognerud, C.L.; Ou, C.N. Accuracy in the alteration of acetaminophen suppositories. *Anesth. Analg.* **2005**, *100*, 1303–1305. [CrossRef] [PubMed]
50. Chen, L.; Zhang, M.; Yung, J.; Chen, J.; McNair, C.; Lee, K.S. Safety of rectal administration of acetaminophen in neonates. *Can. J. Hosp. Pharm.* **2018**, *71*, 364–369. [CrossRef] [PubMed]

51. Demir, N.; Peker, E.; Ece, I.; Balahoroglu, R.; Tuncer, O. Efficacy and safety of rectal ibuprofen for patent ductus arteriosus closure in very low birth weight preterm infants. *J. Matern. Fetal Neonatal Med.* **2017**, *30*, 2119–2125. [CrossRef] [PubMed]
52. Surkov, D.; Obolonskiy, A.; Kapustina, O.; Volkov, D. Use of rectal ibuprofen for pda closure in preterm neonates. *PACCJ* **2014**, *2*, 11–16.
53. Tozan, Y.; Klein, E.Y.; Darley, S.; Panicker, R.; Laxminarayan, R.; Breman, J.G. Prereferral rectal artesunate for treatment of severe childhood malaria: A cost-effectiveness analysis. *Lancet* **2010**, *376*, 1910–1915. [CrossRef]
54. Kauss, T.; Langlois, M.H.; Guyonnet-Duperat, A.; Phoeung, T.; Xie, X.Y.; Cartwright, A.; White, N.; Gomes, M.; Gaudin, K. Development of rectodispersible tablets and granulate capsules for the treatment of serious neonatal sepsis in developing countries. *J. Pharm. Sci.* **2019**. [CrossRef]
55. Jobe, A.H. What is rds in 2012? *Early Hum. Dev.* **2012**, *88*, S42–S44. [CrossRef]
56. Jobe, A. Metabolism of endogenous surfactant and exogenous surfactants for replacement therapy. *Semin. Perinatol.* **1988**, *12*, 231–244. [PubMed]
57. Kandraju, H.; Murki, S.; Subramanian, S.; Gaddam, P.; Deorari, A.; Kumar, P. Early routine versus late selective surfactant in preterm neonates with respiratory distress syndrome on nasal continuous positive airway pressure: A randomized controlled trial. *Neonatology* **2013**, *103*, 148–154. [CrossRef]
58. Verder, H.; Albertsen, P.; Ebbesen, F.; Greisen, G.; Robertson, B.; Bertelsen, A.; Agertoft, L.; Djernes, B.; Nathan, E.; Reinholdt, J. Nasal continuous positive airway pressure and early surfactant therapy for respiratory distress syndrome in newborns of less than 30 weeks' gestation. *Pediatrics* **1999**, *103*, E24. [CrossRef]
59. Kribs, A.; Roll, C.; Gopel, W.; Wieg, C.; Groneck, P.; Laux, R.; Teig, N.; Hoehn, T.; Bohm, W.; Welzing, L.; et al. Nonintubated surfactant application vs conventional therapy in extremely preterm infants: A randomized clinical trial. *JAMA Pediatr.* **2015**, *169*, 723–730. [CrossRef]
60. Fabbri, L.; Klebermass-Schrehof, K.; Aguar, M.; Harrison, C.; Gulczynska, E.; Santoro, D.; Di Castri, M.; Rigo, V. Five-country manikin study found that neonatologists preferred using the lisacath rather than the angiocath for less invasive surfactant administration. *Acta Paediatr.* **2018**, *107*, 780–783. [CrossRef]
61. Pillow, J.J.; Minocchieri, S. Innovation in surfactant therapy ii: Surfactant administration by aerosolization. *Neonatology* **2012**, *101*, 337–344. [CrossRef] [PubMed]
62. Pohlmann, G.; Iwatschenko, P.; Koch, W.; Windt, H.; Rast, M.; de Abreu, M.G.; Taut, F.J.; De Muynck, C. A novel continuous powder aerosolizer (cpa) for inhalative administration of highly concentrated recombinant surfactant protein-c (rsp-c) surfactant to preterm neonates. *J. Aerosol Med. Pulm. Drug Deliv.* **2013**, *26*, 370–379. [CrossRef] [PubMed]
63. Delara, M.; Chauhan, B.F.; Le, M.-L.; Abou-Setta, A.M.; Zarychanski, R.; 'tJong, G.W. Efficacy and safety of pulmonary application of corticosteroids in preterm infants with respiratory distress syndrome: A systematic review and meta-analysis. *Arch. Dis. Child. Fetal Neonatal Ed.* **2019**, *104*, F137–F144. [CrossRef] [PubMed]
64. MacLoughlin, R.; Telfer, C.; Clark, A.; Fink, J. Aerosol: A novel vehicle for pharmacotherapy in neonates. *Curr. Pharm. Des.* **2017**, *23*, 5928–5934. [CrossRef] [PubMed]
65. Ricci, F.; Catozzi, C.; Ravanetti, F.; Murgia, X.; D'Alo, F.; Macchidani, N.; Sgarbi, E.; Di Lallo, V.; Saccani, F.; Pertile, M.; et al. In vitro and in vivo characterization of poractant alfa supplemented with budesonide for safe and effective intratracheal administration. *Pediatr. Res.* **2017**, *82*, 1056–1063. [CrossRef] [PubMed]
66. Barry, P.W.; O'Callaghan, C. An in vitro analysis of the output of budesonide from different nebulizers. *J. Allergy Clin. Immunol.* **1999**, *104*, 1168–1173. [CrossRef]
67. Barry, P.W.; O'Callaghan, C. Drug output from nebulizers is dependent on the method of measurement. *Eur. Respir. J.* **1998**, *12*, 463–466. [CrossRef]
68. Minocchieri, S.; Burren, J.M.; Bachmann, M.A.; Stern, G.; Wildhaber, J.; Buob, S.; Schindel, R.; Kraemer, R.; Frey, U.P.; Nelle, M. Development of the premature infant nose throat-model (print-model): An upper airway replica of a premature neonate for the study of aerosol delivery. *Pediatr. Res.* **2008**, *64*, 141–146. [CrossRef]
69. Cates, C.J.; Crilly, J.A.; Rowe, B.H. Holding chambers (spacers) versus nebulisers for beta-agonist treatment of acute asthma. *Cochrane Database Syst. Rev.* **2006**, CD000052. [CrossRef]
70. Uhlig, T.; Eber, E.; Devadason, S.G.; Pemberton, P.; Badawi, N.; Lesouef, P.N.; Wildhaber, J.H. Aerosol delivery to spontaneously breathing neonates: Spacer or nebulizer? *Pediatr. Asthma Allergy Immunol.* **1997**, *11*, 111–117. [CrossRef]

71. NICE. Guidance on the Use of Inhaler Systems (Devices) in Children under the Age of 5 Years with Chronic Asthma. Technology Appraisal Guidance [ta10]. Available online: https://www.nice.org.uk/guidance/ta10 (accessed on 21 February 2019).
72. Harlos, M.S.; Stenekes, S.; Lambert, D.; Hohl, C.; Chochinov, H.M. Intranasal fentanyl in the palliative care of newborns and infants. *J. Pain Symptom Manage.* **2013**, *46*, 265–274. [CrossRef]
73. Milési, C.; Baleine, J.; Mura, T.; Benito-Castro, F.; Ferragu, F.; Thiriez, G.; Thévenot, P.; Combes, C.; Carbajal, R.; Cambonie, G. Nasal midazolam vs ketamine for neonatal intubation in the delivery room: A randomised trial. *Arch. Dis. Child. Fetal Neonatal Ed.* **2018**, *103*, F221–F226. [CrossRef]
74. Hanson, L.R.; Frey, W.H., 2nd. Intranasal delivery bypasses the blood-brain barrier to target therapeutic agents to the central nervous system and treat neurodegenerative disease. *BMC Neurosci.* **2008**, *9*, S5. [CrossRef]
75. Misra, A.; Kher, G. Drug delivery systems from nose to brain. *Curr. Pharm. Biotechnol.* **2012**, *13*, 2355–2379. [CrossRef]
76. Kozlovskaya, L.; Abou-Kaoud, M.; Stepensky, D. Quantitative analysis of drug delivery to the brain via nasal route. *J. Control. Release* **2014**, *189*, 133–140. [CrossRef]
77. Ainsworth, S.B. *Neonatal Formulary 7: Drug Use in Pregnancy and the First Year of Life*, 7th ed.; John Wiley & Sons Inc.: Chichester, UK, 2014.
78. Demirjian, A.; Levy, O. Safety and efficacy of neonatal vaccination. *Eur. J. Immunol.* **2009**, *39*, 36–46. [CrossRef]
79. Ghersi-Egea, J.-F.; Saudrais, E.; Strazielle, N.J.P.R. Barriers to drug distribution into the perinatal and postnatal brain. *Pharm. Res.* **2018**, *35*, 84. [CrossRef]
80. Saunders, N.R.; Dziegielewska, K.M.; Møllgård, K.; Habgood, M.D. Physiology and molecular biology of barrier mechanisms in the fetal and neonatal brain. *J. Physiol. (Lond.)* **2018**, *596*, 5723–5756. [CrossRef]
81. Webster, R. Presentation-Blood Brain Barrier Maturation: Implications for Drug Development. Available online: https://www.ema.europa.eu/documents/presentation/presentation-blood-brain-barrier-maturation-implications-drug-development_en.pdf (accessed on 13 March 2019).
82. Vecsernyés, M.; Fenyvesi, F.; Bácskay, I.; Deli, M.A.; Szente, L.; Fenyvesi, É. Cyclodextrins, blood–brain barrier, and treatment of neurological diseases. *Arch. Med. Res.* **2014**, *45*, 711–729. [CrossRef]
83. Merkus, F.W.; Verhoef, J.C.; Marttin, E.; Romeijn, S.G.; van der Kuy, P.H.; Hermens, W.A.; Schipper, N.G. Cyclodextrins in nasal drug delivery. *Adv. Drug Deliv. Rev.* **1999**, *36*, 41–57. [CrossRef]
84. Rivers, J.R.; Maggo, S.D.; Ashton, J.C. Neuroprotective effect of hydroxypropyl-beta-cyclodextrin in hypoxia-ischemia. *Neuroreport* **2012**, *23*, 134–138. [CrossRef] [PubMed]
85. Millar, L.J.; Shi, L.; Hoerder-Suabedissen, A.; Molnár, Z. Neonatal hypoxia ischaemia: Mechanisms, models, and therapeutic challenges. *Front. Cell. Neurosci.* **2017**, *11*. [CrossRef] [PubMed]
86. Archambault, J.; Moreira, A.; McDaniel, D.; Winter, L.; Sun, L.; Hornsby, P. Therapeutic potential of mesenchymal stromal cells for hypoxic ischemic encephalopathy: A systematic review and meta-analysis of preclinical studies. *PLoS One* **2017**, *12*, e0189895. [CrossRef] [PubMed]
87. Lee, W.L.A.; Michael-Titus, A.T.; Shah, D.K. Hypoxic-ischaemic encephalopathy and the blood-brain barrier in neonates. *Dev. Neurosci.* **2017**, *39*, 49–58. [CrossRef] [PubMed]
88. van Velthoven, C.T.; Kavelaars, A.; van Bel, F.; Heijnen, C.J. Nasal administration of stem cells: A promising novel route to treat neonatal ischemic brain damage. *Pediatr. Res.* **2010**, *68*, 419–422. [CrossRef]
89. Wolfe, T.R.; Braude, D.A. Intranasal medication delivery for children: A brief review and update. *Pediatrics* **2010**, *126*, 532–537. [CrossRef]
90. Kalia, Y.N.; Nonato, L.B.; Lund, C.H.; Guy, R.H. Development of skin barrier function in premature infants. *J. Invest. Dermatol.* **1998**, *111*, 320–326. [CrossRef]
91. Delgado-Charro, M.B.; Guy, R.H. Effective use of transdermal drug delivery in children. *Adv. Drug Deliv. Rev.* **2014**, *73*, 63–82. [CrossRef]
92. European Medicines Agency. Sodium Laurilsulfate Used as an Excipient. Available online: www.ema.europa.eu/docs/en_GB/document_library/Report/2017/10/WC500235925.pdf (accessed on 17 May 2019).
93. Oranges, T.; Dini, V.; Romanelli, M. Skin physiology of the neonate and infant: Clinical implications. *Adv. Wound Care (New Rochelle)* **2015**, *4*, 587–595. [CrossRef]
94. Chiou, Y.B.; Blume-Peytavi, U. Stratum corneum maturation. A review of neonatal skin function. *Skin Pharmacol. Physiol.* **2004**, *17*, 57–66. [CrossRef]

95. Cuzzolin, L. Off-label drug in the newborn. *J. Pediatr. Neonat. Individual. Med.* **2014**, *3*, e030224.
96. Krzyzaniak, N.; Pawlowska, I.; Bajorek, B. Review of drug utilization patterns in nicus worldwide. *J. Clin. Pharm. Ther.* **2016**, *41*, 612–620. [CrossRef] [PubMed]
97. Arenas-Lopez, S.; Stanley, I.M.; Tunstell, P.; Aguado-Lorenzo, V.; Philip, J.; Perkins, J.; Durward, A.; Calleja-Hernandez, M.A.; Tibby, S.M. Safe implementation of standard concentration infusions in paediatric intensive care. *J. Pharm. Pharmacol.* **2017**, *69*, 529–536. [CrossRef] [PubMed]
98. Kaushal, R.; Bates, D.W.; Landrigan, C.; McKenna, K.J.; Clapp, M.D.; Federico, F.; Goldmann, D.A. Medication errors and adverse drug events in pediatric inpatients. *JAMA* **2001**, *285*, 2114–2120. [CrossRef] [PubMed]
99. Larsen, G.Y.; Parker, H.B.; Cash, J.; O'Connell, M.; Grant, M.C. Standard drug concentrations and smart-pump technology reduce continuous-medication-infusion errors in pediatric patients. *Pediatrics* **2005**, *116*, e21–e25. [CrossRef] [PubMed]
100. The Joint Commission. Preventing Pediatric Medication Errors. Available online: https://www.jointcommission.org/sentinel_event_alert_issue_39_preventing_pediatric_medication_errors/ (accessed on 19 March 2019).
101. Christie-Taylor, S.A.; Tait, P.A. Implementation of standard concentration medication infusions for preterm infants. *Infant* **2012**, *8*, 155–159.
102. ISMP. High-Alert Medications in Acute Care Settings. Available online: https://www.ismp.org/recommendations/high-alert-medications-acute-list (accessed on 22 February 2019).
103. Perkins, J.; Aguado-Lorenzo, V.; Arenas-Lopez, S. Standard concentration infusions in paediatric intensive care: The clinical approach. *J. Pharm. Pharmacol.* **2017**, *69*, 537–543. [CrossRef]
104. Lala, A.C.; Broadbent, R.S.; Medlicott, N.J.; Sherwin, C.M.; Reith, D.M. Illustrative neonatal cases regarding drug delivery issues. *J. Paediatr. Child Health* **2015**, *51*, 478–481. [CrossRef]
105. Medlicott, N.J.; Reith, D.M.; McCaffrey, F.; Krittaphol, W.; Broadbent, R.S. Delayed delivery of intravenous gentamicin in neonates: Impact of infusion variables. *J. Pharm. Pharmacol.* **2013**, *65*, 370–378. [CrossRef]
106. Sherwin, C.M.; McCaffrey, F.; Broadbent, R.S.; Reith, D.M.; Medlicott, N.J. Discrepancies between predicted and observed rates of intravenous gentamicin delivery for neonates. *J. Pharm. Pharmacol.* **2009**, *61*, 465–471. [CrossRef]
107. RCPCH; NPPG. Position Statement 18-01: Using Standardised Strengths of Unlicensed Liquid Medicines in Children. Available online: http://nppg.org.uk/wp-content/uploads/2019/03/NPPG-Position-Statement-18-01-Dec-2018.pdf (accessed on 19 March 2019).
108. Allen, L.V., Jr.; Levinson, R.S.; Phisutsinthop, D. Compatibility of various admixtures with secondary additives at y-injection sites of intravenous administration sets. *Am. J. Hosp. Pharm.* **1977**, *34*, 939–943. [CrossRef] [PubMed]
109. Staven, V.; Iqbal, H.; Wang, S.; Gronlie, I.; Tho, I. Physical compatibility of total parenteral nutrition and drugs in y-site administration to children from neonates to adolescents. *J. Pharm. Pharmacol.* **2017**, *69*, 448–462. [CrossRef]
110. Tonnesen, H.H. Formulation and stability testing of photolabile drugs. *Int. J. Pharm.* **2001**, *225*, 1–14. [CrossRef]
111. Baertschi, S.W.; Clapham, D.; Foti, C.; Jansen, P.J.; Kristensen, S.; Reed, R.A.; Templeton, A.C.; Tonnesen, H.H. Implications of in-use photostability: Proposed guidance for photostability testing and labeling to support the administration of photosensitive pharmaceutical products, part 1: Drug products administered by injection. *J. Pharm. Sci.* **2013**, *102*, 3888–3899. [CrossRef]
112. Baertschi, S.W.; Clapham, D.; Foti, C.; Kleinman, M.H.; Kristensen, S.; Reed, R.A.; Templeton, A.C.; Tonnesen, H.H. Implications of in-use photostability: Proposed guidance for photostability testing and labeling to support the administration of photosensitive pharmaceutical products, part 2: Topical drug product. *J. Pharm. Sci.* **2015**, *104*, 2688–2701. [CrossRef] [PubMed]
113. Brustugun, J.; Tønnesen, H.H.; Edge, R.; Navaratnam, S. Formation and reactivity of free radicals in 5-hydroxymethyl-2-furaldehyde—The effect on isoprenaline photostability. *J. Photochem. Photobiol. B Biol.* **2005**, *79*, 109–119. [CrossRef]
114. Kogermann, K.; Lass, J.; Nellis, G.; Metsvaht, T.; Lutsar, I. Age-appropriate formulations including pharmaceutical excipients in neonatal medicines. *Curr. Pharm. Des.* **2017**, *23*, 5779–5789. [CrossRef] [PubMed]

115. Nellis, G.; Metsvaht, T.; Varendi, H.; Toompere, K.; Lass, J.; Mesek, I.; Nunn, A.J.; Turner, M.A.; Lutsar, I.; Esnee Consortium. Potentially harmful excipients in neonatal medicines: A pan-european observational study. *Arch. Dis. Child.* **2015**, *100*, 694–699. [CrossRef]
116. Whittaker, A.; Currie, A.E.; Turner, M.A.; Field, D.J.; Mulla, H.; Pandya, H.C. Toxic additives in medication for preterm infants. *Arch. Dis. Child. Fetal Neonatal Ed.* **2009**, *94*, F236–F240. [CrossRef] [PubMed]
117. Graham, S.; Turner, M. European study of neonatal exposure to excipients (esnee). *Infant* **2011**, *7*, 196–199.
118. Turner, M.A.; Duncan, J.C.; Shah, U.; Metsvaht, T.; Varendi, H.; Nellis, G.; Lutsar, I.; Yakkundi, S.; McElnay, J.C.; Pandya, H.; et al. Risk assessment of neonatal excipient exposure: Lessons from food safety and other areas. *Adv. Drug Deliv. Rev.* **2014**, *73*, 89–101. [CrossRef]
119. EuPFI. The Step Database. Available online: http://www.eupfi.org/step-database-info/ (accessed on 22 February 2019).
120. European Medicines Agency. Excipients Labelling. Available online: https://www.ema.europa.eu/en/human-regulatory/marketing-authorisation/product-information/reference-guidelines/excipients-labelling (accessed on 22 February 2019).
121. European Medicines Agency. Propylene Glycol and Esters. Available online: https://www.ema.europa.eu/en/propylene-glycol-esters (accessed on 22 February 2019).
122. Tuleu, C.; Breitkreutz, J. Educational paper: Formulation-related issues in pediatric clinical pharmacology. *Eur. J. Pediatr.* **2013**, *172*, 717–720. [CrossRef]
123. Walker, N. Trends in Excipient Demand. Available online: https://www.pharmamanufacturing.com/articles/2017/trends-in-excipient-demand (accessed on 17 May 2019).
124. Reker, D.; Blum, S.M.; Steiger, C.; Anger, K.E.; Sommer, J.M.; Fanikos, J.; Traverso, G. "Inactive" ingredients in oral medications. *Sci. Transl. Med.* **2019**, *11*. [CrossRef]
125. Richey, R.H.; Craig, J.V.; Shah, U.U.; Nunn, A.J.; Turner, M.A.; Barker, C.E.; Ford, J.L.; Peak, M. Modric—Manipulation of drugs in children. *Int. J. Pharm.* **2013**, *457*, 339–341. [CrossRef]
126. Orubu, E.S.; Tuleu, C. Medicines for children: Flexible solid oral formulations. *Bull. World Health Organ.* **2017**, *95*, 238–240. [CrossRef]
127. Kayitare, E.; Vervaet, C.; Ntawukulilyayo, J.D.; Seminega, B.; Bortel, V.; Remon, J.P. Development of fixed dose combination tablets containing zidovudine and lamivudine for paediatric applications. *Int. J. Pharm.* **2009**, *370*, 41–46. [CrossRef]
128. Broadhurst, E.C.; Ford, J.L.; Nunn, A.J.; Rowe, P.H.; Roberts, M. Dose uniformity of samples prepared from dispersible aspirin tablets for paediatric use. *Eur. J. Hosp. Pharm. Sci.* **2008**, *14*, 27–31.
129. JustMilk. Available online: http://www.justmilk.org/ (accessed on 22 February 2019).
130. Orubu, S.E.; Hobson, N.J.; Basit, A.W.; Tuleu, C. The milky way: Paediatric milk-based dispersible tablets prepared by direct compression—A proof-of-concept study. *J. Pharm. Pharmacol.* **2017**, *69*, 417–431. [CrossRef]
131. Binte Abu Bakar, S.Y.; Salim, M.; Clulow, A.J.; Hawley, A.; Boyd, B.J. Revisiting dispersible milk-drug tablets as a solid lipid formulation in the context of digestion. *Int. J. Pharm.* **2019**, *554*, 179–189. [CrossRef]
132. Thabet, Y.; Klingmann, V.; Breitkreutz, J. Drug formulations: Standards and novel strategies for drug administration in pediatrics. *J. Clin. Pharmacol.* **2018**, *58*, S26–S35. [CrossRef]
133. Thabet, Y.; Walsh, J.; Breitkreutz, J. Flexible and precise dosing of enalapril maleate for all paediatric age groups utilizing orodispersible minitablets. *Int. J. Pharm.* **2018**, *541*, 136–142. [CrossRef] [PubMed]
134. Orlu, M.; Ranmal, S.R.; Sheng, Y.; Tuleu, C.; Seddon, P. Acceptability of orodispersible films for delivery of medicines to infants and preschool children. *Drug Deliv.* **2017**, *24*, 1243–1248. [CrossRef]
135. Parshuram, C.S.; To, T.; Seto, W.; Trope, A.; Koren, G.; Laupacis, A. Systematic evaluation of errors occurring during the preparation of intravenous medication. *CMAJ* **2008**, *178*, 42–48. [CrossRef]
136. Bhambhani, V.; Beri, R.S.; Puliyel, J.M. Inadvertent overdosing of neonates as a result of the dead space of the syringe hub and needle. *Arch. Dis. Child. Fetal Neonatal Ed.* **2005**, *90*, F444–F445. [CrossRef]
137. Stucki, C.; Sautter, A.M.; Wolff, A.; Fleury-Souverain, S.; Bonnabry, P. Accuracy of preparation of i.V. Medication syringes for anesthesiology. *Am. J. Health Syst. Pharm.* **2013**, *70*, 137–142. [CrossRef]
138. Aguado-Lorenzo, V.; Weeks, K.; Tunstell, P.; Turnock, K.; Watts, T.; Arenas-Lopez, S. Accuracy of the concentration of morphine infusions prepared for patients in a neonatal intensive care unit. *Arch. Dis. Child.* **2013**, *98*, 975–979. [CrossRef]

139. Isaac, R.E.; Duncan, H.; Marriott, J.F.; Ng, A. The Effect of Different Manipulation Techniques on the Accuracy and Reproducibility of Small Dose Volume i.V. Measurements. In Proceedings of the 6th World Congress on Pediatric Critical Care: One World Sharing Knowledge, Sydney, Australia, 13–17 March 2011; p. A99.
140. Howard, C.; Macken, W.L.; Connolly, A.; Keegan, M.; Coghlan, D.; Webb, D.W. Percutaneous endoscopic gastrostomy for refractory epilepsy and medication refusal. *Arch. Dis. Child.* **2019**. Published Online First: 4 March 2019. [CrossRef]
141. European Medicines Agency. Quality of Medicines Questions and Answers: Part 2: Administration of Oral Immediate Release Medicinal Products through Enteral Feeding Tubes New December 2018. Available online: https://www.ema.europa.eu/en/human-regulatory/research-development/scientific-guidelines/qa-quality/quality-medicines-questions-answers-part-2#administration-of-oral-immediate-release-medicinal-products-through-enteral-feeding-tubes-new-december-2018-section (accessed on 22 February 2019).
142. Duesing, L.A.; Fawley, J.A.; Wagner, A.J. Central venous access in the pediatric population with emphasis on complications and prevention strategies. *Nutr. Clin. Pract.* **2016**, *31*, 490–501. [CrossRef]
143. Zahid, N.; Taylor, K.M.; Gill, H.; Maguire, F.; Shulman, R. Adsorption of insulin onto infusion sets used in adult intensive care unit and neonatal care settings. *Diabetes Res. Clin. Pract.* **2008**, *80*, e11–e13. [CrossRef]
144. Hooymans, P.M.; Janknegt, R.; Lohman, J.J. Comparison of clonazepam sorption to polyvinyl chloride-coated and polyethylene-coated tubings. *Pharm. Weekbl. Sci.* **1990**, *12*, 188–189. [CrossRef] [PubMed]
145. Fuloria, M.; Friedberg, M.A.; DuRant, R.H.; Aschner, J.L. Effect of flow rate and insulin priming on the recovery of insulin from microbore infusion tubing. *Pediatrics* **1998**, *102*, 1401–1406. [CrossRef] [PubMed]
146. Simeon, P.S.; Geffner, M.E.; Levin, S.R.; Lindsey, A.M. Continuous insulin infusions in neonates: Pharmacologic availability of insulin in intravenous solutions. *J. Pediatr.* **1994**, *124*, 818–820. [CrossRef]
147. Trissel, L.A. *Handbook on Injectable Drugs*, 17th ed.; American Society of Health-System Pharmacists: Bethesda, MD, USA, 2013.
148. United States Pharmacopeial Convention (Ed.) USP 41 NF 36. [1664] assessment of drug product leachables associated with pharmaceutical pacakaging/delivery systems. In *The United States Pharmacopoeia|the National Formulary*, 41|36 ed.; United States Pharmacopeial Convention: Rockville, MD, USA, 2017; Volume 5, pp. 7924–7937.
149. WHO; Children's Health and the Environment. Children are not Little Adults. Available online: http://www.who.int/ceh/capacity/Children_are_not_little_adults.pdf (accessed on 3 April 2019).
150. ISO. *Iso 10993-17:2002 Biological Evaluation of Medical Devices-Part 17: Establishment of Allowable Limits for Leachable Substances*; ISO: Geneva, Switzerland, 2016.
151. Bagel-Boithias, S.; Sautou-Miranda, V.; Bourdeaux, D.; Tramier, V.; Boyer, A.; Chopineau, J. Leaching of diethylhexyl phthalate from multilayer tubing into etoposide infusion solutions. *Am. J. Health Syst. Pharm.* **2005**, *62*, 182–188. [CrossRef]
152. Malarvannan, G.; Onghena, M.; Verstraete, S.; van Puffelen, E.; Jacobs, A.; Vanhorebeek, I.; Verbruggen, S.C.A.T.; Joosten, K.F.M.; Van den Berghe, G.; Jorens, P.G.; et al. Phthalate and alternative plasticizers in indwelling medical devices in pediatric intensive care units. *J. Hazard. Mater.* **2019**, *363*, 64–72. [CrossRef]
153. Demirel, A.; et al. Hidden toxicity in neonatal intensive care units: Phthalate exposure in very low birth weight infants. *J. Clin. Res. Pediatr. Endocrinol.* **2016**, *8*, 298–304. [CrossRef]
154. CERHR. *Ntp-cerhr Monograph on the Potential Human Reproductive and Developmental Effects of di(2-ethylhexyl) Phthalate (dehp)*; NIH Publication No. 06-4476; National Toxicology Program: Bethesda, MD, USA, 2006.
155. European Chemicals Bureau. *European Union Risk Assessment Report: Bis(2-ethylhexyl)phthalate (dehp): Risk Assessment*; Institute of Health and Consumer Protection (IHCP): Ispra, Italy, 2008.
156. Wedekind, C.A.; Fidler, B.D. Compatibility of commonly used intravenous infusions in a pediatric intensive care unit. *Crit. Care Nurse* **2001**, *21*, 45–51.
157. Staven, V.; Wang, S.; Gronlie, I.; Tho, I. Development and evaluation of a test program for y-site compatibility testing of total parenteral nutrition and intravenous drugs. *Nutr. J.* **2016**, *15*, 29. [CrossRef]
158. Fink, J.B. Delivery of inhaled drugs for infants and small children: A commentary on present and future needs. *Clin. Ther.* **2012**, *34*, S36–S45. [CrossRef]
159. Amirav, I.; Halamish, A.; Gorenberg, M.; Omar, H.; Newhouse, M.T. More realistic face model surface improves relevance of pediatric in-vitro aerosol studies. *PLoS One* **2015**, *10*, e0128538. [CrossRef]

160. Amirav, I.; Newhouse, M.T.; Luder, A.; Halamish, A.; Omar, H.; Gorenberg, M. Feasibility of aerosol drug delivery to sleeping infants: A prospective observational study. *BMJ Open* **2014**, *4*, e004124. [CrossRef] [PubMed]
161. Batchelor, H.K.; Fotaki, N.; Klein, S. Paediatric oral biopharmaceutics:Key considerations and current challenges. *Adv. Drug Deliv. Rev.* **2014**, *73*, 102–126. [CrossRef] [PubMed]
162. Turner, C.; Aye Mya Thein, N.; Turner, P.; Nosten, F.; White, N.J. Rectal ph in well and unwell infants. *J. Trop. Pediatr.* **2012**, *58*, 311–313. [CrossRef] [PubMed]
163. Kamstrup, D.; Berthelsen, R.; Sassene, P.J.; Selen, A.; Müllertz, A. In vitro model simulating gastro-intestinal digestion in the pediatric population (neonates and young infants). *AAPS PharmSciTech.* **2017**, *18*, 317–329. [CrossRef] [PubMed]
164. Nicolas, J.M.; Bouzom, F.; Hugues, C.; Ungell, A.L. Oral drug absorption in pediatrics: The intestinal wall, its developmental changes and current tools for predictions. *Biopharm. Drug Dispos.* **2017**, *38*, 209–230. [CrossRef] [PubMed]
165. Villiger, A.; Stillhart, C.; Parrott, N.; Kuentz, M. Using physiologically based pharmacokinetic (pbpk) modelling to gain insights into the effect of physiological factors on oral absorption in paediatric populations. *AAPS J.* **2016**, *18*, 933–947. [CrossRef] [PubMed]
166. Batchelor, H.; Kaukonen, A.M.; Klein, S.; Davit, B.; Ju, R.; Ternik, R.; Heimbach, T.; Lin, W.; Wang, J.; Storey, D. Food effects in paediatric medicines development for products co-administered with food. *Int. J. Pharm.* **2018**, *536*, 530–535. [CrossRef] [PubMed]
167. Martir, J.; Flanagan, T.; Mann, J.; Fotaki, N. Recommended strategies for the oral administration of paediatric medicines with food and drinks in the context of their biopharmaceutical properties: A review. *J. Pharm. Pharmacol.* **2017**, *69*, 384–397. [CrossRef]
168. Fernandez Polo, A.; Cabanas Poy, M.J.; Clemente Bautista, S.; Oliveras Arenas, M.; Castillo Salinas, F.; Hidalgo Albert, E. [osmolality of oral liquid dosage forms to be administered to newborns in a hospital]. *Farm. Hosp.* **2007**, *31*, 311–314.
169. White, K.C.; Harkavy, K.L. Hypertonic formula resulting from added oral medications. *Am. J. Dis. Child.* **1982**, *136*, 931–933. [CrossRef]
170. Ellis, Z.-M.; Tan, H.S.G.; Embleton, N.D.; Sangild, P.T.; van Elburg, R.M. Milk feed osmolality and adverse events in newborn infants and animals: A systematic review. *Arch. Dis. Child. Fetal Neonatal Ed.* **2019**, *104*, F333–F340. [CrossRef]
171. Akram, G.; Mullen, A.B. Paediatric nurses' knowledge and practice of mixing medication into foodstuff. *Int. J. Pharm. Pract.* **2012**, *20*, 191–198. [CrossRef]
172. Wollmer, E.; Neal, G.; Whitaker, M.J.; Margetson, D.; Klein, S. Biorelevant in vitro assessment of dissolution and compatibility properties of a novel paediatric hydrocortisone drug product following exposure of the drug product to child-appropriate administration fluids. *Eur. J. Pharm. Biopharm.* **2018**, *133*, 277–284. [CrossRef]
173. Karkossa, F.; Krueger, A.; Urbaniak, J.; Klein, S. Simulating different dosing scenarios for a child-appropriate valproate er formulation in a new pediatric two-stage dissolution model. *AAPS PharmSciTech.* **2017**, *18*, 309–316. [CrossRef]
174. Boberg, M.; Vrana, M.; Mehrotra, A.; Pearce, R.E.; Gaedigk, A.; Bhatt, D.K.; Leeder, J.S.; Prasad, B. Age-dependent absolute abundance of hepatic carboxylesterases (ces1 and ces2) by lc-ms/ms proteomics: Application to pbpk modeling of oseltamivir in vivo pharmacokinetics in infants. *Drug Metab. Dispos.* **2017**, *45*, 216–223. [CrossRef]
175. Duan, P.; Wu, F.; Moore, J.N.; Fisher, J.; Crentsil, V.; Gonzalez, D.; Zhang, L.; Burckart, G.J.; Wang, J. Assessing cyp2c19 ontogeny in neonates and infants using physiologically based pharmacokinetic models: Impact of enzyme maturation versus inhibition. *CPT: Pharmacometrics Syst. Pharmacol.* **2019**, *8*, 158–166.
176. Emoto, C.; Johnson, T.N.; Neuhoff, S.; Hahn, D.; Vinks, A.A.; Fukuda, T. Pbpk model of morphine incorporating developmental changes in hepatic oct1 and ugt2b7 proteins to explain the variability in clearances in neonates and small infants. *CPT:Pharmacometrics Syst. Pharmacol.* **2018**, *7*, 464–473. [CrossRef]
177. Hahn, D.; Emoto, C.; Euteneuer, J.C.; Mizuno, T.; Vinks, A.A.; Fukuda, T. Influence of oct1 ontogeny and genetic variation on morphine disposition in critically ill neonates: Lessons from pbpk modeling and clinical study. *Clin. Pharmacol. Ther.* **2019**, *105*, 761–768. [CrossRef]

178. Johnson, T.N.; Bonner, J.J.; Tucker, G.T.; Turner, D.B.; Jamei, M. Development and applications of a physiologically-based model of paediatric oral drug absorption. *Eur. J. Pharm. Sci.* **2018**, *115*, 57–67. [CrossRef]
179. Mahmood, I.; Tegenge, M.A. A comparative study between allometric scaling and physiologically based pharmacokinetic modeling for the prediction of drug clearance from neonates to adolescents. *J. Clin. Pharmacol.* **2019**, *59*, 189–197. [CrossRef]
180. Michelet, R.; Van Bocxlaer, J.; Allegaert, K.; Vermeulen, A. The use of pbpk modeling across the pediatric age range using propofol as a case. *J. Pharmacokinet. Pharmacodyn.* **2018**, *45*, 765–785. [CrossRef]
181. T'jollyn, H.; Vermeulen, A.; Van Bocxlaer, J. Pbpk and its virtual populations: The impact of physiology on pediatric pharmacokinetic predictions of tramadol. *AAPS J.* **2019**, *21*, 8. [CrossRef]
182. Troutman, J.A.; Sullivan, M.C.; Carr, G.J.; Fisher, J. Development of growth equations from longitudinal studies of body weight and height in the full term and preterm neonate: From birth to four years postnatal age. *Birth Defects Res.* **2018**, *110*, 916–932. [CrossRef]
183. Allegaert, K.; Simons, S.H.P.; Tibboel, D.; Krekels, E.H.; Knibbe, C.A.; van den Anker, J.N. Non-maturational covariates for dynamic systems pharmacology models in neonates, infants, and children: Filling the gaps beyond developmental pharmacology. *Eur. J. Pharm. Sci.* **2017**, *109*, S27–S31. [CrossRef]
184. Smits, A.; De Cock, P.; Vermeulen, A.; Allegaert, K. Physiologically based pharmacokinetic (pbpk) modeling and simulation in neonatal drug development: How clinicians can contribute. *Expert Opin. Drug Metab. Toxicol.* **2019**, *15*, 25–34. [CrossRef]
185. European Medicines Agency. 10-Year Report to the European Commission: General Report on the Experience Acquired as a Result of the Application of the Paediatric Regulation. Available online: https://ec.europa.eu/health/sites/health/files/files/paediatrics/2016_pc_report_2017/ema_10_year_report_for_consultation.pdf (accessed on 19 March 2019).
186. Zisowsky, J.; Krause, A.; Dingemanse, J. Drug development for pediatric populations: Regulatory aspects. *Pharmaceutics* **2010**, *2*, 364–388. [CrossRef] [PubMed]
187. Ivanovska, V.; Rademaker, C.M.A.; van Dijk, L.; Mantel-Teeuwisse, A.K. Pediatric drug formulations: A review of challenges and progress. *Pediatrics* **2014**, *134*, 361–372. [CrossRef] [PubMed]
188. Ernest, T.B.; Craig, J.; Nunn, A.; Salunke, S.; Tuleu, C.; Breitkreutz, J.; Alex, R.; Hempenstall, J. Preparation of medicines for children—A hierarchy of classification. *Int. J. Pharm.* **2012**, *435*, 124–130. [CrossRef] [PubMed]
189. Richey, R.H.; Hughes, C.; Craig, J.V.; Shah, U.U.; Ford, J.L.; Barker, C.E.; Peak, M.; Nunn, A.J.; Turner, M.A. A systematic review of the use of dosage form manipulation to obtain required doses to inform use of manipulation in paediatric practice. *Int. J. Pharm.* **2017**, *518*, 155–166. [CrossRef] [PubMed]
190. European Medicines Agency. Paediatric-Use Marketing Authorisations. Available online: https://www.ema.europa.eu/en/human-regulatory/marketing-authorisation/paediatric-medicines/paediatric-use-marketing-authorisations (accessed on 22 February 2019).
191. EDQM; Council of Europe. European Paediatric Formulary. Available online: https://paedform.edqm.eu/home (accessed on 22 February 2019).

International Journal of
Molecular Sciences

MDPI

Review

Unveiling the Efficacy, Safety, and Tolerability of Anti-Interleukin-1 Treatment in Monogenic and Multifactorial Autoinflammatory Diseases

Alessandra Bettiol [1,2,†], Giuseppe Lopalco [3,†,*], Giacomo Emmi [2], Luca Cantarini [4], Maria Letizia Urban [2], Antonio Vitale [4], Nunzio Denora [5], Antonio Lopalco [5], Annalisa Cutrignelli [5], Angela Lopedota [5], Vincenzo Venerito [3], Marco Fornaro [3], Alfredo Vannacci [1], Donato Rigante [6,7], Rolando Cimaz [8,†] and Florenzo Iannone [3,†]

1 Department of Neurosciences, Psychology, Pharmacology and Child Health (NEUROFARBA), University of Florence, 50139 Florence, Italy; alessandra.bettiol@unifi.it (A.B.); alfredo.vannacci@unifi.it (A.V.)
2 Department of Experimental and Clinical Medicine, University of Florence, 50134 Florence, Italy; giacomaci@yahoo.it (G.E.); marialetiziaurban@hotmail.it (M.L.U.)
3 Department of Emergency and Organ Transplantation, Rheumatology Unit, University of Bari, 70121 Bari, Italy; vincenzo.venerito@gmail.com (V.V.); marco3987@hotmail.it (M.F.); florenzo.iannone@uniba.it (F.I.)
4 Department of Medical Sciences, Surgery and Neurosciences, University of Siena, 53100 Siena, Italy; cantariniluca@hotmail.com (L.C.); espositogenn@gmail.com (A.V.)
5 Department of Pharmacy-Drug Sciences, University of Bari "A. Moro", 70125 Bari, Italy; nunzio.denora@uniba.it (N.D.); antonio.lopalco@uniba.it (A.L.); annalisa.cutrignelli@uniba.it (A.C.); angelaassunta.lopedota@uniba.it (A.L.)
6 Institute of Pediatrics, Fondazione Policlinico Universitario "A. Gemelli" IRCCS, 00168 Rome, Italy; drigante@gmail.com
7 Institute of Pediatrics, Università Cattolica Sacro Cuore, 00198 Rome, Italy
8 Department of Neurosciences, Psychology, Drug Research and Child Health, Rheumatology Unit, Meyer Children's Hospital, University of Florence, 50139 Florence, Italy; r.cimaz@gmail.com
* Correspondence: glopalco@hotmail.it
† These authors equally contributed to this work.

Received: 25 March 2019; Accepted: 15 April 2019; Published: 17 April 2019

Abstract: Autoinflammatory diseases (AIDs) are heterogeneous disorders characterized by dysregulation in the inflammasome, a large intracellular multiprotein platform, leading to overproduction of interleukin-1(IL-1)β that plays a predominant pathogenic role in such diseases. Appropriate treatment is crucial, also considering that AIDs may persist into adulthood with negative consequences on patients' quality of life. IL-1β blockade results in a sustained reduction of disease severity in most AIDs. A growing experience with the human IL-1 receptor antagonist, Anakinra (ANA), and the monoclonal anti IL-1β antibody, Canakinumab (CANA), has also been engendered, highlighting their efficacy upon protean clinical manifestations of AIDs. Safety and tolerability have been confirmed by several clinical trials and observational studies on both large and small cohorts of AID patients. The same treatment has been proposed in refractory Kawasaki disease, an acute inflammatory vasculitis occurring in children before 5 years, which has been postulated to be autoinflammatory for its phenotypical and immunological similarity with systemic juvenile idiopathic arthritis. Nevertheless, minor concerns about IL-1 antagonists have been raised regarding their employment in children, and the development of novel pharmacological formulations is aimed at minimizing side effects that may affect adherence to treatment. The present review summarizes current findings on the efficacy, safety, and tolerability of ANA and CANA for treatment of AIDs and Kawasaki vasculitis with a specific focus on the pediatric setting.

Keywords: Interleukin-1; anakinra; canakinumab; innovative biotechnologies; autoinflammatory disease; Kawasaki disease; systemic juvenile idiopathic arthritis; personalized medicine; child; pediatrics

1. Introduction

Autoinflammatory diseases (AIDs) are a heterogeneous group of monogenic and multifactorial diseases characterized by recurrent or chronic inflammation caused by dysregulation of the innate immune system [1,2]. Most of these disorders have an early onset, ranging from the first h to the first decade of life. However, due to their rarity, a diagnostic delay is frequently observed [3]. The main subgroup of AIDs includes different hereditary periodic fever syndromes, such as cryopyrin-associated periodic syndrome (CAPS), tumor necrosis factor receptor-associated periodic fever syndrome (TRAPS), hyperimmunoglobulin D syndrome/mevalonate kinase deficiency (HIDS/MKD), and familial Mediterranean fever (FMF) [2]. These conditions follow an autosomal dominant (CAPS and TRAPS) or recessive (HIDS/MKD and FMF) hereditary pattern and share a common clinical background marked by recurrent fever attacks and inflammation involving different sites, such as skin, serosal membranes, joints, gastrointestinal tract, or central nervous system [4,5]. AA amyloidosis is their most serious complication in the long-term, with an overall prevalence ranging from 2% to 25% of cases [6]. Many multifactorial disorders manifest with the same clinical features observed in inherited AIDs and even share the same pathogenic pathway [7]. In this regard, recurrent fevers with arthritis and cutaneous rashes are also typical features of systemic juvenile idiopathic arthritis (SJIA), which has been classified up to now as a category of juvenile idiopathic arthritis, the most common rheumatic disease in childhood. SJIA can lead to growth retardation, osteoporosis, and life-threatening complications, such as macrophage activation syndrome (MAS), and is now considered an autoinflammatory condition [8,9]. Recently, an autoinflammatory origin has also been suggested for Kawasaki disease (KD), an acute self-limited systemic vasculitis usually occurring in children before 5 years and involving medium-sized vessels, especially coronary arteries, which represents the first cause of childhood-acquired heart disease in developed countries [10]. Irrespective of the specific underlying pathways, these syndromes are characterized by an overproduction of interleukin (IL)-1, which initiates the inflammatory cascade and leads to tissue damage of variable degrees. Therefore, appropriate and effective treatment is crucial, considering that these conditions may persist into adulthood with negative consequences on both the patient's future health and quality of life [11]. Monotherapy blocking IL-1 activity results in a sustained reduction of disease severity, but chronic treatment is often required to control inflammatory flares for the lifetime and prevent long-term complications. Therefore, monitoring the safety profile and tolerability of therapy in children affected by these disorders is of paramount importance since it may affect adherence to treatment and overall clinical efficacy. In this work, we aimed at providing current findings on the efficacy, safety, and tolerability of Anakinra (ANA) and Canakinumab (CANA) for treatment of AIDs and Kawasaki vasculitis with a specific focus on the pediatric setting.

2. IL-1 Blockade in Autoinflammatory Diseases

The IL-1 family includes key cytokines involved in the innate immune response as well as in the mechanisms of fever and inflammation [12,13]. Namely, IL-1 induces the synthesis of major inflammatory mediators, such as cyclooxygenase type 2 (COX-2), type 2 phospholipase A, and inducible nitric oxide synthase, which in turn account for the production of prostaglandin-E2, platelet activating factor, and nitric oxide [12]. The two major cytokines, IL-1α and IL-1β, exert their pro-inflammatory effects by binding the IL-1 family receptors, which are characterized by the presence of the Toll-IL-1 receptor (TIR) domain in the cytoplasmic portion [14,15]. In physiological conditions, both IL-1α and IL-1β bind to type 1 IL-1 receptor (IL-1R1) and to the adaptor protein, IL-1RAcP, in order to trigger signal transduction [16]. On the contrary, the IL-1 receptor antagonist (IL-1RA) is a competitive inhibitor that prevents IL-1α and IL-1β from interacting with the IL-1 receptor 1 (IL-1R1). Much less is instead known about the type 2 IL-1 receptor (IL-1R2), a decoy receptor for IL-1β, lacking a cytoplasmic domain, which does not have a signal role, but rather sequesters IL-1β [17]. Most AIDs are caused by a lacking regulation in the inflammasome, a large intracellular multiprotein platform, leading to an overproduction of IL-1β that plays a predominant pathogenic role in such disorders [18].

Four biologic drugs blocking IL-1 are currently available [19,20]. Of them, ANA and CANA have been approved for clinical use in Europe, whereas the soluble decoy IL-1-receptor, rilonacept, and the human-engineered monoclonal anti-IL-1β, gevokizumab, are not authorized in European countries. ANA is a human IL-1 receptor antagonist that acts by competitively inhibiting the binding of IL-1 with the IL-1 type 1 receptor [21]. ANA (Kineret®) is currently approved for the treatment of rheumatoid arthritis (RA), CAPS, and Still's disease [21]. In adult, adolescent, and pediatric patients aged 8 months or older affected by CAPS, ANA is administered at the starting dose of 1 to 2 mg/kg/day by subcutaneous (s.c.) injection. For the maintenance of response, a regimen of 1 to 2 mg/kg/day is indicated in the case of a milder disease, whereas in more severe cases, the usual maintenance dose is 3 to 4 mg/kg/day, which can be adjusted to a maximum of 8 mg/kg/day [21]. The absolute bioavailability of ANA after 70 mg s.c. injection in healthy subjects is around 95%. In RA patients, maximum plasma concentrations of ANA have been reported at 3 to 7 h after s.c. administration (1 to 2 mg/kg), whereas the half-life ranges from 4 to 6 h. The clearance of ANA is mainly mediated by the kidney, and increases along with creatinine clearance [21]. In children with CAPS, the pharmacokinetics of ANA is significantly influenced by body weight [22]. On the other hand, CANA is a human monoclonal antibody that specifically binds IL-1β, blocking its interaction with IL-1 receptor and preventing the consequent inflammatory response [23]. CANA (Ilaris®) is currently approved for the treatment of periodic fever syndromes in adults, adolescents, and children aged at least 2 years, including CAPS, TRAPS, FMF, and HIDS/MKD, as well as in the treatment of Still's disease and gouty arthritis [23]. The recommended starting dose of CANA in adults, adolescents, and children aged 2 years (and older) is 150 mg for patients with body weight > 40 kg and 2 mg/kg for patients with body weight ≥ 7.5 kg and ≤ 40 kg. CANA is administered every four weeks as a single dose via s.c. injection. In CAPS children aged 2 to 4 years with body weight ≥ 7.5 kg and in adolescents and children older than 4 years with body weight between 7.5 and 15 kg, the starting dose of CANA is 4 mg/kg. In patients with body weight between 15 kg and 40 kg, the indicated starting dose is 2 mg/kg, whereas in patients weighing more than 40 kg CANA should be initially administered at the dose of 150 mg [23]. CANA should be administered every 8 weeks, as a single dose via s.c. injection. Maintenance dose should be defined based on the initial response [23]. In adults, the peak of serum CANA concentration (Cmax) occurs approximately 7 days after a single s.c. administration of 150 mg, whereas the mean half-life is around 26 days. The absolute bioavailability of CANA is estimated to be 66%. Body weight significantly influences both CANA distribution and elimination [23]. In very young children, a modest increase in the turnover rate of IL-1β has been observed. In pediatric patients, no age-related variations of CANA clearance and volume of distribution can be found after correction for body weight [24]. A schematic representation of IL-1 inhibition with CANA and ANA is depicted in Figure 1.

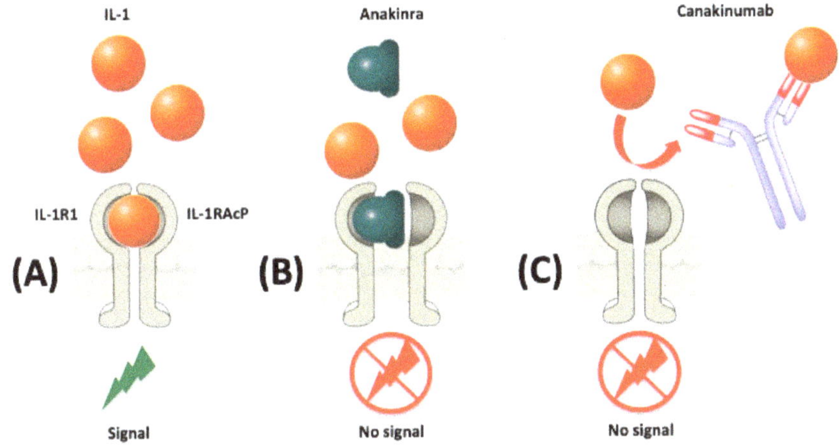

Figure 1. Strategies for IL-1 blockade with Anakinra and Canakinumab. (**A**) Interleukin (IL)-1 binds to type 1 IL-1 receptor (IL-1R1) and to the adaptor protein, IL-1RAcP, in order to trigger signal transduction. (**B**) The recombinant human IL-1R1 antagonist, Anakinra, directly competes with IL-1 for binding to the IL-1R1, blocking the biological activity of IL-1. (**C**) The human monoclonal IgG1 antibody, Canakinumab, selectively neutralizes IL-1β and inhibits its binding to the IL-1 receptor.

3. Cryopyrin-Associated Periodic Syndrome

CAPS includes a spectrum of apparently distinct inflammatory disorders of increasing severity: Familial cold autoinflammatory syndrome (FCAS), Muckle-Wells syndrome (MWS), and chronic infantile neurological, cutaneous, articular (CINCA) syndrome/neonatal-onset multisystem inflammatory disease (NOMID), all caused by mutations in *NLRP3*, the gene encoding cryopyrin, a component of the IL-1 inflammasome that regulates the production of IL-1β [25]. Therapeutic strategies specifically aimed at blocking IL-1 have been widely evaluated in CAPS patients (Table 1). Namely, the efficacy and safety profile of CANA in CAPS have been extensively assessed, both in clinical trials as well as in the real clinical practice. In a 48-week double-blind placebo-controlled randomized withdrawal study, the use of CANA (administered s.c. at the dose of 150 mg or 2 mg/kg for patients weighing 40 kg or less) every 8 weeks was evaluated in a cohort of 35 CAPS patients [26]. Namely, 4 patients were aged 4 to 16 years, whereas the remaining 31 patients were aged 17 to 75 years. Of them, 26 had history of a previous anti-IL1 treatment. A single dose of CANA accounted for complete response in 34 patients (97%); of note, CAPS symptoms significantly improved within 24 h in patients who had a response. Regarding patients entering the double-blind phase, all the 15 patients continuing CANA treatment remained in remission, whereas 81% of patients (13/16) in the placebo group had a disease flare, after a median time of 100 days from the start of placebo. As for safety, the proportion of patients experiencing at least one adverse event ranged from 77% in the third phase of the study to 100% in the second phase. Interestingly, during the same phase, adverse events were also recorded in 88% of patients treated with placebo. Most frequent adverse events included nasopharyngitis, rhinitis, diarrhea, nausea, influenza, bronchitis, headache, and vertigo. Overall, 9 patients experienced serious adverse events. There were no reports of severe injection-site reactions. Furthermore, no safety issues emerged regarding blood monitoring and urinalysis. In a phase II open-label study on 7 pediatric patients with CAPS (5 children with MWS and 2 adolescents with NOMID) [27], CANA (2 mg/kg s.c. for patients weighing ≤ 40 kg or 150 mg s.c. for patients weighing > 40 kg) led to a complete response within 7 days after the first dose in all cases. According to physicians' assessments, a relevant improvement in symptoms occurred within 24 h after the first dose. Six patients were retreated on relapse, and 4 achieved a second complete response within 7 days following retreatment. CANA was generally well tolerated. Only one severe adverse event (vertigo) was reported. Most frequent adverse events were upper respiratory tract infection ($n = 5$), rash ($n = 4$), pharyngitis ($n = 3$), nasopharyngitis ($n = 3$), and vomiting ($n = 3$). In another open-label multicentre phase III study conducted on 109 CANA-naïve adult and pediatric patients and 57 patients with previous history of CANA treatment, CANA was administered at the dose of 150 mg or 2 mg/kg every 8 weeks for up to 2 years [28]. Among CANA-naïve patients, complete response was achieved in 85 cases (78%), 79 of which occurred within the first 8 days of treatment. The other 23 patients who did not achieve complete response showed variable disease improvement. Data of the relapse assessment were available for 141 patients; 90% of them did not relapse, whereas 14 had a clinical relapse on at least one occasion. Overall, the median duration of treatment was 414 days (29–687 days) for the entire cohort, and 290 days (29–625 days) for pediatric patients. In order to control disease, higher doses were required in pediatric cases (≤ 40 kg) compared with adults, and in patients with NOMID compared with other phenotypes. Overall, 40 patients needed dose (or dose frequency) adjustments to control the disease. Overall, 90.4% of patients ($n = 150$) experienced at least one adverse event. Most frequent adverse events included headache ($n = 34$), rhinitis ($n = 27$), arthralgia ($n = 24$), bronchitis ($n = 18$), diarrhea ($n = 18$), and upper respiratory tract infections ($n = 17$). Eighteen patients experienced at least one severe adverse event. In the pediatric cohort, six serious adverse events were reported, related to tonsillitis ($n = 3$), severe intra-abdominal abscess following appendicitis, severe bronchitis, and pneumonia. In a double-blind placebo-controlled randomized withdrawal study by Koné-Paut et al. [29], 35 CAPS patients (of whom 5 were pediatric) received CANA 150 mg s.c. every 8 weeks. According to the study protocol, a double-blind placebo-controlled withdrawal phase was performed from week 9 to 24, whereas in the open-label phase from week 24 to 48 all patients resumed CANA. On day 8, 89% of patients had minimal or no disease activity. By day 8, clinical response was associated with a decrease of inflammatory markers, and considerable improvement in all 36-item Short-Form Health Survey (SF-36) domain scores. Response was sustained in patients treated with 8-weekly CANA, whereas it was rapidly lost during the placebo-controlled phase in patients receiving placebo and regained following CANA resumption. The 48-week treatment with CANA was generally well tolerated. Only two patients experienced serious adverse events. Namely, one patient had recurrent antibiotic-resistant lower urinary tract infection and sepsis, which required CANA discontinuation; vertigo and increased intraocular pressure, acute glaucoma, and unilateral blindness (complications of CAPS) were observed in the second patient. In an open-label study on 19 Japanese patients aged 2 to 48 years affected by NOMID ($n = 12$) or MWS ($n = 7$), CANA was administered every 8 weeks for 24 weeks, at the dose of 150 mg s.c. or 2 mg/kg for patients with a body weight over and under 40 kg, respectively [30]. Complete response was achieved in 18 out of 19 patients, though with dose and/or frequency adjustments. At day 24, relapse occurred in four patients, whereas one patient discontinued CANA before week 24. AID activity scores and inflammatory markers significantly decreased following CANA treatment. Namely, mean C-reactive protein levels decreased by 2.94 ± 2.99 mg/dL, dropping to 1.19 mg/dL at the end of the study compared to 4.52 mg/dL at baseline. A similar trend was observed for serum amyloid-A (SAA) levels. Interestingly, anti-CANA antibodies were detected in 3/19 patients, but the presence of these antibodies was not confirmed in subsequent evaluations. As for the CANA safety profile, 18 patients (95%) experienced one or more adverse events. Specifically, most common adverse events included nasopharyngitis ($n = 7$), gastroenteritis ($n = 6$), upper respiratory tract infection ($n = 3$), and rhinorrhea ($n = 3$). Most adverse events were mild, and only 3 were considered of moderate severity. Severe adverse events related to diffuse vasculitis ($n = 1$) and pneumonia ($n = 1$) were also reported. Higher CANA doses (> 150 mg or 2 mg/kg every 8 weeks) did not appear to be associated with a differential safety profile. In the extension phase of this trial [31], both the efficacy and safety of CANA were evaluated over a 22 month-period. After 48 weeks of treatment, as well as at the end of the study period, all patients had a complete response. All patients experienced at least one adverse event during the study, with the most common event being upper respiratory tract infections ($n = 14$). Serious adverse events were recorded in five patients, and included multiple infections, pneumonia, sinoatrial block, headache, asthma, and appendicitis. No permanent discontinuation of CANA due to adverse events was reported. Regarding ANA, a 5-year prospective open-label cohort study evaluated the safety profile of ANA in 43 CAPS patients (of whom 36 were children) [32]. The ANA starting dose ranged from 0.5 to 1.5 mg/kg/day, but was adjusted to 1.5 to 2.5 mg/kg/day during the follow-up period. The number of adverse events reported during the 5-year study period was 1233, giving an overall reporting rate of 7.7 events per patient per year. Most frequent adverse events included headache ($n = 21$), arthralgia ($n = 18$), and upper respiratory tract infections ($n = 17$). Serious adverse events occurred in 14 patients, mostly within the first year of treatment and mostly related to infections. There were no deaths and all adverse events resolved during the study period.

Table 1. Main clinical trials evaluating the use of Anakinra (ANA) and Canakinumab (CAN) for the treatment of cryopyrin-associated periodic syndrome (CAPS), tumor necrosis factor receptor-associated periodic syndrome (TRAPS), familial Mediterranean fever (FMF), and hyperimmunoglobulin D syndrome (HIDS)/mevalonate kinase deficiency (MKD) in the pediatric population.

Authors	Title	Study Design	Population	Drug
Cryopyrin-Associated Periodic Syndrome (Including CINCA/NOMID and MWS)				
Goldbach-Mansky et al., 2006 [33]	Neonatal-onset multisystem inflammatory disease responsive to interleukin-1beta inhibition.	Clinical trial	Pediatric + adults (total $n = 18$)	ANA (1 to 2 mg/kg/day s.c.)
Goldbach-Mansky et al., 2006 [33]	Neonatal-onset multisystem inflammatory disease responsive to interleukin-1beta inhibition.	Clinical trial	Pediatric ($n = 15$) + adults ($n = 3$)	ANA (1 to 2 mg/kg/day s.c.)
Imagawa et al., 2013 [30]	Safety and efficacy of Canakinumab in Japanese patients with phenotypes of cryopyrin-associated periodic syndrome as established in the first open-label, phase-3 pivotal study (24-week results).	Phase-3 pivotal study	Pediatric + adults (total $n = 19$)	CANA (150 mg s.c. (or 2 mg/kg for patients with a body weight ≤ 40 kg) every 8 weeks for 24 weeks)
Koné-Paut et al., 2011 [29]	Sustained remission of symptoms and improved health-related quality of life in patients with cryopyrin-associated periodic syndrome treated with Canakinumab: Results of a double-blind placebo-controlled randomized withdrawal study.	Double-blind placebo-controlled randomized withdrawal study	Pediatric + adults (total $n = 35$)	CANA (150 mg s.c. at 8-week intervals)
Kuemmerle-Deschner et al., 2011 [27]	Canakinumab (ACZ885, a fully human IgG1 anti-IL-1β mAb) induces sustained remission in pediatric patients with cryopyrin-associated periodic syndrome (CAPS).	Phase II, open-label study	Pediatric ($n = 7$)	CANA (2 mg/kg s.c. for patients ≤ 40 kg or 150 mg s.c. for patients > 40 kg; re-dosing upon each relapse)
Kuemmerle-Deschner et al., 2011 [28]	Two-year results from an open-label multicentre phase III study evaluating the safety and efficacy of Canakinumab in patients with cryopyrin-associated periodic syndrome across different severity phenotypes.	Open-label, multicentre, phase III study	Pediatric ($n = 47$) + adults ($n = 119$)	CANA (s.c. 150 mg or 2 mg/kg for patients ≤ 40 kg every 8 weeks for up to 2 years)
Kullenberg et al., 2016 [32]	Long-term safety profile of Anakinra in patients with severe cryopyrin-associated periodic syndrome.	Prospective, open-label, single center trial	Pediatric ($n = 36$) + adults ($n = 7$)	ANA (starting dose 0.5 to 2.5 mg/kg/day)
Lachmann et al., 2009 [26]	Use of Canakinumab in the cryopyrin-associated periodic syndrome.	Double-blind, placebo-controlled, randomized withdrawal study	Pediatric ($n = 4$) + adults ($n = 31$)	CANA (part 1: 150 mg s.c.; part 2: Either 150 mg of CANA or placebo every 8 weeks for up to 24 week)
Sibley et al., 2017 [34]	A 24-month open-label study of Canakinumab in neonatal-onset multisystem inflammatory disease.	Open-label study	Pediatric ($n = 2$) + adults ($n = 4$)	CANA (150 mg (or 2 mg/kg in patients ≤ 40 kg) or 300 mg (or 4 mg/kg) with escalation up to 600 mg (or 8 mg/kg) every 4 weeks)
Wikén et al., 2018 [35]	Development and effect of antibodies to Anakinra during treatment of severe CAPS: Sub-analysis of a long-term safety and efficacy study.	Post hoc analysis on data from a prospective, open-label, single-center, clinical cohort study	Pediatric ($n = 36$) + adults ($n = 7$)	ANA (1.0 to 2.4 mg/kg/day s.c., increased to 2.0 to 5.0 mg/kg/day according to clinical need (median 3.1 mg/kg/day))
Yokota et al., 2017 [31]	Long-term safety and efficacy of Canakinumab in cryopyrin-associated periodic syndrome: results from an open-label, phase III pivotal study in Japanese patients.	Open-label, phase III pivotal study	Pediatric + adults ($n = 35$)	CANA (2 to 8 mg/kg s.c. every 8 weeks)

Table 1. *Cont.*

Authors	Title	Study Design	Population	Drug
Tumor Necrosis Factor Receptor-Associated Periodic Syndrome				
De Benedetti et al., 2018 [36]	Canakinumab for the treatment of autoinflammatory recurrent fever syndromes.	Clinical trial	Pediatric (TRAPS $n = 27$) + adult (TRAPS $n = 19$)	CANA (initial phase 150 mg/4 weeks)
Gattorno et al., 2008 [37]	Persistent efficacy of Anakinra in patients with tumor necrosis factor receptor-associated periodic syndrome.	Clinical trial	Pediatric ($n = 4$) + adult ($n = 1$)	ANA (1.5 mg/kg/day)
Gattorno et al., 2017 [38]	Canakinumab treatment for patients with active recurrent or chronic TNF receptor-associated periodic syndrome (TRAPS): An open-label, phase II study.	Open-label, phase II study	Pediatric ($n = 6$) + adults ($n = 14$)	CANA (150 mg every 4 weeks for 4 months (2 mg/kg for patients ≤ 40 kg))
Torene et al., 2017 [39]	Canakinumab reverses overexpression of inflammatory response genes in tumor necrosis factor receptor-associated periodic syndrome.	Open-label, multicentre, proof-of-concept study (gene analysis)	Pediatric (n.a.) + adults (n.a.)	CANA (150 mg every 4 weeks for 4 months)
Familial Mediterranean Fever				
Brik et al., 2014 [40]	Canakinumab for the treatment of children with colchicine-resistant familial Mediterranean fever: A 6-month open-label, single-arm pilot study.	Open-label, single-arm pilot study	Pediatric ($n = 7$)	CANA (2 mg/kg (maximum 150 mg); the dose was doubled to 4 mg/kg (maximum 300 mg) if an attack occurred between day 1 and day 29)
De Benedetti et al., 2018 [36]	Canakinumab for the treatment of autoinflammatory recurrent fever syndromes.	Clinical trial	Pediatric (crFMF $n = 29$) + adult (crFMF $n = 34$)	CANA (initial phase 150 mg every 4 weeks)
Gül et al., 2015 [41]	Efficacy and safety of Canakinumab in adolescents and adults with colchicine-resistant familial Mediterranean fever.	Open-label pilot study	Pediatric (n.a.) + adults (n.a.)	CANA (3 injections 150 mg s.c. injections every 4 weeks)
Hyperimmunoglobulin D Syndrome/Mevalonate Kinase Deficiency				
Arostegui et al., 2017 [42]	Open-label phase II study to assess the efficacy and safety of Canakinumab treatment in active hyperimmunoglobulinemia D with periodic fever syndrome.	Clinical trial	Pediatric ($n = 6$) + adults ($n = 3$)	CANA (300 mg or 4 mg/kg every 6 weeks)
De Benedetti et al., 2018 [36]	Canakinumab for the treatment of autoinflammatory recurrent fever syndromes.	Clinical trial	Pediatric (MKD $n = 54$) + adult (MKD $n = 18$)	CANA (initial phase 150 mg/4 weeks)

ANA: Anakinra; CANA: Canakinumab; n.a.: not applicable; s.c.: subcutaneous.

We report below, distinctly, the most severe forms of CAPS, including CINCA/NOMID and MWS, focusing on the main studies in which ANA and CANA were employed.

CINCA/NOMID represents the most severe form of CAPS and is characterized by the triad of cutaneous urticarial-like rash, arthropathy, and central nervous system (CNS) involvement, in association with typical dysmorphic features, including frontal bossing and saddle back nose. CNS manifestations encompass chronic aseptic meningitis leading to brain atrophy and sensorineural hearing loss. Anti-IL-1 treatment represents the standard therapy for this condition. A trial by Goldbach-Mansky et al. evaluated both the efficacy and safety of ANA (1 to 2 mg/kg s.c.) in 18 NOMID patients [33]. All patients had a prompt clinical response to ANA. In detail, rash and conjunctivitis disappeared within three days and all inflammatory markers, including SAA, rapidly dropped. After 3 months of treatment, 11 patients underwent a withdrawal phase of a maximum of 7 days. Disease flare occurred in all except one patient, after a median time of 5 days (2.5 to 7 days). ANA resumption was associated with a rapid response, and improvements were maintained over the 6 month follow-up. As for specific CNS response, ANA treatment was associated with a significant decrease from 0.5 to 0.1 ($p < 0.001$) in the median daily headache scores (classified from 0 to 4 for increasing severity). In 12 patients in whom cerebrospinal fluid was evaluated, intracranial pressure, protein levels, and white-cell counts also decreased significantly. Furthermore, brain magnetic resonance imaging showed a relevant improvement in cochlear and leptomeningeal lesions as compared with baseline. Adverse events reported during ANA treatment included upper respiratory infections ($n = 15$), urinary tract infections ($n = 2$), and dehydration from nonbacterial diarrhea requiring hospitalization ($n = 1$). Localized, injection-site erythematous reactions were reported in eight cases. However, none of the patients discontinued drug treatment. The use of ANA in NOMID is further supported by evidence coming from four observational studies [43–45] as well as from different isolated experiences (see details in Table 1).

In a 24-month open-label phase I/II study [34], six patients aged 11 to 34 years were treated with CANA 150 mg (or 2 mg/kg in patients ≤ 40 kg) or 300 mg (or 4 mg/kg) with escalation up to 600 mg (or 8 mg/kg) every 4 weeks, after discontinuation of previous ANA treatment. CANA led to a significant improvement in all patients, with four patients having inflammatory remission at month 6. CNS remission was not achieved by any of the six patients. However, five patients had a significant improvement in their headache diary scores, and the other patient had a normal CSF leucocyte count, albeit showing a persistent headache. Overall, CANA was well tolerated. Only one severe adverse event related to a methicillin-resistant *Staphylococcus aureus* abscess was recorded. Twelve infection-related adverse events occurred in six patients.

MWS is a rare autosomal dominant disease belonging to the family of CAPS: Its manifestations include urticaria-like rashes, arthralgia, progressive sensorineural deafness, episodic fever, and renal amyloidosis. IL-1 inhibitors have been shown to be revolutionary in MWS [46]. The use of ANA in MWS was evaluated in a single-center observational study on 12 patients (5 children and 7 adults) with severe MWS [47] (see Table 1). After 2 weeks of treatment, a significant decrease in disease activity was reported, leading to significant improvement of organ manifestations as well as improved inflammatory parameters. Treatment was well tolerated, and no severe adverse events were reported. In another single-center open-label prospective observational study on patients diagnosed with active MWS between 2004 and 2008, ANA was started in five pediatric and seven adult patients, whereas CANA was initiated in six children and eight adults [48]. Both treatments led to a significant reduction of disease activity and inflammatory parameters. After a mean time of 12 months (range 5 to 14 months) for ANA and 11 months (6 to 15 months) for CANA, disease remission was achieved in 75% and 93% of patients, respectively. No detectable difference in treatment efficacy was found when comparing anti-IL-1 naïve versus rollover patients (i.e., treated with CANA after ANA discontinuation). In the ANA cohort, no serious adverse events were observed. Mild adverse events included injection-site reactions in five patients and mild upper respiratory infections in four patients, respectively. In the CANA cohort, vertigo occurred in one patient and required hospitalization. No injection-site reactions were observed. Other adverse events included mild upper respiratory tract symptoms in four patients and transient headache in two patients. The use of ANA in MWS is further supported by evidence deriving from five case reports [49–53].

4. Tumor Necrosis Factor Receptor-Associated Periodic Syndrome

TRAPS is the most frequent autosomal dominant autoinflammatory disorder, which is caused by mutations in the *TNFRSF1A* gene, encoding the 55-kD type-1 receptor of tumor necrosis factor (TNF)-α [54]. The average age at disease onset is around 3 years; however, adult onset has been described up to the sixth decade of life. This disease is clinically characterized by recurrent episodes of long-lasting fever and inflammation involving different organs, such as the skin, gastrointestinal tract, serous membranes, joints, muscles, and eyes. Inflammatory attacks are initially responsive to corticosteroids, but the progressive loss of response and recurrence of uncontrolled attacks is further associated with the development of secondary amyloidosis [55]. As IL-1-mediated inflammation is directly involved in the pathogenesis of this syndrome, anti-IL-1 treatments are gaining a relevant role in the management of this disorder (Table 1) [56,57]. Namely, Gattorno et al. evaluated the use of ANA in four children and one adult with TRAPS [37]: All patients had a rapid response to ANA, with a disappearance of symptoms and normalization of inflammatory parameters, including SAA. According to the study protocol, ANA was discontinued after 15 days of treatment. In all cases, a disease relapse was observed, within a mean time of 6 days (range, 3 to 8 days), whereas reintroduction of ANA was associated with a regain of disease control. Over a mean follow-up of 11 months, no fever episodes as well as no disease-related clinical manifestations were observed. The only adverse events reported were cutaneous reactions, including rash, pain, and itching, all occurring during the first week of treatment. Use of CANA in TRAPS has been evaluated in an open-label phase II study on 20 patients aged 7 to 78 years, with active recurrent or chronic TRAPS [38]. CANA, at the dose of 150 mg every 4 weeks for 4 months or 2 mg/kg for patients weighing 40 kg or less, induced clinical remission (Physician's Global Assessment score ≤ 1) and full or partial serological remission within day 15 in 95% of patients ($n = 19$). Responses to CANA occurred rapidly, with a median time to clinical remission of 4 days (3–8 days). Furthermore, a significant improvement in the quality of life was also reported. According to the study protocol, CAN was withdrawn after 4 months of treatment: All patients relapsed following CANA discontinuation after a median time of 91.5 days (range of 65 to 117 days). However, CANA was well tolerated. All patients reported at least one adverse event, and most common events were nasopharyngitis ($n = 12$), abdominal pain ($n = 11$), headache ($n = 11$), and oropharyngeal pain ($n = 10$). Serious adverse events were reported in seven patients, and included pericarditis, abdominal pain, diarrhea, intestinal obstruction, vomiting, upper respiratory tract infection, meniscus injury, hypertriglyceridemia, and hyperkalemia. No significant changes in laboratory, clinical, and vital parameters were reported, and no patients developed anti-drug antibodies.

5. Familial Mediterranean Fever

FMF is the most frequent autosomal recessive autoinflammatory disorder, characterized by self-limited episodes of fever and polyserositis, which may also lead to long-term complications, such as renal amyloidosis [58,59]. Although its pathogenesis is not fully understood, the *MEFV* gene encodes mutant pyrins, crucial players in the regulation of innate immunity, which lead to uncontrolled production of IL-1 [59]. Use of CANA in colchicine-resistant FMF (crFMF) is supported by three clinical trials (Table 1). In the 6-month open-label, single-arm pilot study by Brik et al. [40], seven children with crFMF were treated with subcutaneous injections of CANA at the dose of 2 mg/kg (maximum 150 mg), 4 weeks apart. Six patients experienced a reduction of 76% to 100% of FMF attacks, and three did not experience any attacks during the treatment phase. After the last CANA injection, five participants developed an attack, after a median time of 25 days (range of 5 to 34 days). Overall, 11 adverse events were reported in four patients, including two cases of infections. All adverse events were mild, except a moderate streptococcal throat infection. No significant laboratory abnormalities and formation of neutralizing anti-CANA antibodies were reported. In another open-label trial on nine adolescents and adults with crFMF [41], CANA accounted for a reduction of 50% or more in attack frequency in all treated patients, with only one patient experiencing an attack during the treatment period. Furthermore, following CANA administration, a significant improvement was observed in both the physical and mental component assessed by SF-36. Following the last CANA injection, five patients had an attack, after a median time of 71 days (range of 31 to 78 days). Eight patients (89%) experienced one or more adverse events. Namely, the most frequent adverse events were headache ($n = 4$), and respiratory tract infections. Other adverse events included, anxiety, hidradenitis, pruritus, tooth infection, and vomiting, all reported in one case each. Effectiveness and safety of CANA in crFMF is further supported by two long-term retrospective observational studies on 15 children [60] and 14 adolescent or adult patients [61], respectively. On the other hand, the use of ANA in FMF is mainly supported by observational evidence. In a study by Başaran et al. [62], eight children and adolescents with crFMF were treated with ANA (1 mg/kg/day to 3 mg/kg/day). A switch to CANA was required in four cases, due to noncompliance to daily injections in three cases and clinical and laboratory worsening in one case. Overall, the use of anti-IL-1 drugs was beneficial to all patients, both in terms of a reduction of attack frequency and normalization of inflammatory parameters. No severe adverse events were reported, and only one patient experienced a local injection-site erythema with ANA.

Yet, in another retrospective study by Cetin et al. [63] conducted on 20 patients with crFMF, 4 pediatric patients were treated with ANA or CAN (in two cases each). In two children treated with ANA, the number of monthly attacks decreased from 4 to 0 in one case and from 1 to 0 in the other one. ANA was continued for 12 and 7 months, respectively, and was still ongoing at the end of the study. Similarly, CANA also accounted for a significant reduction both in the rate of monthly and annual attacks. No adverse skin or allergic events was reported. Biopsy-proven amyloidosis was present in one child treated with CANA, with a significant decrease of urinary protein excretion (from 25.6 mg/m^2/h to 12 mg/m^2/h before versus after CANA treatment, respectively). In the retrospective study by Özçakar et al. [64], three pediatric patients suffering from crFMF were treated with ANA and one case with CAN. Three of them (two treated with ANA and one with CANA) had concomitant FMF-related amyloidosis. Among the three patients treated with ANA, the occurrence of attacks was significantly reduced. However, for the two patients with amyloidosis, one needed renal transplantation and the other had end-stage renal disease. In the patient treated with CANA, the frequency of disease attacks was significantly reduced (from 24 attacks/year to 0), and partial remission of nephrotic syndrome was achieved. None of these patients experienced drug-related adverse events.

6. Hyperimmunoglobulin D Syndrome/Mevalonate Kinase Deficiency

HIDS/MKD is an autosomal recessive disorder caused by mutations in the gene encoding the enzyme mevalonate kinase, directly involved in cholesterol and isoprenoid biosynthesis. This autoinflammatory disease usually starts in the first year of life and is characterized by lifelong recurrent fever episodes (every 4 to 6 weeks), typically lasting from 3 to 7 days [65,66]. The clinical signs range from the milder HIDS to its most severe expression, named "mevalonic aciduria". The most frequent symptoms during febrile attacks, sometimes precipitated by vaccinations, infections, emotional stress, trauma, or surgery, are abdominal pain, diarrhea, vomiting, arthralgia, lymphadenopathy, heterogeneous skin lesions, and aphthous ulcers [4]. An increase in mevalonic acid and activation of small GTPases result in IL-1 overexpression via caspase-1 activation. Therefore, short-term IL-1 blockade may be effective for stopping inflammatory attacks [2]. To date, two clinical trials have evaluated the use of the IL-1 inhibitor, CANA, in this condition, whereas no controlled clinical trial on ANA has been conducted (Table 1). Namely, in the open-label phase II study by Arostegui et al. [42], CANA was administered subcutaneously at the dose of 300 mg (or 4 mg/kg for patients weighing ≤ 40 kg) every 6 weeks to six pediatric patients and three adults with active HIDS. In this cohort, the first CANA injection led to good or excellent control of the disease, with a median time to disease resolution of

3 days. Furthermore, CANA accounted for a significant reduction in the severity or disappearance of HIDS features, including fever, lymphadenopathy, abdominal pain, and aphthous ulcers. According to the study protocol, CANA was discontinued after 6 months. Seven out of nine patients experienced a relapse following CANA discontinuation, the median time to relapse being 110 days (range of 62 to 196 days). As for its safety profile, all nine patients experienced at least one adverse event during the study period, albeit only one was judged as drug-related (i.e., non-serious fungal vaginitis). Overall, 98 adverse events (of whom 14 serious) were observed. Most frequent non-serious adverse events were related to infections and required systemic antibiotics in most cases. As for serious adverse events, 8 out of 14 events included acute peritonitis, anemia, bacteremia due to *Streptococcus pneumoniae*, gastrointestinal bleeding, hypertensive crisis, pneumonia, and severe anemia, which occurred in the same patient. Another patient was hospitalized due to hidradenitis suppurativa, and one patient developed cellulitis in the left arm. No death was reported, and no patient required drug discontinuation because of adverse events. Furthermore, no relevant change in clinical laboratory parameters and vital signs was observed.

Use of ANA in the treatment of pediatric HIDS/MKD is supported only by observational evidence. Specifically, two studies evaluated both the efficacy and safety of ANA for this condition. In the prospective observational study by Bodar et al. [67], 11 patients with HIDS (4 pediatric patients aged 5 to 17) were treated with either continuous (n = 3) or on-demand (n = 8) ANA (starting at first symptoms of an attack, 100 mg/day or 1 mg/kg/day for 5 to 7 days). In the two pediatric patients with mevalonic aciduria, continuous ANA treatment induced partial remission only in one case, and no response in the other one. Among the other nine patients, continuous ANA treatment induced complete remission, but was further discontinued for safety reasons. Among the nine patients with on-demand therapy, ANA induced a clinical response in 8 out of 12 attacks, but did not impact on its frequency. No major adverse events were observed; local injection-site reactions as well as mild upper respiratory tract infections were the only adverse events. Of note, treatment discontinuation was required in one patient. In the cohort of 103 adult and pediatric HIDS patients investigated in the observational study by van der Hilst et al., the use of ANA was reported in 11 cases: Among them, four achieved a good response to treatment, three a partial response, whereas the remaining four had no response to therapy [68]. Use of both ANA and CANA has been further evaluated in an observational study by Galeotti et al. [69]: Eleven French adult and pediatric patients were treated with either ANA (n = 9) or CANA (n = 6, of whom four following ANA treatment), reaching a complete and partial remission in four and seven cases, respectively. Anti-IL-1 treatment was also associated with a decrease in a 12-item clinical score, in the number of days with fever during attacks as well as in the level of inflammatory parameters. Both drugs were well tolerated. During ANA treatment, four patients experienced injection-site reactions, whereas shivers and hypothermia after the first injection and bacterial pneumonia were reported in one patient each. As for CANA, adverse events included injection-site reaction (n = 1), recurrent pharyngitis (n = 1), and transient hepatitis (n = 2, one of whom was without confirmation of viral or autoimmune etiology). No alterations in hematological and urinary parameters were reported. Yet, in a national Japanese survey conducted on 10 pediatric patients suffering from MKD, the use of anti-IL-1 treatment was reported in two patients [70]: In both patients, initial ANA treatment accounted for partial response, whereas a switch to CANA led to a complete response. During anti-IL-1 treatment, transaminase elevation and arthritis were reported in one patient each.

7. Additional Evidence on IL-1 Inhibition in Autoinflammatory Diseases

Beside the above summarized evidence for specific AIDs, the use of anti-IL-1 has been evaluated also in mixed cohorts of patients with different conditions.

A retrospective chart review by Ozen et al. evaluated the treatment pattern of 134 patients with FMF (n = 49), TRAPS (n = 47), or HIDS/MKD (n = 38), highlighting the central role of anti-IL-1 agents in the management of these conditions [71]. Similarly, data derived from the Eurofever Registry and related to 496 patients with FMF, CAPS, TRAPS, MKD, pyogenic arthritis-pyoderma gangrenosum-acne

(PAPA) syndrome, deficiency of IL-1 receptor antagonist (DIRA), NLRP12-related autoinflammatory disorder, and periodic fever-aphthosis-pharyngitis-adenitis (PFAPA) syndrome pointed out the key-role of IL-1 blockade for DIRA and CAPS, as well as for cases of poorly controlled MKD, TRAPS, PAPA, and crFMF [72]. According to an Italian study aimed at evaluating the use of IL-1 inhibitors among 475 patients (of whom 111 were aged 16 or less), 86% and 56% of all treatments with ANA and CANA, respectively, were mainly related to adult onset Still's disease (18.5%), SJIA (13.5%), Behçet's disease (9.7%), FMF (7.6%), idiopathic recurrent acute pericarditis (5.6%), and TRAPS (5.0%) in the ANA group, and to Behçet's disease (14.3%), TRAPS (13.3%), FMF (5.7%), and HIDS (3.8%) in the CANA one [73].

Efficacy and safety of ANA in patients with AIDs is further supported by the result of a French nationwide survey on 189 patients [74]. On the other hand, damage caused by amyloid deposits in AIDs seems not to improve with anti-IL 1 treatment [75].

In a recent clinical trial by De Benedetti et al. [36], 63 patients with crFMF (29 were children), 72 with MKD (54 children), and 46 with TRAPS (27 children) were randomized to receive CANA 150 mg s.c. or placebo every 4 weeks, with an add-on injection of 150 mg of CANA in the case of no flare resolution. At week 16, the proportion of patients with complete response was significantly higher in the CANA group if compared with the placebo group: Sixty-one percent versus 6% for patients with crFMF ($p < 0.001$), 35% versus 6% for MKD ($p = 0.003$), and 45% versus 8% for TRAPS ($p = 0.006$). Considering also patients who required an increase in the CANA dose, a complete response was achieved in 71% of patients with crFMF, in 57% of patients with MKD, and in 73% of those with TRAPS. After 16 weeks, disease control was maintained in 46%, 23%, and 53% of patients with crFMF, MKD, and TRAPS, respectively. Numbers of adverse events observed during CANA treatment were 332, 613, and 265 among patients with crFMF, MKD, and TRAPS, respectively. Namely, numbers of adverse events related to infections were 79, 160, and 58 in the three disease groups. As for events unrelated to infections, most frequent adverse events included abdominal pain, headache, arthralgia, and injection-site reactions. This study led to market authorization of CANA for the three conditions.

8. Systemic Juvenile Idiopathic Arthritis

SJIA is an inflammatory disease associated with dysregulation of the innate immune system [7,9]. The disease is characterized by fever, lymphadenopathy, arthritis, rash, and serositis. Furthermore, complications of SJIA include invalidating arthritis and MAS, a condition characterized by unremitting high fever, pancytopenia, hepatosplenomegaly, hepatic dysfunction, encephalopathy, coagulation abnormalities, and increased levels of ferritin. As IL-1 and IL-6 have been shown to play a primary role in the pathogenesis of SJIA, anti-IL-1 treatments as well as anti-IL-6 drugs represent promising therapeutic strategies for the control of this disease [7,9]. The use of IL-1 inhibitors, ANA and CANA, in SJIA has been extensively evaluated, both in clinical trials and observational studies (Table 2). In a study by Gattorno et al. [76], ANA (at the starting dosage of 1 mg/kg/day, for a maximum of 100 mg) was administered to 22 patients aged 1 to 19 years. Within the first week of treatment, two distinct patterns of response to ANA could be distinguished. One group of 10 patients achieved prompt improvement of systemic and articular manifestations as well as improved inflammatory parameters, maintaining complete disease control during a mean follow-up of 1.36 years (range of 0.3 to 2.59 years). On the other hand, a second group of 11 patients experienced a variable response to ANA, with an improvement soon after treatment began, but with a general tendency towards relapses, particularly at the articular level. These two clusters appeared to be characterized by different clinical features, in particular patients with complete response had a significantly lower number of active joints and an increased absolute neutrophil count.

Table 2. Main clinical trials evaluating the use of Anakinra (ANA) and Canakinumab (CANA) for the treatment of systemic juvenile idiopathic arthritis (SJIA) in the pediatric population.

Systemic Juvenile Idiopathic Arthritis				
Authors	**Title**	**Study Design**	**Population**	**Drug**
Brachat et al., 2017 [77]	Early changes in gene expression and inflammatory proteins in systemic juvenile idiopathic arthritis patients on Canakinumab therapy.	Gene expression analysis (data from the two phase-3 trials evaluating CANA for SJIA)	Pediatric (n.a.)	CANA (Trial 1: Patients were randomly assigned to a single s.c. dose of CANA (4 mg/kg) or placebo. Trial 2: Open-label phase (s.c. CANA 4 mg/kg every 4 weeks for up to 32 weeks) + withdrawal phase)
Feist et al., 2018 [78]	Efficacy and safety of Canakinumab in patients with Still's disease: Exposure-response analysis of pooled systemic juvenile idiopathic arthritis data by age groups.	Pooled results of clinical trials	Pediatric (n = 216) + adolescents (n = 56) + adult (n = 29)	CANA (Study 1: Single s.c. CANA at 4 mg/kg (maximum of 300 mg) or placebo; Study 2: CANA 4 mg/kg (maximum dose of 300 mg) every 4 weeks for up to 8 months + second double-blind randomized placebo-controlled phase; Study 3: S.c. CANA 4 mg/kg every 4 weeks for 12 weeks (patients had received CANA in either Study 1 or Study 2, with an additional cohort of CANA-naïve patients); Study 4: Dose-ranging study (0.5–9 mg/kg)
Gattorno et al., 2008 [76]	The pattern of response to anti-interleukin-1 treatment distinguishes two subsets of patients with systemic-onset juvenile idiopathic arthritis.	Clinical study	Pediatric (n = 22)	ANA (starting dosage of 1 mg/kg/day, s.c. (maximum 100 mg))
Grom et al., 2016 [79]	Rate and clinical presentation of macrophage activation syndrome in patients with systemic juvenile idiopathic arthritis treated with Canakinumab.	Pooled analysis	Pediatric (n = 21)	CANA (n.a.)
Ilowite et al., 2009 [80]	Anakinra in the treatment of polyarticular-course juvenile rheumatoid arthritis: Safety and preliminary efficacy results of a randomized multicenter study.	2-week open-label run-in phase	Pediatric (n = 86)	ANA (1 mg/kg daily, maximum 100 mg/day)
Kimura et al., 2017 [81]	Pilot study comparing the Childhood Arthritis & Rheumatology Research Alliance (CARRA) systemic juvenile idiopathic arthritis consensus treatment plans.	Pilot interventional study	Pediatric (n = 30; IL-1 inhibitors n = 12)	ANA (CANA) (median initial dose of ANA 2.93 (IQR 2–3.6))
Quartier et al., 2011 [82]	A multicentre randomized double-blind placebo-controlled trial with the interleukin-1 receptor antagonist Anakinra in patients with systemic-onset juvenile idiopathic arthritis (ANAJIS trial).	Multicentre, randomized, double-blind, placebo-controlled trial	Pediatric (n = 24)	ANA (2 mg/kg s.c. daily, maximum 100 mg)
Ruperto et al. (trial 1), 2012 [83]	Two randomized trials of Canakinumab in systemic juvenile idiopathic arthritis.	2 phase III trials	Pediatric (n = 84 + 177)	CANA (s.c., 4 mg/kg per month (maximum dose, 300 mg))
Ruperto et al. (trial 2), 2012 [84]	A phase II, multicenter, open-label study evaluating dosing and preliminary safety and efficacy of Canakinumab in systemic juvenile idiopathic arthritis with active systemic features.	Phase II, multicenter, open-label, dosage-escalation study	Pediatric (n = 23)	CANA (single s.c. dose of 0.5 to 9 mg/kg)
Ruperto et al., 2018 [85]	Canakinumab in patients with systemic juvenile idiopathic arthritis and active systemic features: Results from the 5-year long-term extension of the phase III pivotal trials.	5-year long-term extension of the phase III pivotal trials.	Pediatric (n = 177; 144 in the long-term extension phase)	CANA (4 mg/kg s.c. every 4 weeks (maximum dose 300 mg); in the long-term extension, tapered to 2 mg/kg every 4 weeks in patients who were glucocorticoid free as per physicians' judgement)

ANA: Anakinra; CANA: Canakinumab; IQR: interquartile range; n.a.: not applicable; s.c.: subcutaneous.

Another multicentre randomized double-blind placebo-controlled trial compared the efficacy of ANA (2 mg/kg s.c. daily, maximum 100 mg) versus placebo in 24 SJIA patients aged 8.5 ± 4.5 years [82]. After 1 month of treatment, response (defined as a 30% improvement of the pediatric American College of Rheumatology (ACR) criteria for JIA, resolution of systemic symptoms, and a decrease of at least 50% of inflammatory parameters compared with baseline) was achieved in 8 out of 12 patients treated with ANA and only in 1 patient receiving placebo ($p = 0.003$). Ten patients were switched from placebo to ANA; of them, nine achieved response within month 2. Fourteen adverse events were recorded in the ANA group and 13 in the placebo group. No serious adverse event was reported. Namely, in the ANA group, adverse events, including injection-site pain ($n = 8$), post-injection erythema ($n = 3$), and infections involving gastrointestinal and upper respiratory tract ($n = 2$), were recorded.

In a multicenter double-blind trial on 50 patients aged 3 to 17 years with polyarticular-course JIA (11 of whom had systemic onset), randomized to ANA (1 mg/kg/day, for a maximum of 100 mg) or placebo [80], the three most common reported adverse events were injection-site reactions (12% in each group), upper respiratory infections (16% versus 20% in the ANA and placebo groups, respectively), and headache (24% versus 4%). No case of adverse event-related discontinuation of the blinded phase was observed. In the subsequent extension phase of ANA treatment, the most common adverse events reported were arthralgia (23%), fever (21%), and abdominal pain (16%), with three patients (7%) requiring discontinuation of ANA because of safety issues.

In another pilot study evaluating different therapeutic strategies for treatment of SJIA [81], 12 out of 30 patients received IL-1 inhibitors (ANA as initial treatment, eventually switched to CANA). Of them, two patients needed to add methotrexate, whereas two needed to switch to an IL-6 inhibitor. Overall, clinical disease inactivity was reached in 42% of patients treated with IL-1 inhibitors. There were four serious adverse events: Two resulted in hospitalization for intravenous antimicrobial therapy (cellulitis in a child taking CANA and glucocorticoids, and varicella in a child taking ANA), one hospitalization for appendicitis and appendectomy in a child on methotrexate and glucocorticoid, and one for MAS in a child on tocilizumab. As for CANA, its efficacy and safety in SJIA has been extensively evaluated in a pooled analysis of data coming from four trials [78,83–85]. CANA (mostly administered at the dose of 4 mg/kg every 4 weeks) was given to a total population of 233 children, 60 young adolescents and 31 older adolescents or young adults. Within day 15 of treatment, at least 50% of patients in each age group had absence of fever as well as ≥ 70% improvement according to the adapted ACR pediatric response criteria. Responses were stable and maintained or improved over the 85 days of follow-up. Similarly, clinical and laboratory findings also markedly improved in all age groups. Regarding safety, adverse events were reported in 86.7% to 88.3% of patients in the different age groups, with 11% to 19% of patients who experienced adverse events leading to treatment discontinuation. In all age groups, most common adverse events were infections (70–76%), gastrointestinal disorders (52–58%), and musculoskeletal or connective tissue disorders (51–55%). Other adverse events included disorders of the skin, subcutaneous tissue, and respiratory tract. Serious adverse events were reported in 29% to 42% of patients, and included JIA reactivation, MAS, gastroenteritis, and *Cytomegalovirus* infection. Beside the above mentioned evidence from clinical trials, the use of ANA and CANA in SJIA is also supported by growing evidence coming from observational studies [79,86–93].

9. Kawasaki Disease

KD is an acute vasculitis of unknown etiology, which is typically observed in the pediatric age. If untreated, patients with KD are at significantly higher risk of developing coronary artery abnormalities, thromboembolic occlusions, and myocardial infarction, with subsequent increased risk of mortality [94]. Many shortcomings still exist in studies aimed at defining the etiology of KD, though different levels of evidence support the hypothesis that it is a complex disease with a unique pathogenesis [95]. Intravenous immunoglobulins (IVIG) in association with aspirin represent the main treatment for KD and their administration within the first 10 days following fever onset has been associated with a 5-fold reduction in the risk of coronary artery aneurysms [96].

The extent of acute phase response and a younger age at onset may be related to patients' responsiveness to IVIG [97]. In particular, 10% to 15% of patients develop resistance to this treatment: The prediction of IVIG resistance is a crucial issue for managing these children, as recognizing high-risk patients should allow the administration of an intensified initial treatment in combination with IVIG, and prevent coronary injuries [98]. Limited and local experiences suggest the possible benefit of IL-1 inhibition in children with KD [99–104]. A retrospective case series by Koné-Paut et al. [105] evaluated the use of ANA (2 to 8 mg/kg) in 11 children with KD aged 4 months to 9 years refractory to standard treatment. Specifically, the main reasons for starting ANA were persistent inflammation, progression of coronary dilations, and severe myocarditis with cardiac failure. ANA proved to be effective in controlling KD. Namely, all patients had fever resolution and a decrease of inflammatory parameters. Furthermore, 10 out of 11 patients had a decrease in coronary artery dilations. The other patient died suddenly due to massive pericardial effusion secondary to giant aneurysm rupture while on anticoagulant treatment. To date, two trials with ANA in children with Kawasaki disease are ongoing (NCT02179853 and NCT02390596).

10. Conclusions

Recent evidence from both observational studies and clinical trials have clarified the efficacy of ANA and CANA in the main AIDs, also revealing a good safety profile with minor concerns regarding tolerability. In particular, the major treatment-related side effects of ANA are skin reactions at the injection-site. This high rate of injection-site reactions can become so irritating for pediatric patients that they require treatment withdrawal. In this regard, convincing patients, especially children, to continue therapy can be challenging. Reactions can be mitigated by the application of topical hydrocortisone or anti-histamine cream, but it may not be enough [106]. On the other hand, the overall safety of CANA has shown an excellent tolerability [107], as highlighted by very few discontinuation rates and few injection-site reactions. However, a slightly increased rate of non-serious infections related to the upper respiratory tract has been observed [26]. Although these two anti-IL1 agents represent the most effective treatments available in AIDs and also a promising tool in refractory KD, the development of novel pharmacological formulations that further reduce side effects in pediatric sceneries is expected.

Author Contributions: Conceptualization, A.B., G.L., G.E., L.C., D.R., R.C. and F.I.; Literature review, A.B., M.L.U. and G.L.; Writing—Original Draft Preparation, A.B. and G.L.; Writing—Review & Editing, A.B., G.L., G.E., M.L.U., L.C., A.V. (Antonio Vitale), A.V. (Alfredo Vannacci), N.D., A.L. (Antonio Lopalco), A.C., A.L. (Angela Lopedota), V.V., M.F., D.R., R.C. and F.I.

Funding: This research received no external funding.

Acknowledgments: We would like to thank Giovanni Lapadula for his valuable insights.

Conflicts of Interest: The authors declare no conflict of interest.

References

1. Teague, M. Pediatric rheumatologic diseases: A review for primary care NPs. *Nurse Pract.* **2017**, *42*, 43–47. [CrossRef]
2. Rigante, D. A systematic approach to autoinflammatory syndromes: A spelling booklet for the beginner. *Expert Rev. Clin. Immunol.* **2017**, *13*, 571–597. [CrossRef] [PubMed]
3. Cantarini, L.; Vitale, A.; Lucherini, O.M.; De Clemente, C.; Caso, F.; Costa, L.; Emmi, G.; Silvestri, E.; Magnotti, F.; Maggio, M.C.; et al. The labyrinth of autoinflammatory disorders: A snapshot on the activity of a third-level center in Italy. *Clin. Rheumatol.* **2015**, *34*, 17–28. [CrossRef]
4. Rigante, D.; Lopalco, G.; Vitale, A.; Lucherini, O.M.; Caso, F.; De Clemente, C.; Molinaro, F.; Messina, M.; Costa, L.; Atteno, M.; et al. Untangling the web of systemic autoinflammatory diseases. *Mediat. Inflamm.* **2014**, *2014*, 948154. [CrossRef] [PubMed]
5. Cattalini, M.; Soliani, M.; Lopalco, G.; Rigante, D.; Cantarini, L. Systemic and organ involvement in monogenic autoinflammatory disorders: A global review filtered through internists' lens. *Intern. Emerg. Med.* **2016**, *11*, 781–791. [CrossRef]

6. Obici, L.; Merlini, G. Amyloidosis in autoinflammatory syndromes. *Autoimmun. Rev.* **2012**, *12*, 14–17. [CrossRef]
7. Cimaz, R. Systemic-onset juvenile idiopathic arthritis. *Autoimmun. Rev.* **2016**, *15*, 931–934. [CrossRef] [PubMed]
8. Lee, J.J.Y.; Schneider, R. Systemic juvenile idiopathic arthritis. *Pediatr. Clin. N. Am.* **2018**, *65*, 691–709. [CrossRef]
9. Barut, K.; Adrovic, A.; Şahin, S.; Kasapçopu, Ö. Juvenile idiopathic arthritis. *Balk. Med. J.* **2017**, *34*, 90–101. [CrossRef]
10. Patel, R.M.; Shulman, S.T. Kawasaki disease: A comprehensive review of treatment options. *J. Clin. Pharm.* **2015**, *40*, 620–625. [CrossRef] [PubMed]
11. Hersh, A.; von Scheven, E.; Yelin, E. Adult outcomes of childhood-onset rheumatic diseases. *Nat. Rev. Rheumatol.* **2011**, *7*, 290–295. [CrossRef]
12. Weber, A.; Wasiliew, P.; Kracht, M. Interleukin-1 (IL-1) pathway. *Sci. Signal.* **2010**, *3*, cm1. [CrossRef] [PubMed]
13. Cantarini, L.; Lopalco, G.; Cattalini, M.; Vitale, A.; Galeazzi, M.; Rigante, D. Interleukin-1: Ariadne's Thread in autoinflammatory and autoimmune disorders. *Isr. Med. Assoc. J.* **2015**, *17*, 93–97. [PubMed]
14. O'Neill, L.A.J. The interleukin-1 receptor/Toll-like receptor superfamily: 10 years of progress. *Immunol. Rev.* **2008**, *226*, 10–18. [CrossRef] [PubMed]
15. Beesu, M.; Caruso, G.; Salyer, A.C.D.; Shukla, N.M.; Khetani, K.K.; Smith, L.J.; Fox, L.M.; Tanji, H.; Ohto, U.; Shimizu, T.; et al. Identification of a human Toll-Like Receptor (TLR) 8-specific agonist and a functional pan-TLR inhibitor in 2-aminoimidazoles. *J. Med. Chem.* **2016**, *59*, 3311–3330. [CrossRef] [PubMed]
16. Federici, S.; Martini, A.; Gattorno, M. The Central Role of Anti-IL-1 Blockade in the Treatment of Monogenic and Multi-Factorial Autoinflammatory Diseases. *Front. Immunol.* **2013**, *4*, 351. [CrossRef] [PubMed]
17. Dinarello, C.A. Overview of the IL-1 family in innate inflammation and acquired immunity. *Immunol. Rev.* **2018**, *281*, 8–27. [CrossRef] [PubMed]
18. Masters, S.L.; Simon, A.; Aksentijevich, I.; Kastner, D.L. Horror autoinflammaticus: The molecular pathophysiology of autoinflammatory disease (*). *Annu. Rev. Immunol.* **2009**, *27*, 621–668. [CrossRef]
19. Toplak, N.; Blazina, Š.; Avčin, T. The role of IL-1 inhibition in systemic juvenile idiopathic arthritis: Current status and future perspectives. *Drug Des. Dev. Ther.* **2018**, *12*, 1633–1643. [CrossRef]
20. Bettiol, A.; Silvestri, E.; Di Scala, G.; Amedei, A.; Becatti, M.; Fiorillo, C.; Lopalco, G.; Salvarani, C.; Cantarini, L.; Soriano, A.; et al. The right place of interleukin-1 inhibitors in the treatment of Behçet's syndrome: A systematic review. *Rheumatol. Int.* **2019**. [CrossRef]
21. European Medicines Agency. Summary of Product Characteristics—Kineret. Available online: https://www.ema.europa.eu/documents/product-information/kineret-epar-product-information_en.pdf (accessed on 8 March 2019).
22. Urien, S.; Bardin, C.; Bader-Meunier, B.; Mouy, R.; Compeyrot-Lacassagne, S.; Foissac, F.; Florkin, B.; Wouters, C.; Neven, B.; Treluyer, J.-M.; et al. Anakinra pharmacokinetics in children and adolescents with systemic-onset juvenile idiopathic arthritis and autoinflammatory syndromes. *BMC Pharmacol. Toxicol.* **2013**, *14*, 40. [CrossRef]
23. European Medicines Agency. Summary of Product Characteristics—Ilaris. Available online: https://www.ema.europa.eu/en/medicines/human/EPAR/ilaris (accessed on 17 April 2019).
24. Sun, H.; Van, L.M.; Floch, D.; Jiang, X.; Klein, U.R.; Abrams, K.; Sunkara, G. Pharmacokinetics and pharmacodynamics of canakinumab in patients with systemic juvenile idiopathic arthritis. *J. Clin. Pharmacol.* **2016**, *56*, 1516–1527. [CrossRef]
25. Cantarini, L.; Lucherini, O.M.; Frediani, B.; Brizi, M.G.; Bartolomei, B.; Cimaz, R.; Galeazzi, M.; Rigante, D. Bridging the gap between the clinician and the patient with cryopyrin-associated periodic syndromes. *Int. J. Immunopathol. Pharmacol.* **2011**, *24*, 827–836. [CrossRef]
26. Lachmann, H.J.; Kone-Paut, I.; Kuemmerle-Deschner, J.B.; Leslie, K.S.; Hachulla, E.; Quartier, P.; Gitton, X.; Widmer, A.; Patel, N.; Hawkins, P.N.; et al. Use of canakinumab in the cryopyrin-associated periodic syndrome. *N. Engl. J. Med.* **2009**, *360*, 2416–2425. [CrossRef] [PubMed]

27. Kuemmerle-Deschner, J.B.; Ramos, E.; Blank, N.; Roesler, J.; Felix, S.D.; Jung, T.; Stricker, K.; Chakraborty, A.; Tannenbaum, S.; Wright, A.M.; et al. Canakinumab (ACZ885, a fully human IgG1 anti-IL-1β mAb) induces sustained remission in pediatric patients with cryopyrin-associated periodic syndrome (CAPS). *Arthritis Res. Ther.* **2011**, *13*, R34. [CrossRef]
28. Kuemmerle-Deschner, J.B.; Hachulla, E.; Cartwright, R.; Hawkins, P.N.; Tran, T.A.; Bader-Meunier, B.; Hoyer, J.; Gattorno, M.; Gul, A.; Smith, J.; et al. Two-year results from an open-label, multicentre, phase III study evaluating the safety and efficacy of canakinumab in patients with cryopyrin-associated periodic syndrome across different severity phenotypes. *Ann. Rheum. Dis.* **2011**, *70*, 2095–2102. [CrossRef]
29. Koné-Paut, I.; Lachmann, H.J.; Kuemmerle-Deschner, J.B.; Hachulla, E.; Leslie, K.S.; Mouy, R.; Ferreira, A.; Lheritier, K.; Patel, N.; Preiss, R.; et al. Sustained remission of symptoms and improved health-related quality of life in patients with cryopyrin-associated periodic syndrome treated with canakinumab: Results of a double-blind placebo-controlled randomized withdrawal study. *Arthritis Res. Ther.* **2011**, *13*, R202. [CrossRef] [PubMed]
30. Imagawa, T.; Nishikomori, R.; Takada, H.; Takeshita, S.; Patel, N.; Kim, D.; Lheritier, K.; Heike, T.; Hara, T.; Yokota, S. Safety and efficacy of canakinumab in Japanese patients with phenotypes of cryopyrin-associated periodic syndrome as established in the first open-label, phase-3 pivotal study (24-week results). *Clin. Exp. Rheumatol.* **2013**, *31*, 302–309. [PubMed]
31. Yokota, S.; Imagawa, T.; Nishikomori, R.; Takada, H.; Abrams, K.; Lheritier, K.; Heike, T.; Hara, T. Long-term safety and efficacy of canakinumab in cryopyrin-associated periodic syndrome: Results from an open-label, phase III pivotal study in Japanese patients. *Clin. Exp. Rheumatol.* **2016**, *35* (Suppl. 108), 19–26.
32. Kullenberg, T.; Löfqvist, M.; Leinonen, M.; Goldbach-Mansky, R.; Olivecrona, H. Long-term safety profile of anakinra in patients with severe cryopyrin-associated periodic syndromes. *Rheumatology* **2016**, *55*, 1499–1506. [CrossRef]
33. Goldbach-Mansky, R.; Dailey, N.J.; Canna, S.W.; Gelabert, A.; Jones, J.; Rubin, B.I.; Kim, H.J.; Brewer, C.; Zalewski, C.; Wiggs, E.; et al. Neonatal-onset multisystem inflammatory disease responsive to interleukin-1beta inhibition. *N. Engl. J. Med.* **2006**, *355*, 581–592. [CrossRef]
34. Sibley, C.H.; Chioato, A.; Felix, S.; Colin, L.; Chakraborty, A.; Plass, N.; Rodriguez-Smith, J.; Brewer, C.; King, K.; Zalewski, C.; et al. A 24-month open-label study of canakinumab in neonatal-onset multisystem inflammatory disease. *Ann. Rheum. Dis.* **2015**, *74*, 1714–1719. [CrossRef] [PubMed]
35. Wikén, M.; Hallén, B.; Kullenberg, T.; Koskinen, L.O. Development and effect of antibodies to anakinra during treatment of severe CAPS: Sub-analysis of a long-term safety and efficacy study. *Clin. Rheumatol.* **2018**, *37*, 3381–3386. [CrossRef] [PubMed]
36. De Benedetti, F.; Gattorno, M.; Anton, J.; Ben-Chetrit, E.; Frenkel, J.; Hoffman, H.M.; Koné-Paut, I.; Lachmann, H.J.; Ozen, S.; Simon, A.; et al. Canakinumab for the Treatment of Autoinflammatory Recurrent Fever Syndromes. *N. Engl. J. Med.* **2018**, *378*, 1908–1919. [CrossRef] [PubMed]
37. Gattorno, M.; Pelagatti, M.A.; Meini, A.; Obici, L.; Barcellona, R.; Federici, S.; Buoncompagni, A.; Plebani, A.; Merlini, G.; Martini, A. Persistent efficacy of anakinra in patients with tumor necrosis factor receptor-associated periodic syndrome. *Arthritis Rheum.* **2008**, *58*, 1516–1520. [CrossRef]
38. Gattorno, M.; Obici, L.; Cattalini, M.; Tormey, V.; Abrams, K.; Davis, N.; Speziale, A.; Bhansali, S.G.; Martini, A.; Lachmann, H.J. Canakinumab treatment for patients with active recurrent or chronic TNF receptor-associated periodic syndrome (TRAPS): An open-label, phase II study. *Ann. Rheum. Dis.* **2017**, *76*, 173–178. [CrossRef] [PubMed]
39. Torene, R.; Nirmala, N.; Obici, L.; Cattalini, M.; Tormey, V.; Caorsi, R.; Starck-Schwertz, S.; Letzkus, M.; Hartmann, N.; Abrams, K.; et al. Canakinumab reverses overexpression of inflammatory response genes in tumour necrosis factor receptor-associated periodic syndrome. *Ann. Rheum. Dis.* **2017**, *76*, 303–309. [CrossRef]
40. Brik, R.; Butbul-Aviel, Y.; Lubin, S.; Ben Dayan, E.; Rachmilewitz-Minei, T.; Tseng, L.; Hashkes, P.J. Canakinumab for the treatment of children with colchicine-resistant familial Mediterranean fever: A 6-month open-label, single-arm pilot study. *Arthritis Rheumatol.* **2014**, *66*, 3241–3243. [CrossRef] [PubMed]
41. Gül, A.; Ozdogan, H.; Erer, B.; Ugurlu, S.; Kasapcopur, O.; Davis, N.; Sevgi, S. Efficacy and safety of canakinumab in adolescents and adults with colchicine-resistant familial Mediterranean fever. *Arthritis Res. Ther.* **2015**, *17*, 243. [CrossRef]

42. Arostegui, J.I.; Anton, J.; Calvo, I.; Robles, A.; Iglesias, E.; López-Montesinos, B.; Banchereau, R.; Hong, S.; Joubert, Y.; Junge, G.; et al. Open-label phase II Study to assess the efficacy and safety of canakinumab treatment in active hyperimmunoglobulinemia D with periodic fever syndrome. *Arthritis Rheumatol.* **2017**, *69*, 1679–1688. [CrossRef]
43. Caroli, F.; Pontillo, A.; D'Osualdo, A.; Travan, L.; Ceccherini, I.; Crovella, S.; Alessio, M.; Stabile, A.; Gattorno, M.; Tommasini, A.; et al. Clinical and genetic characterization of Italian patients affected by CINCA syndrome. *Rheumatology* **2007**, *46*, 473–478. [CrossRef] [PubMed]
44. Neven, B.; Marvillet, I.; Terrada, C.; Ferster, A.; Boddaert, N.; Couloignier, V.; Pinto, G.; Pagnier, A.; Bodemer, C.; Bodaghi, B.; et al. Long-term efficacy of the interleukin-1 receptor antagonist anakinra in ten patients with neonatal-onset multisystem inflammatory disease/chronic infantile neurologic, cutaneous, articular syndrome. *Arthritis Rheum.* **2010**, *62*, 258–267. [CrossRef] [PubMed]
45. Sibley, C.H.; Plass, N.; Snow, J.; Wiggs, E.A.; Brewer, C.C.; King, K.A.; Zalewski, C.; Kim, H.J.; Bishop, R.; Hill, S.; et al. Sustained response and prevention of damage progression in patients with neonatal-onset multisystem inflammatory disease treated with anakinra: A cohort study to determine three- and five-year outcomes. *Arthritis Rheum.* **2012**, *64*, 2375–2386. [CrossRef] [PubMed]
46. Tran, T. Muckle–Wells syndrome: Clinical perspectives. *Open Access Rheumatol. Res. Rev.* **2017**, *9*, 123–129. [CrossRef] [PubMed]
47. Kuemmerle-Deschner, J.B.; Tyrrell, P.N.; Koetter, I.; Wittkowski, H.; Bialkowski, A.; Tzaribachev, N.; Lohse, P.; Koitchev, A.; Deuter, C.; Foell, D.; et al. Efficacy and safety of anakinra therapy in pediatric and adult patients with the autoinflammatory Muckle-Wells syndrome. *Arthritis Rheum.* **2011**, *63*, 840–849. [CrossRef] [PubMed]
48. Kuemmerle-Deschner, J.B.; Wittkowski, H.; Tyrrell, P.N.; Koetter, I.; Lohse, P.; Ummenhofer, K.; Reess, F.; Hansmann, S.; Koitschev, A.; Deuter, C.; et al. Treatment of Muckle-Wells syndrome: Analysis of two IL-1-blocking regimens. *Arthritis Res. Ther.* **2013**, *15*, R64. [CrossRef]
49. Dalgic, B.; Egritas, O.; Sari, S.; Cuisset, L. A variant Muckle-Wells syndrome with a novel mutation in CIAS1 gene responding to anakinra. *Pediatr. Nephrol.* **2007**, *22*, 1391–1394. [CrossRef]
50. Maksimovic, L.; Stirnemann, J.; Caux, F.; Ravet, N.; Rouaghe, S.; Cuisset, L.; Letellier, E.; Grateau, G.; Morin, A.-S.; Fain, O. New *CIAS1* mutation and anakinra efficacy in overlapping of Muckle-Wells and familial cold autoinflammatory syndromes. *Rheumatology* **2008**, *47*, 309–310. [CrossRef]
51. Marchica, C.; Zawawi, F.; Basodan, D.; Scuccimarri, R.; Daniel, S.J. Resolution of unilateral sensorineural hearing loss in a pediatric patient with a severe phenotype of Muckle-Wells syndrome treated with Anakinra: A case report and review of the literature. *J. Otolaryngol. Head Neck Surg.* **2018**, *47*, 9. [CrossRef]
52. Stew, B.T.; Fishpool, S.J.C.; Owens, D.; Quine, S. Muckle-Wells syndrome: A treatable cause of congenital sensorineural hearing loss. *B-Ent* **2013**, *9*, 161–163.
53. Yamazaki, T.; Masumoto, J.; Agematsu, K.; Sawai, N.; Kobayashi, S.; Shigemura, T.; Yasui, K.; Koike, K. Anakinra improves sensory deafness in a Japanese patient with Muckle-Wells syndrome, possibly by inhibiting the cryopyrin inflammasome. *Arthritis Rheum.* **2008**, *58*, 864–868. [CrossRef] [PubMed]
54. Rigante, D.; Lopalco, G.; Vitale, A.; Lucherini, O.M.; De Clemente, C.; Caso, F.; Emmi, G.; Costa, L.; Silvestri, E.; Andreozzi, L.; et al. Key facts and hot spots on tumor necrosis factor receptor-associated periodic syndrome. *Clin. Rheumatol.* **2014**, *33*, 1197–1207. [CrossRef] [PubMed]
55. Magnotti, F.; Vitale, A.; Rigante, D.; Lucherini, O.M.; Cimaz, R.; Muscari, I.; Granados Afonso de Faria, A.; Frediani, B.; Galeazzi, M.; Cantarini, L. The most recent advances in pathophysiology and management of tumour necrosis factor receptor-associated periodic syndrome (TRAPS): Personal experience and literature review. *Clin. Exp. Rheumatol.* **2013**, *31*, 141–149. [PubMed]
56. Lopalco, G.; Rigante, D.; Vitale, A.; Frediani, B.; Iannone, F.; Cantarini, L. Tumor necrosis factor receptor-associated periodic syndrome managed with the couple canakinumab-alendronate. *Clin. Rheumatol.* **2015**, *34*, 807–809. [CrossRef]
57. Cantarini, L.; Lopalco, G.; Vitale, A.; Caso, F.; Lapadula, G.; Iannone, F.; Galeazzi, M.; Rigante, D. Delights and let-downs in the management of tumor necrosis factor receptor-associated periodic syndrome: The canakinumab experience in a patient with a high-penetrance T50M *TNFRSF1A* variant. *Int. J. Rheum. Dis.* **2015**, *18*, 473–475. [CrossRef] [PubMed]
58. Rigante, D.; Lopalco, G.; Tarantino, G.; Compagnone, A.; Fastiggi, M.; Cantarini, L. Non-canonical manifestations of familial Mediterranean fever: A changing paradigm. *Clin. Rheumatol.* **2015**, *34*, 1503–1511. [CrossRef] [PubMed]

59. Alghamdi, M. Familial Mediterranean fever, review of the literature. *Clin. Rheumatol.* **2017**, *36*, 1707–1713. [CrossRef] [PubMed]
60. Gülez, N.; Makay, B.; Sözeri, B. Long-Term Effectıveness and safety of canakınumab ın pedıatrıc famılıal Medıterranean fever patıents. *Mod. Rheumatol.* **2018**, 1–13. [CrossRef] [PubMed]
61. Laskari, K.; Boura, P.; Dalekos, G.N.; Garyfallos, A.; Karokis, D.; Pikazis, D.; Settas, L.; Skarantavos, G.; Tsitsami, E.; Sfikakis, P.P. Long-term beneficial effect of canakinumab in colchicine-resistant familial Mediterranean fever. *J. Rheumatol.* **2017**, *44*, 102–109. [CrossRef]
62. Başaran, Ö.; Uncu, N.; Çelikel, B.A.; Taktak, A.; Gür, G.; Cakar, N. Interleukin-1 targeting treatment in familial Mediterranean fever: An experience of pediatric patients. *Mod. Rheumatol.* **2015**, *25*, 621–624. [CrossRef]
63. Cetin, P.; Sari, I.; Sozeri, B.; Cam, O.; Birlik, M.; Akkoc, N.; Onen, F.; Akar, S. Efficacy of Interleukin-1 Targeting Treatments in Patients with Familial Mediterranean Fever. *Inflammation* **2015**, *38*, 27–31. [CrossRef] [PubMed]
64. Özçakar, Z.B.; Özdel, S.; Yılmaz, S.; Kurt-Şükür, E.D.; Ekim, M.; Yalçınkaya, F. Anti-IL-1 treatment in familial Mediterranean fever and related amyloidosis. *Clin. Rheumatol.* **2016**, *35*, 441–446. [CrossRef] [PubMed]
65. Houten, S.M.; Kuis, W.; Duran, M.; de Koning, T.J.; van Royen-Kerkhof, A.; Romeijn, G.J.; Frenkel, J.; Dorland, L.; de Barse, M.M.J.; Huijbers, W.A.R.; et al. Mutations in *MVK*, encoding mevalonate kinase, cause hyperimmunoglobulinaemia D and periodic fever syndrome. *Nat. Genet.* **1999**, *22*, 175–177. [CrossRef] [PubMed]
66. Ter Haar, N.M.; Jeyaratnam, J.; Lachmann, H.J.; Simon, A.; Brogan, P.A.; Doglio, M.; Cattalini, M.; Anton, J.; Modesto, C.; Quartier, P.; et al. The phenotype and genotype of mevalonate kinase deficiency: A series of 114 cases from the Eurofever registry. *Arthritis Rheumatol.* **2016**, *68*, 2795–2805. [CrossRef] [PubMed]
67. Bodar, E.J.; Kuijk, L.M.; Drenth, J.P.H.; van der Meer, J.W.M.; Simon, A.; Frenkel, J. On-demand anakinra treatment is effective in mevalonate kinase deficiency. *Ann. Rheum. Dis.* **2011**, *70*, 2155–2158. [CrossRef] [PubMed]
68. van der Hilst, J.C.H.; Bodar, E.J.; Barron, K.S.; Frenkel, J.; Drenth, J.P.H.; van der Meer, J.W.M.; Simon, A. International HIDS Study Group Long-term follow-up, clinical features, and quality of life in a series of 103 patients with hyperimmunoglobulinemia D syndrome. *Medicine* **2008**, *87*, 301–310. [CrossRef]
69. Galeotti, C.; Meinzer, U.; Quartier, P.; Rossi-Semerano, L.; Bader-Meunier, B.; Pillet, P.; Koné-Paut, I. Efficacy of interleukin-1-targeting drugs in mevalonate kinase deficiency. *Rheumatology* **2012**, *51*, 1855–1859. [CrossRef]
70. Tanaka, T.; Yoshioka, K.; Nishikomori, R.; Sakai, H.; Abe, J.; Yamashita, Y.; Hiramoto, R.; Morimoto, A.; Ishii, E.; Arakawa, H.; et al. National survey of Japanese patients with mevalonate kinase deficiency reveals distinctive genetic and clinical characteristics. *Mod. Rheumatol.* **2018**, *29*, 181–187. [CrossRef]
71. Ozen, S.; Kuemmerle-Deschner, J.B.; Cimaz, R.; Livneh, A.; Quartier, P.; Kone-Paut, I.; Zeft, A.; Spalding, S.; Gul, A.; Hentgen, V.; et al. International retrospective chart review of treatment patterns in severe familial Mediterranean fever, tumor necrosis factor receptor-associated periodic syndrome, and mevalonate kinase deficiency/hyperimmunoglobulinemia D syndrome. *Arthritis Care Res.* **2017**, *69*, 578–586. [CrossRef]
72. Ter Haar, N.; Lachmann, H.; Özen, S.; Woo, P.; Uziel, Y.; Modesto, C.; Koné-Paut, I.; Cantarini, L.; Insalaco, A.; Neven, B.; et al. Treatment of autoinflammatory diseases: Results from the Eurofever Registry and a literature review. *Ann. Rheum. Dis.* **2013**, *72*, 678–685. [CrossRef]
73. Vitale, A.; Insalaco, A.; Sfriso, P.; Lopalco, G.; Emmi, G.; Cattalini, M.; Manna, R.; Cimaz, R.; Priori, R.; Talarico, R.; et al. A snapshot on the on-label and off-label use of the interleukin-1 inhibitors in Italy among rheumatologists and pediatric rheumatologists: A nationwide multi-center retrospective observational study. *Front. Pharmacol.* **2016**, *7*, 380. [CrossRef]
74. Rossi-Semerano, L.; Fautrel, B.; Wendling, D.; Hachulla, E.; Galeotti, C.; Semerano, L.; Touitou, I.; Koné-Paut, I. Tolerance and efficacy of off-label anti-interleukin-1 treatments in France: A nationwide survey. *Orphanet J. Rare Dis.* **2015**, *10*, 19. [CrossRef]
75. Topaloglu, R.; Batu, E.D.; Orhan, D.; Ozen, S.; Besbas, N. Anti-interleukin 1 treatment in secondary amyloidosis associated with autoinflammatory diseases. *Pediatr. Nephrol.* **2016**, *31*, 633–640. [CrossRef]
76. Gattorno, M.; Piccini, A.; Lasiglìè, D.; Tassi, S.; Brisca, G.; Carta, S.; Delfino, L.; Ferlito, F.; Pelagatti, M.A.; Caroli, F.; et al. The pattern of response to anti-interleukin-1 treatment distinguishes two subsets of patients with systemic-onset juvenile idiopathic arthritis. *Arthritis Rheum.* **2008**, *58*, 1505–1515. [CrossRef]

77. Brachat, A.H.; Grom, A.A.; Wulffraat, N.; Brunner, H.I.; Quartier, P.; Brik, R.; McCann, L.; Ozdogan, H.; Rutkowska-Sak, L.; Schneider, R.; et al. Early changes in gene expression and inflammatory proteins in systemic juvenile idiopathic arthritis patients on canakinumab therapy. *Arthritis Res. Ther.* **2017**, *19*, 13. [CrossRef]
78. Feist, E.; Quartier, P.; Fautrel, B.; Schneider, R.; Sfriso, P.; Efthimiou, P.; Cantarini, L.; Lheritier, K.; Leon, K.; Karyekar, C.S.; et al. Efficacy and safety of canakinumab in patients with Still's disease: Exposure-response analysis of pooled systemic juvenile idiopathic arthritis data by age groups. *Clin. Exp. Rheumatol.* **2018**, *36*, 668–675.
79. Grom, A.A.; Ilowite, N.T.; Pascual, V.; Brunner, H.I.; Martini, A.; Lovell, D.; Ruperto, N.; Abrams, K.; Leon, K.; Lheritier, K.; et al. Rate and clinical presentation of macrophage activation syndrome in patients with systemic juvenile idiopathic arthritis treated with canakinumab. *Arthritis Rheumatol.* **2016**, *68*, 218–228. [CrossRef]
80. Ilowite, N.; Porras, O.; Reiff, A.; Rudge, S.; Punaro, M.; Martin, A.; Allen, R.; Harville, T.; Sun, Y.-N.; Bevirt, T.; et al. Anakinra in the treatment of polyarticular-course juvenile rheumatoid arthritis: Safety and preliminary efficacy results of a randomized multicenter study. *Clin. Rheumatol.* **2009**, *28*, 129–137. [CrossRef]
81. Kimura, Y.; Grevich, S.; Beukelman, T.; Morgan, E.; Nigrovic, P.A.; Mieszkalski, K.; Graham, T.B.; Ibarra, M.; Ilowite, N.; Klein-Gitelman, M.; et al. Pilot study comparing the Childhood Arthritis & Rheumatology Research Alliance (CARRA) systemic Juvenile Idiopathic Arthritis consensus treatment plans. *Pediatr. Rheumatol. Online J.* **2017**, *15*, 23.
82. Quartier, P.; Allantaz, F.; Cimaz, R.; Pillet, P.; Messiaen, C.; Bardin, C.; Bossuyt, X.; Boutten, A.; Bienvenu, J.; Duquesne, A.; et al. A multicentre, randomised, double-blind, placebo-controlled trial with the interleukin-1 receptor antagonist anakinra in patients with systemic-onset juvenile idiopathic arthritis (ANAJIS trial). *Ann. Rheum. Dis.* **2011**, *70*, 747–754. [CrossRef]
83. Ruperto, N.; Brunner, H.I.; Quartier, P.; Constantin, T.; Wulffraat, N.; Horneff, G.; Brik, R.; McCann, L.; Kasapcopur, O.; Rutkowska-Sak, L.; et al. Two randomized trials of canakinumab in systemic juvenile idiopathic arthritis. *N. Engl. J. Med.* **2012**, *367*, 2396–2406. [CrossRef] [PubMed]
84. Ruperto, N.; Quartier, P.; Wulffraat, N.; Woo, P.; Ravelli, A.; Mouy, R.; Bader-Meunier, B.; Vastert, S.J.; Noseda, E.; D'Ambrosio, D.; et al. A phase II, multicenter, open-label study evaluating dosing and preliminary safety and efficacy of canakinumab in systemic juvenile idiopathic arthritis with active systemic features. *Arthritis Rheum.* **2012**, *64*, 557–567. [CrossRef]
85. Ruperto, N.; Brunner, H.I.; Quartier, P.; Constantin, T.; Wulffraat, N.M.; Horneff, G.; Kasapcopur, O.; Schneider, R.; Anton, J.; Barash, J.; et al. Canakinumab in patients with systemic juvenile idiopathic arthritis and active systemic features: Results from the 5-year long-term extension of the phase III pivotal trials. *Ann. Rheum. Dis.* **2018**, *77*, 1710–1719. [CrossRef] [PubMed]
86. Hedrich, C.M.; Bruck, N.; Fiebig, B.; Gahr, M. Anakinra: A safe and effective first-line treatment in systemic onset juvenile idiopathic arthritis (SoJIA). *Rheumatol. Int.* **2012**, *32*, 3525–3530. [CrossRef]
87. Horneff, G.; Schulz, A.C.; Klotsche, J.; Hospach, A.; Minden, K.; Foeldvari, I.; Trauzeddel, R.; Ganser, G.; Weller-Heinemann, F.; Haas, J.P. Experience with etanercept, tocilizumab and interleukin-1 inhibitors in systemic onset juvenile idiopathic arthritis patients from the BIKER registry. *Arthritis Res. Ther.* **2017**, *19*, 256. [CrossRef] [PubMed]
88. Kearsley-Fleet, L.; Beresford, M.W.; Davies, R.; De Cock, D.; Baildam, E.; Foster, H.E.; Southwood, T.R.; Thomson, W.; Hyrich, K.L. Short-term outcomes in patients with systemic juvenile idiopathic arthritis treated with either tocilizumab or anakinra. *Rheumatology* **2019**, *58*, 94–102. [CrossRef] [PubMed]
89. Lequerré, T.; Quartier, P.; Rosellini, D.; Alaoui, F.; De Bandt, M.; Mejjad, O.; Koné-Paut, I.; Michel, M.; Dernis, E.; Khellaf, M.; et al. Interleukin-1 receptor antagonist (anakinra) treatment in patients with systemic-onset juvenile idiopathic arthritis or adult onset Still disease: Preliminary experience in France. *Ann. Rheum. Dis.* **2008**, *67*, 302–308. [CrossRef] [PubMed]
90. Nigrovic, P.A.; Mannion, M.; Prince, F.H.M.; Zeft, A.; Rabinovich, C.E.; van Rossum, M.A.J.; Cortis, E.; Pardeo, M.; Miettunen, P.M.; Janow, G.; et al. Anakinra as first-line disease-modifying therapy in systemic juvenile idiopathic arthritis: Report of forty-six patients from an international multicenter series. *Arthritis Rheum.* **2011**, *63*, 545–555. [CrossRef] [PubMed]
91. Pardeo, M.; Pires Marafon, D.; Insalaco, A.; Bracaglia, C.; Nicolai, R.; Messia, V.; De Benedetti, F. Anakinra in systemic juvenile idiopathic arthritis: A single-center experience. *J. Rheumatol.* **2015**, *42*, 1523–1527. [CrossRef] [PubMed]

92. Romano, M.; Pontikaki, I.; Gattinara, M.; Ardoino, I.; Donati, C.; Boracchi, P.; Meroni, P.L.; Gerloni, V. Drug survival and reasons for discontinuation of the first course of biological therapy in 301 juvenile idiopathic arthritis patients. *Reumatismo* **2014**, *65*, 278–285. [CrossRef]
93. Vastert, S.J.; de Jager, W.; Noordman, B.J.; Holzinger, D.; Kuis, W.; Prakken, B.J.; Wulffraat, N.M. Effectiveness of first-line treatment with recombinant interleukin-1 receptor antagonist in steroid-naive patients with new-onset systemic juvenile idiopathic arthritis: Results of a prospective cohort study. *Arthritis Rheumatol.* **2014**, *66*, 1034–1043. [CrossRef]
94. Marrani, E.; Burns, J.C.; Cimaz, R. How should we classify Kawasaki disease? *Front. Immunol.* **2018**, *9*, 2974. [CrossRef]
95. Principi, N.; Rigante, D.; Esposito, S. The role of infection in Kawasaki syndrome. *J. Infect.* **2013**, *67*, 1–10. [CrossRef]
96. De Rosa, G.; Pardeo, M.; Rigante, D. Current recommendations for the pharmacologic therapy in Kawasaki syndrome and management of its cardiovascular complications. *Eur. Rev. Med. Pharmacol. Sci.* **2007**, *11*, 301–308.
97. Rigante, D.; Valentini, P.; Rizzo, D.; Leo, A.; De Rosa, G.; Onesimo, R.; De Nisco, A.; Angelone, D.F.; Compagnone, A.; Delogu, A.B. Responsiveness to intravenous immunoglobulins and occurrence of coronary artery abnormalities in a single-center cohort of Italian patients with Kawasaki syndrome. *Rheumatol. Int.* **2010**, *30*, 841–846. [CrossRef]
98. Rigante, D.; Andreozzi, L.; Fastiggi, M.; Bracci, B.; Natale, M.F.; Esposito, S. Critical overview of the risk scoring systems to predict non-responsiveness to intravenous immunoglobulin in Kawasaki syndrome. *Int. J. Mol. Sci.* **2016**, *17*, 278. [CrossRef] [PubMed]
99. Agarwal, S.; Agrawal, D.K. Kawasaki disease: Etiopathogenesis and novel treatment strategies. *Expert Rev. Clin. Immunol.* **2017**, *13*, 247–258. [CrossRef] [PubMed]
100. Blonz, G.; Lacroix, S.; Benbrik, N.; Warin-Fresse, K.; Masseau, A.; Trewick, D.; Hamidou, M.; Stephan, J.-L.; Néel, A. Severe late-onset Kawasaki disease successfully treated with anakinra. *J. Clin. Rheumatol.* **2018**, 1. [CrossRef]
101. Cohen, S.; Tacke, C.E.; Straver, B.; Meijer, N.; Kuipers, I.M.; Kuijpers, T.W. A child with severe relapsing Kawasaki disease rescued by IL-1 receptor blockade and extracorporeal membrane oxygenation. *Ann. Rheum. Dis.* **2012**, *71*, 2059–2061. [CrossRef] [PubMed]
102. Guillaume, M.-P.; Reumaux, H.; Dubos, F. Usefulness and safety of anakinra in refractory Kawasaki disease complicated by coronary artery aneurysm. *Cardiol. Young* **2018**, *28*, 739–742. [CrossRef]
103. Sánchez-Manubens, J.; Gelman, A.; Franch, N.; Teodoro, S.; Palacios, J.R.; Rudi, N.; Rivera, J.; Antón, J. A child with resistant Kawasaki disease successfully treated with anakinra: A case report. *BMC Pediatr.* **2017**, *17*, 102. [CrossRef] [PubMed]
104. Shafferman, A.; Birmingham, J.D.; Cron, R.Q. High dose Anakinra for treatment of severe neonatal Kawasaki disease: A case report. *Pediatr. Rheumatol. Online J.* **2014**, *12*, 26. [CrossRef]
105. Koné-Paut, I.; Cimaz, R.; Herberg, J.; Bates, O.; Carbasse, A.; Saulnier, J.P.; Maggio, M.C.; Anton, J.; Piram, M. The use of interleukin 1 receptor antagonist (anakinra) in Kawasaki disease: A retrospective cases series. *Autoimmun. Rev.* **2018**, *17*, 768–774. [CrossRef] [PubMed]
106. Kaiser, C.; Knight, A.; Nordström, D.; Pettersson, T.; Fransson, J.; Florin-Robertsson, E.; Pilström, B. Injection-site reactions upon Kineret (anakinra) administration: Experiences and explanations. *Rheumatol. Int.* **2012**, *32*, 295–299. [CrossRef] [PubMed]
107. Sota, J.; Vitale, A.; Insalaco, A.; Sfriso, P.; Lopalco, G.; Emmi, G.; Cattalini, M.; Manna, R.; Cimaz, R.; Priori, R.; et al. Safety profile of the interleukin-1 inhibitors anakinra and canakinumab in real-life clinical practice: A nationwide multicenter retrospective observational study. *Clin. Rheumatol.* **2018**, *37*, 2233–2240. [CrossRef]

MDPI
St. Alban-Anlage 66
4052 Basel
Switzerland
Tel. +41 61 683 77 34
Fax +41 61 302 89 18
www.mdpi.com

International Journal of Molecular Sciences Editorial Office
E-mail: ijms@mdpi.com
www.mdpi.com/journal/ijms

www.ingramcontent.com/pod-product-compliance
Lightning Source LLC
Chambersburg PA
CBHW041947260926
51850CB00018B/1817
* 9 7 8 3 0 3 6 5 0 7 4 0 8 *